基于 Abaqus 的复合材料有限元分析

Finite Element Analysis of Composite
Materials Using Abaqus

[美] 埃弗·J. 巴尔贝罗（Ever J. Barbero） 著

郑 权　杨颜志　宋林郁　陈鸣亮　译

国防工业出版社

·北京·

著作权合同登记　图字:01-2022-5648号

图书在版编目(CIP)数据

基于Abaqus的复合材料有限元分析/(美)埃弗·J.巴尔贝罗(Ever J. Barbero)著;郑权等译. —北京:国防工业出版社,2023.1

书名原文:Finite Element Analysis of Composite Materials Using Abaqus

ISBN 978-7-118-12816-1

Ⅰ.①基… Ⅱ.①埃… ②郑… Ⅲ.①复合材料—有限元分析—应用软件 Ⅳ.①TB33-39

中国国家版本馆CIP数据核字(2023)第007994号

Finite Element Analysis of Composite Materials using Abaqus^{TM}/by Ever J. Barbero/9781466516618
Copyright@ 2013 by Taylor & Francis Group, LLC
Authorized translation from English language edition published by CRC Press, part of Taylor & Francis Group LLC; All rights reserved.

本书原版由Taylor & Francis出版集团旗下，CRC出版公司出版,并经其授权翻译出版。版权所有,侵权必究。

National Defense Industry Press is authorized to publish and distribute exclusively the Chinese (Simplified Characters) language edition. This edition is authorized for sale throughout Mainland of China. No part of the publication may be reproduced or distributed by any means, or stored in a database or retrieval system, without the prior written permission of the publisher.

本书中文简体翻译版经授权由国防工业出版社独家出版,并限在中国大陆地区销售。未经出版者书面许可,不得以任何方式复制或发行本书的任何部分。

Copies of this book sold without a Taylor & Francis sticker on the cover are unauthorized and illegal.
本书封面贴有Taylor & Francis公司防伪标签,无标签者不得销售。

※

*国防工业出版社*出版发行

(北京市海淀区紫竹院南路23号　邮政编码100048)
三河市腾飞印务有限公司印刷
新华书店经售

*

开本710×1000　1/16　印张26½　字数478千字
2023年1月第1版第1次印刷　印数1—3000册　定价98.00元

(本书如有印装错误,我社负责调换)

国防书店:(010)88540777　　书店传真:(010)88540776
发行业务:(010)88540717　　发行传真:(010)88540762

译者序

目前，复合材料尤其是连续纤维增强树脂基复合材料已在国内外很多领域中广泛应用，如航空、航天、汽车、建筑、体育用品等。与常规各向同性的金属材料相比，复合材料结构设计更为复杂，不仅需要进行结构构型设计仿真，对于复合材料本身还要开展设计仿真工作。复合材料力学作为已经发展数十年的固体力学分支，已形成独立的学科体系，且国内外大部分高等院校已经将复合材料设置为必修科目。在国家航空航天等领域大力发展的背景下，了解并掌握复合材料力学及其分析方法是每一位相关行业工作人员必备的知识和能力。

本书为英文书籍 *Finite Element Analysis of Composite Materials Using Abaqus* 的译著。该书主要分析对象为复合材料及结构，从细观到宏观全面涵盖了复合材料力学的基础理论。同时本书将有限元分析方法引入到复合材料及结构的分析中，并且以 Abaqus 软件为平台开展了大量工程实例分析，使本书不仅具有较高的理论学习价值，同时具有很好的工程应用借鉴作用。本书可作为复合材料相关专业设计人员以及高校复合材料相关专业学生的参考书籍。

参加翻译的人员分工如下：郑权负责第 1、2、5、6、8 章；杨颜志负责序言，第 3、4、7 章；宋林郁负责第 9 章；陈鸣亮负责第 10 章、附录；全书的统稿工作由郑权完成。在本书的翻译过程中，得到了上海宇航系统工程研究所复合材料领域相关专家的指导，他们提出了大量宝贵意见，对于本书的完善起到了很大的帮助，在此表示由衷的感谢！

由于译者知识的局限，本书翻译难免有不妥之处，希望读者提出宝贵意见和建议。

译 者

丛书序

复合材料的商业化应用半个世纪以来,其在许多行业中已得到广泛应用。例如航空航天、风车叶片、高速公路桥梁改造以及需要20年或更长时间安全可靠使用的其他设计领域。利用复合材料时,用户可以通过选择不同的材料组成、材料比例以及材料的几何分布等,获得不同的刚度、强度、热导率和耐火性等材料性能。换句话说,工程师在设计结构时可以同时设计材料。但是,复合材料的破坏模式要比常规材料更复杂。对复合材料性能、安全性和可靠性的需求使得工程师在设计过程中需要考虑多种的破坏现象。因此,*Composite Materials:Analysis and Design*丛书旨在给设计工程师提供一系列相关成果,其由不同领域的复合材料专家根据自身实际设计工作过程总结而成。

由于可应用的复合材料数量和类型不断增加,导致复合材料体系和制造技术的多样性、复杂性呈指数增长。鉴于复合材料的多样性以及复合材料性能不断地提升和变化,对复合材料的研究还远未完成。该系列丛书不仅适用于工程师的实践应用,而且对于期望了解最先进的复合材料特性和结构响应以及开发新的工具用于建模、预测结构响应的研究人员和学生也同样适用。

因此,该系列丛书着重于向读者介绍现有的和正在研究的相关知识,包括材料-性能关系、工艺-性能关系以及复合材料的材料和结构响应。该系列丛书包含了复合材料的分析方法、试验方法和数值方法,为复合材料结构设计提供了相关参考。

埃弗·J. 巴尔贝罗
西弗吉尼亚大学,摩根敦,西弗吉尼亚

序

　　复合材料有限元分析的主要分析对象为复合材料结构,也称 Composites。本书对复合材料的分析包括细观层面的材料本身的分析以及由复合材料制备的结构的分析。本书包含了我在西弗吉尼亚大学教授"MAE 646 Advanced Mechanics of Composite Materials"研究生课程的笔记。虽然本书也是关于复合材料力学的教科书,但是有限元方法的应用对于解决复合材料高级分析中的复杂边值问题至关重要,这也是本书命名缘由。

　　关于先进的复合材料力学图书很多,但是像本书一样,能在解决实际问题过程中将理论引入实践的却很少。有一些专门致力于有限元分析方法的图书,其包含了一些关于复合材料建模的实例,但还远不足以指导实际工程中复合材料的材料、结构设计与分析。本书对复合材料详细分析的相关概念进行了解释,对需要转换为数学表达的相关力学概念也进行了深入阐释,同时使用商业有限元分析软件(如 Abaqus)详细阐述了边界值问题求解过程。此外,本书中将 50 多个详细分析的实例贯穿于相关理论,在章节的末尾提供了多达 75 个练习题,以及 50 多段独立的 Abaqus 操作步骤代码用于详细说明实例的求解过程。读者可以复制实例并完成练习。当需要进行有限元分析时,读者可以使用商业软件或其他计算程序来完成。本书的官网设置了下载必要软件的链接,读者可以链接下载。利用 Abaqus 和 MATLAB 的分析实例有很多,相关代码可以从官网中下载。此外,读者可以通过编写用户材料子例程和 Python 脚本来扩展 Abaqus 的分析能力,具体可参考本书中的相关实例。

　　本书的第 1 章至第 7 章可以涵盖第一学期的研究生课程。第 2 章(有限元分析介绍)主要介绍了有限元分析方法,其针对那些没有系统学习过有限元方法或没有相关知识的读者。第 4 章(屈曲)的内容在本书的后续章节中并未涉及,因此为便于更详尽地了解后续章节内容,此章节可以省略。第 7 章(黏弹性)、第 8 章(连续介质损伤力学)和第 9 章(离散损伤力学)3 章内容放在一起介绍以强调它们内容之间的连贯性。但是,若需要更多强调损伤和分层等内容,那么第一学期课程可以跳过第 7 章。

　　本书中大量使用了归纳法。具体是,通过越来越多复杂的实例引入相关概念,直到读者对涉及的物理现象达到足够的理解,此时再介绍通用的理论就会毫无困难。此方法在学习初期时需要先接受某些事实，建立的模型和相互关系，然后它们会在后续学习中逐步被证明。例如，在第 7 章中，一开始就介绍了黏弹性模型以帮助读者理解黏弹性材料的响应。同时，简略地介绍了叠加原理和拉普拉斯变换，这些在本章的后续部分进行了正式的介绍。对于习惯了演绎法的读者来说，理解起来很奇怪，但是多年的教学经验使我相信学生们可以通过该方法更有效地学习相关知识。

本书假设读者已经熟悉复合材料的基础力学概念，其可以通过入门级的教材进行学习，例如我之前的专著 *Introduction to Composite Material Design*。此外，假设读者掌握了相关知识体系，而这些知识是获得航空航天、力学、土木工程，或类似学科科学学位所必需的。书中对其他书籍和段落的引用以及注脚，用于帮助读者更新这些概念和阐明使用的符号。若读者具有连续介质力学、张量分析和有限元方法方面的学习基础，可以提高本书的学习效率，但其并非学习本书所必需。有限元方法是解决实际问题的工具，大多数情况下，本书均使用 Abaqus 软件。使用 FORTRAN，Python 和 MATLAB 进行编程仅限于建立材料模型和后处理算法。拥有这些编程语言的基础知识是有用的，但也不是必需的。

本书中仅使用了 3 个软件包：Abaqus 用于章节中实例和习题的有限元建模求解；MATLAB 用于章节中实例和习题的符号运算和数值计算。此外，第 4 章中使用的 BMI3，在本书的官网中已免费提供。ANSYS Mechanical、LS-DYNA、MSC-MARC、SolidWorks 等其他软件在本书中也有引用，但未在实例中使用。本书中实例的相关代码可在官网 http://barbero.cadec-online.com/feacm-abaqus/ 中查找。

复合材料现在已在多个领域中广泛应用，如航空航天、汽车、建筑、体育用品等。因为与常规的各向同性的金属材料不同，复合材料的设计特别具有挑战性，在设计复合材料结构时也要对复合材料本身进行设计。复合材料的初步设计一般基于层合板的平面应力假设，并对零件的几何、载荷和支撑条件进行粗略的估算。这样的话，分析方法可以相对简化并可使用简单的代数方法进行计算。但是，初步的分析方法有很多缺点，通过本书中所述的高级的力学方法和有限元分析则可以弥补这些缺点。当前的商业有限元软件具有友好的用户前处理和后处理功能，并且具有强大的用户编程功能，设计人员可以很容易开展复合材料的详细分析。本书旨在将强大的有限元分析工具和复合材料结构的实际问题联系在一起。我期望研究生、工程师和高校教师在深刻理解复合材料高级力学方法后会发现本书是关于复合材料有限元分析的实用教科书。

埃弗·J. 巴尔贝罗

致　谢

　　我要感谢 Elio Sacco、Raimondo Luciano 和 Fritz Campo 对第 6 章所做的工作。感谢 Tom Damiani、Joan Andreu Mayugo 和 Xavier Martinez，他们在 2004 年、2006 年和 2009 年教授过该课程，并对课程笔记做过许多修订和补充，本书也是基于此进行编写的。另外，感谢 Eduardo Sosa 博士在第 4 章中帮助开展了含缺陷结构的屈曲分析，感谢 Fabrizio Greco 教授对第 10 章进行了校对。感谢对本书进行校对的其他人，包括 Enrique Barbero、Guillermo Creus、Luis Godoy、Paolo Lonetti、Severino Marques、Qiao Pizhong、Timothy Norman、Sonia Sanchez 和 Eduardo Sosa。此外，感谢那些帮助我编译求解方案手册的人，包括 Hermann Alcazar、John Sandro Rivas 和 Rajiv Dastane。我还要感谢 John Sandro Rivas 对本书中实例代码进行的手动检查，这些实例代码均需要手动转换为 Abaqus/CAE 6.10 版格式。另外，非常感谢我的同事和学生，他们对本书提出了宝贵的意见和并作出贡献。最后，我要感谢我的妻子 Ana Maria，感谢我的孩子 Margaret 和 Daniel，他们放弃了很多与他们父亲团聚的机会，以便我完成这本书。

符号列表

1. 与正交各向异性材料力学相关的符号

$\boldsymbol{\epsilon}$	应变张量
ε_{ij}	张量符号中的应变分量
ε_α	指标符号缩写形式下的应变分量
ϵ_α^e	弹性应变
ϵ_α^p	塑性应变
λ	拉梅常数
ν	泊松比
ν_{12}	面内泊松比
ν_{23}, ν_{13}	层间泊松比
ν_{xy}	$x-y$ 平面的层合板表观泊松比
$\boldsymbol{\sigma}$	应力张量
σ_{ij}	张量符号中的应力分量
σ_α	指标符号缩写形式下的应力分量
$[a]$	向量的转换矩阵
e_i	全局坐标系下的单位向量分量
e_i'	材料坐标系下的单位向量分量
f_i, f_{ij}	Tsai–Wu 系数
k	体积模量
l, m, n	方向余弦
$\tilde{u}(\varepsilon_{ij})$	单位体积应变能
u_i	位移矢量分量
x_i	全局方向或轴
x_i'	材料方向或轴
$\boldsymbol{C}$	刚度张量
C_{ijkl}	指标符号形式下的刚度分量

$C_{\alpha,\beta}$	指标符号缩写形式下的刚度分量
E	弹性模量
E_1	纵向模量
E_2	横向模量
E_3	厚度方向上的横向模量
E_x	全局坐标系下 x 方向的层合板表观模量
$G = \mu$	剪切模量
G_{12}	面内剪切模量
G_{23}, G_{13}	层间剪切模量
G_{xy}	$x-y$ 平面的层合板表观剪切模量
I_{ij}	2 阶等同张量
I_{ijkl}	4 阶等同张量
Q'_{ij}	单层板坐标系下的单层板刚度分量
$[R]$	Reuter 矩阵
S	柔度张量
S_{ijkl}	指标符号形式下的柔度分量
$S_{\alpha,\beta}$	指标符号缩写形式下的柔度分量
$[T]$	应力的坐标转换矩阵
$[\bar{T}]$	应变的坐标转换矩阵

2. 与有限元分析相关的符号

$\underline{\underline{\partial}}$	矩阵形式的应变－位移方程
$\underline{\epsilon}$	应变分量的六元数组
$\theta_x, \theta_y, \theta_z$	遵循右手法则的旋转角（图 2.19）
$\underline{\sigma}$	应力分量的六元数组
ϕ_x, ϕ_y	板壳理论中使用的旋转角
$\underline{a}$	节点位移数列
u_j^e	离散化计算中的未知参数
$\underline{\underline{B}}$	应变－位移矩阵
$\underline{\underline{C}}$	刚度矩阵
$\underline{\underline{K}}$	组装后的整体刚度矩阵
$\underline{\underline{K}}^e$	单元刚度矩阵

$\underline{N}$	插值函数数列
N_j^e	离散化后单元的插值函数
$\underline{P}^e$	单元力数列
$\underline{P}$	组装后的整体力数列

3. 与层合板弹性和强度相关的符号

γ_{xy}^0	面内剪切应变
γ_{4u}	2–3 平面内的极限层间剪切应变
γ_{5u}	1–3 平面内的极限层间剪切应变
γ_{6u}	面内极限剪切应变
$\varepsilon_x^0, \varepsilon_y^0$	面内应变
ϵ_{1t}	纵向极限拉伸应变
ϵ_{2t}	横向极限拉伸应变
ϵ_{3t}	厚度方向的横向极限拉伸应变
ϵ_{1c}	纵向极限压缩应变
ϵ_{2c}	横向极限压缩应变
ϵ_{3c}	厚度方向的横向极限压缩应变
k_x, k_y	弯曲曲率
k_{xy}	扭曲曲率
ϕ_x, ϕ_y	壳中面旋转角(图 2.19)
c_4, c_5, c_6	Tsai–Wu 耦合系数
t_k	单层板厚度
u_0, v_0, w_0	壳中面位移
z	距壳中面的距离
A_{ij}	拉伸刚度矩阵$[A]$的分量
B_{ij}	拉–弯耦合矩阵$[B]$的分量
D_{ij}	弯曲刚度矩阵$[D]$的分量
$[E_0]$	ANSYS 软件中拉伸刚度矩阵$[A]$的符号
$[E_1]$	ANSYS 软件中拉–弯耦合矩阵$[B]$的符号
$[E_2]$	ANSYS 软件中弯曲刚度矩阵$[D]$的符号
F_{1t}	纵向拉伸强度
F_{2t}	横向拉伸强度

F_{3t}	厚度方向的横向拉伸强度
F_{1c}	纵向压缩强度
F_{2c}	横向压缩强度
F_{3c}	厚度方向的横向压缩强度
F_4	2–3 平面内的层间剪切强度
F_5	1–3 平面内的层间剪切强度
F_6	面内剪切强度
H_{ij}	层间剪切矩阵 $[H]$ 的分量
I_F	失效指数
M_x, M_y, M_{xy}	单位长度的弯矩（图 3.3）
$\hat{M}_n$	单位长度上施加的弯矩
N_x, N_y, N_{xy}	单位长度的面内力（图 3.3）
$\hat{N}_n$	单位长度上施加的面内法向力，边的法向
$\hat{N}_{ns}$	单位长度上施加的面内剪切力，边的切向
$(\overline{Q}_{ij})_k$	第 k 层在层合板坐标系下的单层板刚度分量
V_x, V_y	单位长度的剪切力（图 3.3）

4. 与屈曲问题相关的符号

λ, λ_i	特征值
s	摄动参数
Λ	载荷乘子
$\Lambda^{(cr)}$	分叉乘子或临界载荷乘子
$\Lambda^{(1)}$	后临界路径斜率
$\Lambda^{(2)}$	后临界路径曲率
$\boldsymbol{v}$	特征向量（屈曲模态）
$[K]$	刚度矩阵
$[K_s]$	应力刚度矩阵
P_{CR}	临界载荷

5. 与自由边应力相关的符号

$\eta_{xy,x}, \eta_{xy,y}$	相互影响系数
$\eta_{x,xy}, \eta_{y,xy}$	另一对相互影响系数

F_{yz}	$y-z$ 平面内的层间剪切力
F_{xz}	$x-z$ 平面内的层间剪切力
M_z	层间弯矩

6. 与细观力学相关的符号

$\bar{\epsilon}_\alpha$	平均工程应变分量
$\bar{\varepsilon}_{ij}$	平均应变张量分量
$\epsilon_\alpha^0, \varepsilon_{ij}^0$	远场施加的应变分量
$\bar{\sigma}_\alpha$	平均应力分量
A^i	应变集中张量,第 i 相,指标符号缩写形式
$2a_1, 2a_2, 2a_3$	RVE 的尺寸
A_{ijkl}	应变集中张量分量
B^i	应力集中张量,第 i 相,指标符号缩写形式
B_{ijkl}	应力集中张量分量
I	6×6 单位矩阵
P_{ijkl}	Eshelby 张量
V_f	纤维体积分数
V_m	基体体积分数

7. 与黏弹性相关的符号

$\dot{\varepsilon}$	应变率
η	黏度
θ	老化时间
$\dot{\sigma}$	应力率
τ	材料或系统的时间常数
Γ	Gamma 函数
s	拉普拉斯变量
t	时间
$C_{\alpha,\beta}(t)$	时域内的刚度张量
$C_{\alpha,\beta}(s)$	拉普拉斯域内的刚度张量
$\hat{C}_{\alpha,\beta}(s)$	Carson 域内的刚度张量
$D(t)$	柔量

$D_0, (D_i)_0$	初始柔量
$D_c(t)$	总柔量 $D(t)$ 的蠕变柔量
D', D''	储能柔量和耗散柔量
$E_0, (E_i)_0$	初始模量
E_∞	平衡模量
E, E_0, E_1, E_2	黏弹性模型中的参数(图 7.1)
$E(t)$	松弛模量
E', E''	储能模量和耗散模量
$F[\ ]$	傅里叶变换
$(G_{ij})_0$	初始剪切模量
$H(t-t_0)$	Heaviside 阶跃函数
$H(\theta)$	松弛谱
$L[\ ]$	拉普拉斯变换
$L[\ ]^{-1}$	拉普拉斯逆变换

8. 与损伤相关的符号

α	层合板的热膨胀系数
$\alpha^{(k)}$	层合板中第 k 层的热膨胀系数
α_{cr}	纵向压缩失效时的临界偏折角
α_σ	纤维偏折的标准方差
$\gamma(\delta)$	损伤硬化函数
γ_0	损伤阈值
δ_{ij}	Kronecker 函数
δ	损伤硬化变量
ε	有效应变
$\bar{\varepsilon}$	未损伤应变
ε^p	塑性应变
$\dot{\gamma}$	单位体积散热率
$\dot{\gamma}_s$	内部熵产率
λ	裂纹密度
λ_{lim}	饱和裂纹密度
$\dot{\lambda}, \dot{\lambda}^d$	损伤乘子

$\dot{\lambda}^p$	屈服乘子
ρ	密度
σ	有效应力
$\bar{\sigma}$	未损伤应力
τ_{13}, τ_{23}	层间剪切应力分量
φ, φ^*	应变能密度和应变余能密度
χ	吉布斯能密度
ψ	Helmholtz 自由能密度
ΔT	温度变化量
$\Omega = \Omega_{ij}$	完整性张量
$2a_0$	典型裂纹尺寸
d_i	损伤张量特征值
f^d	损伤流面
f^p	屈服流面
$f(x), F(x)$	概率密度,累积概率
g	伤伤激活函数
g^d	损伤面
g^p	屈服面
h	层合板的厚度
h_k	第 k 层铺层的厚度
m	韦布尔模数
p	屈服强化变量
$\hat{p}$	p 量的量厚度平均值
$\tilde{p}$	p 量的初始值
$\bar{p}$	p 量的体积平均值
$\boldsymbol{q}$	单位面积的热流矢量
r	单位质量的辐射热
s	比熵
$u(\varepsilon_{ij})$	内能密度
A	裂纹面积
$[A]$	层合板面内刚度矩阵

A_{ijkl}	拉-压损伤本构张量
B_{ijkl}	剪切损伤本构张量
B_a	无量纲数(式(8.57))
$\overline{C}_{\alpha,\beta}$	未损伤构型中的刚度矩阵
C^{ed}	切线刚度张量
D_{ij}	损伤张量
D_{1t}^{cr}	纵向拉伸破坏时的临界损伤变量值
D_{1c}^{cr}	纵向压缩破坏时的临界损伤变量值
D_{2t}^{cr}	横向拉伸破坏时的临界损伤变量值
D_2, D_6	损伤变量
$E(D)$	有效模量
$\overline{E}$	未损伤(初始)模量
$G_c = 2\gamma_c$	表面能
G_{Ic}, G_{IIc}	Ⅰ和Ⅱ模式下的临界能量释放率
J_{ijkl}	法向损伤本构张量
M_{ijkl}	损伤效应张量
N	层合板中的铺层数量
$\{N\}$	膜应力的阵列形式
Q	层合板的3×3退化刚度矩阵
$R(p)$	屈服硬化函数
R_0	屈服阈值
S	熵或层合板柔度矩阵,取决于上下文
T	温度
U	应变能
V	RVE的体积
Y_{ij}	热力学力张量

9. 分层相关的符号

α	混合模式的裂纹扩展指数
β_δ, β_G	混合模式比率
δ	界面的CZM分离间距
δ_m	混合模式的分离间距

δ_m^0	损伤初始时的混合模式下分离间距
δ_m^0	断裂时的混合模式下分离间距
σ^0	损伤初始时的 CZM 临界分离间距
l	2D 分层情况下的分层长度
σ^0	界面的 CZM 强度
ψ_{xi}, ψ_{yi}	关于平板中面法向的旋转角
Ω	实体体积
Ω_D	分层区域
Π_e	弹性势能
a_2	总势能
$\dot{\Gamma}$	耗散率
Λ	单位面积的界面应变能密度
$\partial\Omega$	实体边界
d	一维损伤状态变量
k_{xy}, k_z	位移连续性参数
$[A_i], [B_i], [D_i]$	层合板刚度子矩阵
D_I, D_{II}, D_{III}	CZM 在 I, II 和 III 模式下的损伤变量
$G(l)$	二维情况下,总的能量释放率(ERR)
G	三维情况下,总的能量释放率(ERR)
G_I, G_{II}, G_{III}	I, II 和 III 模式下的能量释放率(ERR)
G_c	三维情况下,总的临界能量释放率(ERR)
G_I^c	I 模式的临界能量释放率
$[H_i]$	层合板的层间剪切刚度矩阵
K	罚刚度
$\tilde{K}$	初始罚刚度
K_I, K_{II}, K_{III}	I, II 和 III 模式下的应力强度因子(SIF)
N_i, M_i, T_i	应力合力
U	内能
W	实体对周围环境所做的功
$W_{closure}$	裂纹闭合功

目　录

第1章　正交各向异性材料力学 ·· 1

- 1.1　单层板坐标系 ·· 1
- 1.2　位移 ··· 1
- 1.3　应变 ··· 2
- 1.4　应力 ··· 3
- 1.5　指标符号缩写 ·· 4
- 1.6　平衡方程和虚功原理 ·· 5
- 1.7　边界条件 ··· 7
  - 1.7.1　牵引力边界条件 ··· 7
  - 1.7.2　自由表面边界条件 ·· 7
- 1.8　连续性条件 ·· 8
  - 1.8.1　牵引力的连续性 ··· 8
  - 1.8.2　位移的连续性 ·· 8
- 1.9　相容性 ·· 9
- 1.10　坐标变换 ··· 9
  - 1.10.1　应力变换 ··· 12
  - 1.10.2　应变变换 ··· 14
- 1.11　本构方程变换 ··· 15
- 1.12　3D本构方程 ··· 16
  - 1.12.1　各向异性材料 ··· 17
  - 1.12.2　单对称材料 ·· 18
  - 1.12.3　正交各向异性材料 ··· 19
  - 1.12.4　横观各向同性材料 ··· 21
  - 1.12.5　各向同性材料 ··· 22
- 1.13　工程常数 ··· 23
- 1.14　从3D退化为平面应力状态 ·· 28

1.15　层合板表观性能 ……………………………………………… 30
　　习题 ……………………………………………………………………… 31

第2章　有限元分析介绍 …………………………………………………… 34
　2.1　基本的 FEM 步骤 …………………………………………………… 34
　　2.1.1　离散 ……………………………………………………………… 35
　　2.1.2　单元方程 ………………………………………………………… 35
　　2.1.3　一个单元上的近似解 …………………………………………… 36
　　2.1.4　插值函数 ………………………………………………………… 36
　　2.1.5　一个特定问题的单元方程 ……………………………………… 38
　　2.1.6　单元方程的组装 ………………………………………………… 39
　　2.1.7　边界条件 ………………………………………………………… 39
　　2.1.8　方程的解 ………………………………………………………… 40
　　2.1.9　单元内部的解 …………………………………………………… 40
　　2.1.10　派生结果 ………………………………………………………… 40
　2.2　通用的有限元法过程 ……………………………………………… 41
　2.3　实体建模、分析和可视化 ………………………………………… 44
　　2.3.1　模型几何体 ……………………………………………………… 44
　　2.3.2　材料及截面属性 ………………………………………………… 55
　　2.3.3　装配 ……………………………………………………………… 58
　　2.3.4　分析步 …………………………………………………………… 59
　　2.3.5　载荷 ……………………………………………………………… 60
　　2.3.6　边界条件 ………………………………………………………… 61
　　2.3.7　划分网格和单元类型 …………………………………………… 64
　　2.3.8　求解 ……………………………………………………………… 66
　　2.3.9　后处理和可视化 ………………………………………………… 67
　习题 ……………………………………………………………………… 84

第3章　层合板的刚度和强度 ……………………………………………… 85
　3.1　壳的运动学假设 …………………………………………………… 86
　　3.1.1　一阶剪切变形理论 ……………………………………………… 86
　　3.1.2　Kirchhoff 理论 …………………………………………………… 90
　　3.1.3　简支边界条件 …………………………………………………… 92
　3.2　层合板的有限元分析 ……………………………………………… 92

XVIII

  3.2.1 单元类型和命名规则 ·············· 95
  3.2.2 薄壳单元 ························ 96
  3.2.3 厚壳单元 ························ 97
  3.2.4 通用壳单元 ······················ 97
  3.2.5 连续体壳单元 ···················· 97
  3.2.6 夹层壳 ·························· 98
  3.2.7 节点和曲率 ······················ 98
  3.2.8 面外转动 ························ 98
  3.2.9 层合板 FEA 中 A、B、D、H 参数的输入 ·············· 99
  3.2.10 层合板 FEA 中等效正交各向异性参数的输入 ·········· 106
  3.2.11 多向层合板 FEA 中的 LSS ········ 113
  3.2.12 含丢层层合板的 FEA ············ 124
  3.2.13 夹层壳的 FEA ·················· 134
  3.2.14 单元坐标系 ···················· 147
  3.2.15 约束 ·························· 157
 3.3 失效准则 ····························· 162
  3.3.1 2D 失效准则 ···················· 162
  3.3.2 3D 失效准则 ···················· 165
 3.4 预定义场 ····························· 171
 习题 ··································· 173

第 4 章 屈曲 ································ 176

 4.1 特征值屈曲分析 ························· 176
  4.1.1 缺陷敏感性 ······················ 183
  4.1.2 非对称分叉 ······················ 183
  4.1.3 后临界路径 ······················ 184
 4.2 延续分析方法 ··························· 187
 习题 ··································· 193

第 5 章 自由边应力 ·························· 194

 5.1 泊松比的不匹配 ························· 195
  5.1.1 层间力 ·························· 195
  5.1.2 层间弯矩 ························ 196
 5.2 相互影响系数 ··························· 204

习题 ··· 213

第 6 章 细观力学计算 ··· 216

6.1 均匀化的分析方法 ·· 217
6.1.1 Reuss 模型 ·· 217
6.1.2 Voigt 模型 ··· 217
6.1.3 周期性细观结构模型 ··································· 217
6.1.4 横观各向同性平均化 ··································· 219
6.2 均匀化的数值计算方法 ······································· 220
6.3 局部–整体分析 ·· 239
6.4 层合板的 RVE ·· 242
习题 ··· 248

第 7 章 黏弹性 ··· 249

7.1 黏弹性模型 ·· 251
7.1.1 Maxwell 模型 ··· 251
7.1.2 Kelvin 模型 ·· 252
7.1.3 标准线性固体模型 ······································ 252
7.1.4 Maxwell-Kelvin 模型 ···································· 253
7.1.5 幂指数模型 ·· 254
7.1.6 Prony 级数模型 ·· 254
7.1.7 标准非线性固体模型 ··································· 255
7.1.8 非线性幂指数模型 ······································ 256
7.2 Boltzmann 叠加原理 ·· 257
7.2.1 线性黏弹性材料 ··· 257
7.2.2 耐老化黏弹性材料 ······································ 258
7.3 对应原理 ·· 259
7.4 频域 ··· 260
7.5 频谱表示 ·· 261
7.6 黏弹性复合材料的细观力学 ·································· 261
7.6.1 一维情况 ·· 261
7.6.2 三维情况 ·· 263
7.7 黏弹性复合材料的宏观力学 ·································· 267

7.7.1 平衡对称层合板 ·················· 267
   7.7.2 常规层合板 ····················· 267
  7.8 黏弹性复合材料的FEA ················ 268
  习题 ································· 280

第8章 连续介质损伤力学 ················ 282

  8.1 一维损伤力学 ····················· 282
   8.1.1 损伤变量 ······················ 282
   8.1.2 损伤阈值和激活函数 ················ 285
   8.1.3 动力学方程 ····················· 286
   8.1.4 动力学方程的统计学解释 ·············· 287
   8.1.5 一维随机强度模型 ················· 287
   8.1.6 纤维方向的拉伸损伤 ················ 292
   8.1.7 纤维方向的压缩损伤 ················ 298
  8.2 多维损伤和有效空间 ················· 302
  8.3 热力学公式 ······················· 303
   8.3.1 第一定律 ······················ 304
   8.3.2 第二定律 ······················ 305
  8.4 三维空间中的动力学定律 ··············· 311
  8.5 损伤和塑性 ······················· 321
  习题 ································· 321

第9章 离散损伤力学 ···················· 324

  9.1 概述 ··························· 325
  9.2 近似值 ·························· 328
  9.3 单层板本构方程 ···················· 329
  9.4 位移场 ·························· 330
   9.4.1 $\Delta T=0$ 时的边界条件 ············· 331
   9.4.2 $\Delta T \neq 0$ 时的边界条件 ············ 332
  9.5 退化的层合板刚度和CTE值 ············· 333
  9.6 退化的单层板刚度 ··················· 333
  9.7 断裂能 ·························· 334
  9.8 求解算法 ························ 335
   9.8.1 铺层迭代 ······················ 335

  9.8.2 层合板迭代 ································ 336
 习题 ·· 346

第10章 分层 ·· 348

 10.1 内聚力方法 ··· 350
  10.1.1 单模式的内聚力模型 ···························· 352
  10.1.2 混合模式的内聚力模型 ························ 355
 10.2 虚拟裂纹闭合技术 ······································ 367
 习题 ·· 370

附录A 张量代数 ·· 372

 A.1 应力和应变的主方向 ··································· 372
 A.2 张量对称性 ··· 372
 A.3 张量的矩阵表示 ··· 372
 A.4 双点积 ··· 373
 A.5 张量求逆 ·· 374
 A.6 张量导数 ·· 375

附录B 二阶对角损伤模型 ···································· 377

 B.1 有效和损伤空间 ··· 377
 B.2 热力学力 Y ··· 378
 B.3 损伤面 ··· 381
 B.4 不可恢复应变面 ··· 381

附录C 使用的软件 ·· 383

 C.1 Abaqus ·· 383
 C.2 BMI3 ·· 387

参考文献 ·· 389

第1章　正交各向异性材料力学

本章内容主要为后续章节的学习提供必要的理论基础。本章主要介绍基本的力学概念和坐标变换、本构方程等复合材料力学特有的相关理论知识。采用连续介质力学描述正交各向异性材料的变形和应力，其基本方程参见1.2～1.9节。后续章节会用到张量运算，所以其相关知识在1.10节中进行介绍。在层合板、单层板等复合材料层合结构坐标系中，应力、应变和刚度等需要大量的坐标变换操作，坐标变换相关知识在1.10～1.11节中进行介绍。后续章节很大程度上利用了本章知识，所以读者可以根据自己掌握的连续介质力学相关知识的程度选择性地复习本章知识。

1.1　单层板坐标系

纤维增强复合材料的单层板可以看作是一种正交各向异性材料，即这种材料有3个相互垂直的对称面。这3个对称面的交线定义为材料的3个方向轴：纤维方向（x'_1）、厚度方向（x'_3）和与前两个方向均垂直的第三个方向 $x'_2 = x'_3 \times x'_1$ [1]。

1.2　位移

在载荷作用下，物体内每一点均会发生刚体平移、旋转以及变形，并占据新的区域。物体内部任意一点 P 的位移使用 u_i 表示（图1.1），u_i 为由三个分量组成的矢量（在直角笛卡儿坐标系中），表示为 $u_i = (u_1, u_2, u_3)$，或者也可以表示为 $u_i = (u, v, w)$。位移是一个矢量也是一个一阶张量。

$$\boldsymbol{u} = u_i = (u_1, u_2, u_3); i = 1, \cdots, 3 \tag{1.1}$$

这里黑体字（如 $\boldsymbol{u}$）表示一个张量，这也是张量的惯用表示方式。此处位移是一个矢量，而矢量是一阶张量。本书中所有张量均用黑体字表示（如 $\boldsymbol{\sigma}$），但张量的分量不用黑体字表示（如 σ_{ij}）。张量的阶数（一阶、二阶或四阶等）必须从上下文中推断，或者如式（1.1）一样，在指标符号表示法中，通过看同一实体的下标数量（如 u_i）得到。

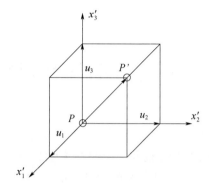

图 1.1 位移分量标记示意图

1.3 应变

对于几何非线性分析,拉格朗日应变张量分量表达形式如下[2]:

$$L_{ij} = \frac{1}{2}(u_{i,j} + u_{j,i} + u_{r,i}u_{r,j}) \qquad (1.2)$$

式中

$$u_{i,j} = \frac{\partial u_i}{\partial x_j} \qquad (1.3)$$

如果位移变化足够小,与 u_i 的线性偏导(一阶)项相比,其两个偏导项的乘积可以忽略不计。那么,在小变形理论中,应变张量 ε_{ij} 可以使用下式表达[2]:

$$\boldsymbol{\varepsilon} = \varepsilon_{ij} = \frac{1}{2}(u_{i,j} + u_{j,i}) \qquad (1.4)$$

再次说明,文中黑体字符表示的是一个张量,其阶数由上下文中推断。例如,ε 表示一个一维标量应变,而 **ε** 表示二阶应变张量。张量大多数情况下均使用指标符号表示法进行表示(如 ε_{ij}),并且变量的张量特征(标量、矢量、二阶张量等)容易由上下文中理解得到。

由式(1.4)可知,应变是一个二阶对称张量($\varepsilon_{ij} = \varepsilon_{ji}$)。展开形式下,应变张量每个分量定义如下:

$$\begin{cases} \varepsilon_{11} = \dfrac{\partial u_1}{\partial x_1} = \epsilon_1; 2\varepsilon_{12} = 2\varepsilon_{21} = \left(\dfrac{\partial u_1}{\partial x_2} + \dfrac{\partial u_2}{\partial x_1}\right) = \gamma_6 = \epsilon_6 \\ \varepsilon_{22} = \dfrac{\partial u_2}{\partial x_2} = \epsilon_2; 2\varepsilon_{13} = 2\varepsilon_{31} = \left(\dfrac{\partial u_1}{\partial x_3} + \dfrac{\partial u_3}{\partial x_1}\right) = \gamma_5 = \epsilon_5 \\ \varepsilon_{33} = \dfrac{\partial u_3}{\partial x_3} = \epsilon_3; 2\varepsilon_{23} = 2\varepsilon_{32} = \left(\dfrac{\partial u_2}{\partial x_3} + \dfrac{\partial u_3}{\partial x_2}\right) = \gamma_4 = \epsilon_4 \end{cases} \qquad (1.5)$$

式中：$\epsilon_\alpha, \alpha = 1, \cdots, 6$ 将在1.5节中定义。正应变分量$(i=j)$代表每单位长度的变化量(图1.2)。剪切应变分量$(i \neq j)$表示相对于初始直角的角度改变量的1/2(图1.3)。通常使用工程剪切应变 $\gamma_\alpha = 2\varepsilon_{ij}, i \neq j$ 替代应变张量中的剪切应变分量，因为在材料力学中剪切模量G由$\tau = G\gamma$定义[3]。由此，二阶应变张量可以使用矩阵的形式表达如下：

$$[\varepsilon] = \begin{bmatrix} \varepsilon_{11} & \varepsilon_{12} & \varepsilon_{13} \\ \varepsilon_{12} & \varepsilon_{22} & \varepsilon_{23} \\ \varepsilon_{13} & \varepsilon_{23} & \varepsilon_{33} \end{bmatrix} = \begin{bmatrix} \epsilon_1 & \epsilon_6/2 & \epsilon_5/2 \\ \epsilon_6/2 & \epsilon_2 & \epsilon_4/2 \\ \epsilon_5/2 & \epsilon_4/2 & \epsilon_3 \end{bmatrix} \quad (1.6)$$

式中：[ ]表示矩阵。

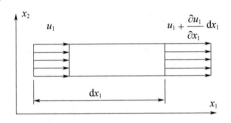

图1.2 正应变

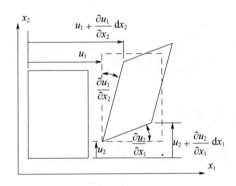

图1.3 工程剪切应变

1.4 应力

平面上一点的应力定义为作用在该平面上通过这一点的单位面积上的载荷。一个二阶的应力张量可以完全描述某一点的应力状态。如图1.4所示，在直角坐标系中，应力张量分量可以在相互垂直的3个平面上表示出来。应力张量的指标符号形式由$\sigma_{ij}(i,j=1,2,3)$表示，第一个下标对应应力所在平面的法向，第二个下标对应应力的方向。当平面法向与应力的方向都是正向或负向时，该点的正应力($i=j$)为正值。图1.4中表示的所有应力分量均为正值。在图1.4中,要保证体积单元的力和力矩平衡,应力张量矩阵必须对称(即$\sigma_{ij}=\sigma_{ji}$)[3]。二阶应力张量的矩阵形式表示如下:

$$[\sigma]=\begin{bmatrix}\sigma_{11} & \sigma_{12} & \sigma_{13}\\ \sigma_{12} & \sigma_{22} & \sigma_{23}\\ \sigma_{13} & \sigma_{23} & \sigma_{33}\end{bmatrix}=\begin{bmatrix}\sigma_1 & \sigma_6 & \sigma_5\\ \sigma_6 & \sigma_2 & \sigma_4\\ \sigma_5 & \sigma_4 & \sigma_3\end{bmatrix} \tag{1.7}$$

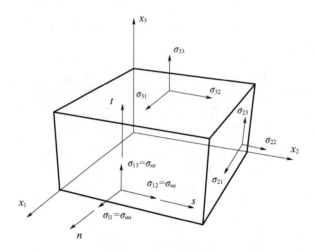

图1.4 应力分量

1.5 指标符号缩写

应力张量矩阵是对称的,因此可以写成Voigt指标符号缩写形式,即

$$\sigma_\alpha=\sigma_{ij}=\sigma_{ji} \tag{1.8}$$

缩写规则如下:

$$\alpha=\begin{cases}i, & i=j\\ 9-i-j, & i\neq j\end{cases} \tag{1.9}$$

这样,缩写后的应力张量分量如式(1.7)所示。

应变张量的缩写规则与应力张量类似,缩写后如式(1.6)所示。注意,应力($\sigma_\alpha,\alpha=1,\cdots,6$)的6个分量可以写成一个列阵并用中括号表示,如式(1.10)所示。但是$\{\sigma\}$不是一个向量,这样做只是为了方便表示一个对称二阶应力张量中的6个单独分量。

不同的有限元软件使用不同的指标缩写规则,如表1.1所列。若要求应力

或应变的标准约定缩写形式变换到 Abaqus 缩写形式,这时就需要一个如下的变换矩阵:

$$\{\sigma_A\}[T]\{\sigma\} \tag{1.10}$$

式中:下标$()_A$代表 Abaqus 软件中的缩写规则表示的量。另外,注意到这里的$\{\}$表示一个拥有6个分量的列阵,$[\ ]$为一个6×6的旋转矩阵,该矩阵表达如下:

$$\begin{bmatrix} 1 & 0 & 0 & 0 & 0 & 0 \\ 0 & 1 & 0 & 0 & 0 & 0 \\ 0 & 0 & 1 & 0 & 0 & 0 \\ 0 & 0 & 0 & 0 & 0 & 1 \\ 0 & 0 & 0 & 0 & 1 & 0 \\ 0 & 0 & 0 & 1 & 0 & 0 \end{bmatrix} \tag{1.11}$$

刚度阵变换如下

$$[C_A] = [T]^T [C][T] \tag{1.12}$$

对于 LS-DYNA 和 ANSYS 软件,变换矩阵为

$$[T] = \begin{bmatrix} 1 & 0 & 0 & 0 & 0 & 0 \\ 0 & 1 & 0 & 0 & 0 & 0 \\ 0 & 0 & 1 & 0 & 0 & 0 \\ 0 & 0 & 0 & 0 & 0 & 1 \\ 0 & 0 & 0 & 1 & 0 & 0 \\ 0 & 0 & 0 & 0 & 1 & 0 \end{bmatrix} \tag{1.13}$$

表 1.1 不同有限元软件中的缩写规则

标准约定	Abaqus/Standard	LS-DYNA 和 Abaqus/Explicit	ANSYS/Mechanical
11→1	11→1	11→1	11→1
22→2	22→2	22→2	22→2
33→3	33→3	33→3	33→3
23→4	12→4	12→4	12→4
13→5	13→5	23→5	23→5
12→6	23→6	13→6	13→6

1.6 平衡方程和虚功原理

物体内每一点的3个平衡方程写成张量形式如下:

$$\sigma_{ij,j} + f_i = 0 \tag{1.14}$$

式中:f_i 为每单位体积的体力;$()_{,j} = \dfrac{\partial}{\partial x_j}$。当体力可以忽略时,在层合板坐标系统 $x-y-z$ 中,平衡方程的展开形式如下:

$$\begin{cases} \dfrac{\partial \sigma_{xx}}{\partial x} + \dfrac{\partial \sigma_{xy}}{\partial y} + \dfrac{\partial \sigma_{xz}}{\partial z} = 0 \\ \dfrac{\partial \sigma_{xy}}{\partial x} + \dfrac{\partial \sigma_{yy}}{\partial y} + \dfrac{\partial \sigma_{yz}}{\partial z} = 0 \\ \dfrac{\partial \sigma_{xz}}{\partial x} + \dfrac{\partial \sigma_{yz}}{\partial y} + \dfrac{\partial \sigma_{zz}}{\partial z} = 0 \end{cases} \tag{1.15}$$

虚功原理(PVW)提供另外一种平衡方程的表达形式[4]。因为虚功原理为积分表达形式,所以比式(1.14)更容易表达为有限元形式。其 PVW 表达式为

$$\int_V \sigma_{ij} \delta\epsilon_{ij} \mathrm{d}V - \int_S t_i \delta u_i \mathrm{d}S - \int_V f_i \delta u_i \mathrm{d}V = 0 \tag{1.16}$$

式中:t_i 为作用在外表面 S 上的单位面积的表面牵引力,式中负号表示该部分功为外力(t_i, f_i)做的功。力和位移遵循相同的符号规定,即指向轴的正向时,其分量为正值。式(1.16)中的第一项是内应力作用的虚功,遵循相同的符号规定时,内应力的虚功为正值。

例 1.1 对于一个细长杆件,顶端吊住垂直放置,在自身重力作用下,求其位移函数 $u(x)$。已知该杆件横截面积为 A,长度为 L,材料弹性模量为 E,密度为 ρ。可以利用原点在杆件顶端并沿着杆件向下指向尾端的坐标 x。

解 假设位移的二次函数表达式为

$$u(x) = C_0 + C_1 x + C_2 x^2$$

在杆件顶端位置利用边界条件(B.C.)容易得到 $C_0 = 0$。由于没有表面牵引力,只有非零应变 ϵ_x,利用胡克定律,虚功原理式(1.16)可以简化为

$$\int_0^L E\epsilon_x \delta\epsilon_x A \mathrm{d}x - \int_0^L \rho g \delta u A \mathrm{d}x = 0$$

由假设的位移函数表达式得到

$$\delta u = x\delta C_1 + x^2 \delta C_2$$

$$\epsilon_x = \dfrac{\mathrm{d}u}{\mathrm{d}x} = C_1 + 2xC_2$$

$$\delta\epsilon_x = \delta C_1 + 2x\delta C_2$$

将其代入后,得

$$EA\int_0^L (C_1 + 2xC_2)(\delta C_1 + 2x\delta C_2)\mathrm{d}x - \rho g A \int_0^L (x\delta C_1 + x^2 \delta C_2)\mathrm{d}x = 0$$

对其进行积分并分别对 δC_1 和 δC_2 合并同类项后,得

$$\left(EC_2L^2 + EC_1L - \frac{\rho g L^2}{2}\right)\delta C_1 + \left(\frac{4}{3}EC_2L^3 + EC_1L^2 - \frac{\rho g L^3}{3}\right)\delta C_2 = 0$$

因为 δC_1 和 δC_2 值的任意性,该方程中 δC_1 和 δC_2 的系数项分别为零,解得

$$C_1 = \frac{L\rho g}{E}; C_2 = -\frac{\rho g}{2E}$$

将 C_1 和 C_2 代回到 $u(x)$,得

$$u(x) = \frac{\rho g}{2E}(2L - x)x$$

与材料力学中得到的精确解一致。

1.7 边界条件

1.7.1 牵引力边界条件

固体力学中问题的求解必须给定边界条件。边界条件可以由位移、应力或两者混合形式指定。对于任意表面上的任一点,其牵引力 T_i 是由 3 个应力分量组成的一个矢量定义。如图 1.4 所示,牵引力矢量由正应力 σ_{nn} 和两个剪切应力 σ_{nt}、σ_{ns} 组成。利用 Cauchy 准则,牵引力矢量可以写成如下形式:

$$T_i = \sigma_{ji}n_j = \sum_{j}^{3} \sigma_{ji}n_j \tag{1.17}$$

式中:n_j 为所考虑的表面上某一点的单位法向量①。对于一个垂直于 x_1 轴的平面来说,$n_i = (1,0,0)$,其牵引力分量为 $T_1 = \sigma_{11}$,$T_2 = \sigma_{12}$,$T_3 = \sigma_{13}$。

1.7.2 自由表面边界条件

自由表面边界条件是指表面上所有牵引应力分量均为 0,即 $T_n = \sigma_{nn} = 0$,$T_t = \sigma_{nt} = 0$,$T_s = \sigma_{ns} = 0$。也指某一特定的分量为 0,而其他分量不为 0。例如纯压强载荷条件下,正应力不为零而剪切应力为 0。

---

① 以式(1.17)为例,介绍爱因斯坦的求和约定。在同一项中,如果同一指标重复出现,表示在其取值范围内求和。此外,每对重复的指标代表一个哑标。也就是说,对于 T_i 来讲,其张量阶数为 1,比运算中所涉及张量的阶数总数少两个。由此产生的张量只保留与哑标无关的自由指标(在该例子中,只剩下 i 了)。

1.8 连续性条件

1.8.1 牵引力的连续性

平衡条件(作用力和反作用力)需要牵引力各分量 T_i 在其通过的任意界面时必须连续。使用数学表达式表示为 $T_i^+ + T_i^- = 0$。根据式(1.17)，$T_i^+ = \sigma_{ji}^+ n_j$。因为 $n_j^+ = -n_j^-$，就得到 $\sigma_{ji}^+ = \sigma_{ji}^-$。就单个应力分量而言，可以得到 $\sigma_{nn}^+ = \sigma_{nn}^-$，$\sigma_{nt}^+ = \sigma_{nt}^-$，$\sigma_{ns}^+ = \sigma_{ns}^-$（图1.5）。由此，作用在界面上的正应力和剪切应力分量在通过界面时一定是连续的。但应力的其他三个分量没有连续性要求，也就是说 $\sigma_{tt}^+ \neq \sigma_{tt}^-$，$\sigma_{ss}^+ \neq \sigma_{ss}^-$，$\sigma_{ts}^+ \neq \sigma_{ts}^-$ 可能成立。由于材料性能的不连续，在通过界面层边界时，两个正应力分量和一个剪切应力分量不连续现象是普遍存在的。

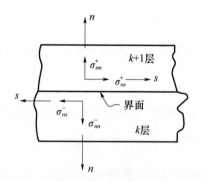

图1.5 通过界面时的牵引力连续性

1.8.2 位移的连续性

在任何理想粘接条件下的连续体中，沿任何界面均必须满足一定的位移连续条件。下面以一个在外压载荷条件下，圆筒结构的屈曲问题为例来具体说明（图1.6）。其 $A'-A$ 剖面两侧的位移必须满足 $u_i^+ = u_i^-$。在理想粘接连续介质中的每一点，位移的连续性条件都必须满足。但是，若材料的不同区域或不同材料相之间发生脱粘、开裂或者滑动等情况时，位移的连续性就不再满足。本例中，其斜率的连续性也必须满足 $\left(\dfrac{\partial w^+}{\partial \theta} = \dfrac{\partial w^-}{\partial \theta}\right)$，这里 w 为径向位移。

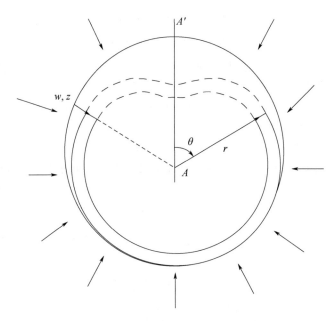

图 1.6 封闭圆管在外压载荷下的失稳模式

1.9 相容性

应变-位移方程组(式(1.5))为 3 个未知位移 u_i 提供了 6 个等式。因此,应变 ε_{ij} 必须满足一定条件,否则对式(1.5)进行积分得不到位移唯一解。若随意给 ε_{ij} 赋值会导致材料中出现不连续,包括出现间隙和(或)重叠等区域。

位移唯一解的必要条件就是相容性条件。虽然式(1.5)中的 6 个等式可用[2],但这里却没有使用它们,因为在本书中使用的位移法不需要它们。也就是说,在求解问题时,位移 u_i 的形式总是被预先假设。然后通过式(1.5)计算得到应变,通过式(1.46)计算得到应力,最后利用 PVW 式(1.16)进行平衡求解。

1.10 坐标变换

一点 P 在主坐标系中的坐标可以通过其在副坐标系中的坐标转化获得。从图 1.7 中可以看出,点 P 的坐标为

$$\begin{cases} x_1' = x_1\cos\theta + x_2\sin\theta \\ x_2' = -x_1\sin\theta + x_2\cos\theta \\ x_3' = x_3 \end{cases} \quad (1.18)$$

或者表示为

$$x_i' = a_{ij}x_j \quad (1.19)$$

又或者表示成矩阵形式

$$\{x'\} = [a]\{x\} \quad (1.20)$$

式中：a_{ij} 为从副坐标系 e_j 转化为主坐标系 e_i' 时产生的方向余弦矩阵分量，按行列形式写为[2]

$$a_{ij} = \cos(e_i', e_j) = \begin{array}{c} \\ e_1' \\ e_2' \\ e_3' \end{array} \begin{array}{ccc} e_1 & e_2 & e_3 \\ a_{11} & a_{12} & a_{13} \\ a_{21} & a_{22} & a_{23} \\ a_{31} & a_{32} & a_{33} \end{array} \quad (1.21)$$

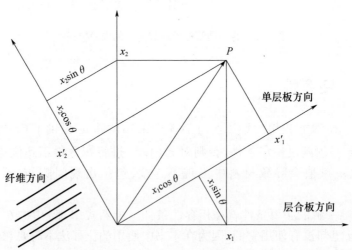

图 1.7 坐标转化

如主坐标系指定为单层板的坐标系，副坐标系指定为层合板的坐标系，那么式(1.19)为层合板坐标系向单层板坐标系转换的表达式。若转换方向相反，一般使用转置矩阵，表达形式如下

$$\{x\} = [a]^T\{x'\} \quad (1.22)$$

例 1.2 对于纤维方向为 $\theta = 30°$ 的复合材料单层板，通过计算单层板的方向余弦（计算层合板坐标系(x_j)向单层板坐标系(x_i')转化时的方向余弦矩阵分量）得到矩阵 $[a]$。

解 从图 1.7 和式(1.19),有

$$\begin{cases} a_{11} = \cos\theta = \dfrac{\sqrt{3}}{2} \\ a_{12} = \sin\theta = \dfrac{1}{2} \\ a_{13} = 0 \\ a_{21} = -\sin\theta = -\dfrac{1}{2} \\ a_{22} = \cos\theta = \dfrac{\sqrt{3}}{2} \\ a_{23} = 0 \\ a_{31} = 0 \\ a_{32} = 0 \\ a_{33} = 1 \end{cases}$$

例 1.3 对于一个纤维增强的复合材料管件,其纤维缠绕方向为环向(1-方向)。该坐标系统中的刚度值(E_1, E_2 等)已给定。但是,当利用通用平面应变单元(Abaqus 中的 CAX4 单元)分析材料横截面性能时,模型一般建立在 X, Y, Z 坐标系统中。因此有必要把刚度值转化为新坐标系下的 E_x, E_y 等。为此创建转换矩阵 $[a]^T$,从单层板坐标系(1-2-3)转换为如图 1.8 所示的 $(X-Y-Z)$ 结构坐标系。

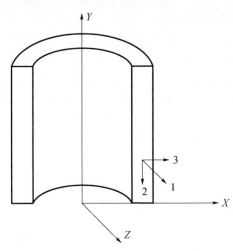

图 1.8 对于轴对称问题分析的坐标转换

解 首先利用式(1.21)按照行的顺序计算建立矩阵 $[a]$。第 i-行包括($i = 1, \cdots, 3$)沿着 $(X-Y-Z)$ 坐标的所有分量。

[a]	X	Y	Z
1	0	0	1
2	0	-1	0
3	1	0	0

本例题中所求的转换矩阵为上述矩阵的转置。

1.10.1 应力变换

二阶张量 σ_{pq} 可以看作为两个矢量 $\boldsymbol{V}_p$ 和 $\boldsymbol{V}_q$ 的外积[①]，即

$$\boldsymbol{\sigma}_{pq} = \boldsymbol{V}_p \otimes \boldsymbol{V}_q \tag{1.23}$$

式中每一项按照式(1.19)转换后如下：

$$\sigma'_{ij} = a_{ip}\boldsymbol{V}_p \otimes a_{jq}\boldsymbol{V}_q \tag{1.24}$$

因此，有

$$\sigma'_{ij} = a_{ip}a_{jq}\sigma_{pq} \tag{1.25}$$

或者使用矩阵表达如下

$$\{\sigma'\} = [a]\{\sigma\}[a]^{\mathrm{T}} \tag{1.26}$$

例如，指标缩写形式下展开 σ'_{11} 为

$$\sigma'_1 = a_{11}^2\sigma_1 + a_{12}^2\sigma_2 + a_{13}^2\sigma_3 + 2a_{11}a_{12}\sigma_6 + 2a_{11}a_{13}\sigma_5 + 2a_{12}a_{13}\sigma_4 \tag{1.27}$$

指标缩写形式下展开 σ'_{12} 为

$$\sigma'_6 = a_{11}a_{21}\sigma_1 + a_{12}a_{22}\sigma_2 + a_{13}a_{23}\sigma_3 + (a_{11}a_{22} + a_{12}a_{21})\sigma_6 \\ + (a_{11}a_{23} + a_{13}a_{21})\sigma_5 + (a_{12}a_{23} + a_{13}a_{22})\sigma_4 \tag{1.28}$$

接下来的计算是为了得到 6×6 坐标转换矩阵 $[T]$，这样式(1.25)可以重新写成如下指标缩写形式：

$$\sigma'_\alpha = T_{\alpha\beta}\sigma_\beta \tag{1.29}$$

如果 $\alpha \leq 3, \beta \leq 3$ 则 $i=j, p=q$，那么，有

$$T_{\alpha\beta} = a_{ip}a_{ip} = a_{ip}^2 (\text{不对 } i,p \text{ 求和}) \tag{1.30}$$

如果 $\alpha \leq 3, \beta > 3$，则 $i=j$，但 $p \neq q$，并且考虑到按照式(1.9)，以 q 交换 p 得到相同的 $\beta = 9-p-q$ 值，得

$$T_{\alpha\beta} = a_{ip}a_{iq} + a_{iq}a_{ip} = 2a_{ip}a_{iq} (\text{不对 } i,p \text{ 求和}) \tag{1.31}$$

若 $\alpha > 3$，则 $i \neq j$，但是由于 σ_{ij} 和 σ_{ji} 数值相等，我们只在式中出现 σ_{ij}，不出现 σ_{ji}。事实上 $i \neq j$ 时，$\sigma_\alpha = \sigma_{ij} = \sigma_{ji}, \alpha = 9-i-j$。另外，若 $\beta \leq 3$ 则 $p=q$，得

[①] 外积保留了所有实体相关的指标，因此产生了一个阶数等于所涉及的实体阶数总和的张量。

$$T_{\alpha\beta} = a_{ip}a_{jq} \text{(不对 } i,p \text{ 求和)} \tag{1.32}$$

当 $\alpha>3$ 且 $\beta>3$, $i\neq j$ 且 $p\neq q$ 时,有

$$T_{\alpha\beta} = a_{ip}a_{jq} + a_{iq}a_{jp} \tag{1.33}$$

由此完成了 $T_{\alpha\beta}$ 的推导。展开式(1.30)~式(1.33)并利用式(1.21),得

$$[T] = \begin{bmatrix} a_{11}^2 & a_{12}^2 & a_{13}^2 & 2a_{12}a_{13} & 2a_{11}a_{13} & 2a_{11}a_{12} \\ a_{21}^2 & a_{22}^2 & a_{23}^2 & 2a_{22}a_{23} & 2a_{21}a_{23} & 2a_{21}a_{22} \\ a_{31}^2 & a_{32}^2 & a_{33}^2 & 2a_{32}a_{33} & 2a_{31}a_{33} & 2a_{31}a_{32} \\ a_{21}a_{31} & a_{22}a_{32} & a_{23}a_{33} & a_{22}a_{33}+a_{23}a_{32} & a_{21}a_{33}+a_{23}a_{31} & a_{21}a_{32}+a_{22}a_{31} \\ a_{11}a_{31} & a_{12}a_{32} & a_{13}a_{33} & a_{12}a_{33}+a_{13}a_{32} & a_{11}a_{33}+a_{13}a_{31} & a_{11}a_{32}+a_{12}a_{31} \\ a_{11}a_{21} & a_{12}a_{22} & a_{13}a_{23} & a_{12}a_{23}+a_{13}a_{22} & a_{11}a_{23}+a_{13}a_{21} & a_{11}a_{22}+a_{12}a_{21} \end{bmatrix}$$

$$\tag{1.34}$$

下面是一个用来生成式(1.34)的 MATLAB 程序(也可以在文献[5]中获取)。

```
% Derivation of the transformation matrix[T]
clear all;
syms T alpha R
syms a a11 a12 a13 a21 a22 a23 a31 a32 a33
a = [a11,a12,a13;
     a21,a22,a23;
     a31,a32,a33];
T(1:6,1:6) = 0;
for i = 1:1:3
for j = 1:1:3
 if i = = j; alpha = j; else alpha = 9-i-j; end
 for p = 1:1:3
 for q = 1:1:3
  if p = = q beta = p; else beta = 9-p-q; end
  T(alpha,beta) = 0;
  if alpha < = 3 & beta < = 3; T(alpha,beta) = a(i,p)* a(i,p); end
  if alpha > 3 & beta < = 3; T(alpha,beta) = a(i,p)* a(j,p); end
  if alpha < = 3 & beta > 3; T(alpha,beta) = a(i,q)* a(i,p) + a(i,p)* a(i,q);end
  if alpha > 3 & beta > 3; T(alpha,beta) = a(i,p)* a(j,q) + a(i,q)* a(j,p);end
 end
 end
end
end
```

T
R = eye(6,6); R(4,4) = 2; R(5,5) = 2; R(6,6) = 2; % Reuter matrix
Tbar = R* T* R^(-1)

1.10.2 应变变换

应变张量分量 ε_{ij} 的转换与应力张量分量的转换方式相同,如下:

$$\varepsilon'_{ij} = a_{ip}a_{jq}\varepsilon_{pq} \tag{1.35}$$

或者表达如下:

$$\varepsilon'_{\alpha} = T_{\alpha\beta}\varepsilon_{\beta} \tag{1.36}$$

其中:$T_{\alpha\beta}$ 通过式(1.34)得到。不同的是,3 个剪切应变 $\varepsilon_{xz},\varepsilon_{yz},\varepsilon_{xy}$ 通常使用工程剪切应变 $\gamma_{xz},\gamma_{yz},\gamma_{xy}$ 替代。工程应变(ϵ 替代 ε)在式(1.5)中已定义,它们可以从下面关系中计算得到

$$\epsilon_{\delta} = R_{\delta\gamma}\varepsilon_{\gamma} \tag{1.37}$$

Reuter 矩阵由下式给出

$$[R] = \begin{bmatrix} 1 & 0 & 0 & 0 & 0 & 0 \\ 0 & 1 & 0 & 0 & 0 & 0 \\ 0 & 0 & 1 & 0 & 0 & 0 \\ 0 & 0 & 0 & 2 & 0 & 0 \\ 0 & 0 & 0 & 0 & 2 & 0 \\ 0 & 0 & 0 & 0 & 0 & 2 \end{bmatrix} \tag{1.38}$$

那么,由式(1.36)和式(1.37),就得到了工程应变的坐标变换公式,即

$$\epsilon'_{\alpha} = \overline{T}_{\alpha\beta}\epsilon_{\beta} \tag{1.39}$$

式中

$$[\overline{T}] = [R][T][R]^{-1} \tag{1.40}$$

$[\overline{T}]$ 仅用于工程应变变换。

展开 $[\overline{T}]$ 得到

$$[\overline{T}] = \begin{bmatrix} a_{11}^2 & a_{12}^2 & a_{13}^2 & a_{12}a_{13} & a_{11}a_{13} & a_{11}a_{12} \\ a_{21}^2 & a_{22}^2 & a_{23}^2 & a_{22}a_{23} & a_{21}a_{23} & a_{21}a_{22} \\ a_{31}^2 & a_{32}^2 & a_{33}^2 & a_{32}a_{33} & a_{31}a_{33} & a_{31}a_{32} \\ 2a_{21}a_{31} & 2a_{22}a_{32} & 2a_{23}a_{33} & a_{22}a_{33} + a_{23}a_{32} & a_{21}a_{33} + a_{23}a_{31} & a_{21}a_{32} + a_{22}a_{31} \\ 2a_{11}a_{31} & 2a_{12}a_{32} & 2a_{13}a_{33} & a_{12}a_{33} + a_{13}a_{32} & a_{11}a_{33} + a_{13}a_{31} & a_{11}a_{32} + a_{12}a_{31} \\ 2a_{11}a_{21} & 2a_{12}a_{22} & 2a_{13}a_{23} & a_{12}a_{23} + a_{13}a_{22} & a_{11}a_{23} + a_{13}a_{21} & a_{11}a_{22} + a_{12}a_{21} \end{bmatrix}$$

$$\tag{1.41}$$

1.11 本构方程变换

使用应变张量($\boldsymbol{\varepsilon}$,不是 $\boldsymbol{\epsilon}$)定义应力 $\boldsymbol{\sigma}$ – 应变 $\boldsymbol{\varepsilon}$ 关系的本构方程为

$$\begin{cases} \boldsymbol{\sigma}' = \boldsymbol{C}' : \boldsymbol{\varepsilon}' \\ \sigma'_{ij} = C'_{ijkl}\varepsilon'_{kl} \end{cases} \quad (1.42)$$

上式中使用了张量的实体和指标符号两种表示法①。

为了简化,考虑一种正交各向异性材料(1.12.3 节)。那么 σ'_{11} 和 σ'_{22} 可表示为

$$\begin{cases} \sigma'_{11} = C'_{1111}\varepsilon'_{11} + C'_{1122}\varepsilon'_{22} + C'_{1133}\varepsilon'_{33} \\ \sigma'_{12} = C'_{1212}\varepsilon'_{12} + C'_{1221}\varepsilon'_{21} = 2C'_{1212}\varepsilon'_{12} \end{cases} \quad (1.43)$$

使用指标缩写形式重写式(1.43),可以明显观察到在指标缩写形式下所有剪切应变项均出现两次,有

$$\begin{cases} \sigma'_1 = C'_{11}\varepsilon'_1 + C'_{12}\varepsilon'_2 + C'_{13}\varepsilon'_3 \\ \sigma'_6 = 2C'_{66}\varepsilon'_6 \end{cases} \quad (1.44)$$

剪切应变前面的系数 2 基于两个事实:①张量 $\boldsymbol{C}$ 和 $\boldsymbol{\varepsilon}$ 的次对称性(the minor symmetry)(见式(1.5)、式(1.55)~式(1.56));②式(1.43)中刚度矩阵 C_{ijkl} 最后两个指标和应变 ε_{kl} 指标的缩并。因此,当张量书写成指标缩写形式时,对于任何具有次对称特征的张量进行双点积运算时均需要使用 Reuter 矩阵(式(1.38))进行修正。接下来,式(1.42)可以写为

$$\sigma'_\alpha = C'_{\alpha\beta} R_{\beta\delta} \varepsilon'_\delta \quad (1.45)$$

注意,利用式(1.37),式(1.45)中的 Reuter 矩阵与应变张量合并,得到由工程应变组成的下式:

$$\sigma'_\alpha = C'_{\alpha\beta} \epsilon'_\beta \quad (1.46)$$

为了得到层合板坐标系下的刚度矩阵 $[C]$,把式(1.29)和式(1.39)代入式(1.46),得

$$T_{\alpha\delta}\sigma_\delta = C'_{\alpha\beta}\overline{T}_{\beta\gamma}\epsilon_\gamma \quad (1.47)$$

由于

$$[T]^{-1} = [\overline{T}]^\mathrm{T} \quad (1.48)$$

因此

$$\{\sigma\} = [C]\{\epsilon\} \quad (1.49)$$

---

① 双点积运算并涉及两个指标的缩并,如式中的下标 k 和 l。另外还要注意,使用黑体字符为张量的实体表示法。

式中

$$[C] = [\bar{T}]^{\mathrm{T}}[C'][\bar{T}] \tag{1.50}$$

$$[C'] = [\bar{T}]^{-\mathrm{T}}[C][\bar{T}]^{-1} = [T][C][T]^{\mathrm{T}} \tag{1.51}$$

柔度矩阵是刚度矩阵的逆矩阵,不是四阶张量 C_{ijkl} 的逆。因此,有

$$[S'] = [C']^{-1} \tag{1.52}$$

考虑到式(1.48)和式(1.50),柔度矩阵转化为

$$[S] = [T]^{\mathrm{T}}[S'][T] \tag{1.53}$$

$$[S'] = [T]^{-\mathrm{T}}[S][T]^{-1} = [\bar{T}][S][\bar{T}]^{\mathrm{T}} \tag{1.54}$$

1.12　3D 本构方程

在三维(3D)问题中,Hooke 定律表达形式见式(1.42)。3D 刚度张量 C_{ijkl} 是一个四阶张量,拥有 81 个分量。对于各向异性材料,只有 21 个独立分量。也就是说,剩下的 60 个分量可以使用这 21 个分量表示。一维情况(1D)下研究材料力学行为也包含在内,其只有 σ_{11} 不为 0,其他项均为 0。而且只有在 1D 情况下,$\sigma_{11} = \sigma, \varepsilon_{11} = \epsilon, C_{1111} = E, \sigma = E\epsilon$。本章节中所有推导均在单层板坐标系中进行,但为了简化起见,仅在本章节中省略(′)记号。

在式(1.42)中,把哑标 i 和 j,k 和 l 相互交换,得

$$\sigma_{ji} = C_{jilk}\varepsilon_{lk} \tag{1.55}$$

因为应力和应变张量是对称的,即 $\sigma_{ij} = \sigma_{ji}, \varepsilon_{kl} = \varepsilon_{lk}$,那么

$$C_{ijkl} = C_{jikl} = C_{ijlk} = C_{jilk} \tag{1.56}$$

由此独立分量的个数由 81 个有效降低为 36 个,例如 $C_{1213} = C_{2131}$ 等。那么 36 个独立分量可以写成一个 6×6 矩阵。

此外,对于无耗散能的弹性材料,受载时储存弹性能,卸载时弹性能完全释放。因此,在应力-应变曲线上的任何点的弹性能与到达该点的路径无关,这个与路径无关的函数称为势函数。这种情况下,这个势函数就是应变能密度函数 $\tilde{u}(\varepsilon_{ij})$。使用泰勒幂级数对其展开,得

$$\tilde{u} = \tilde{u}_0 + \frac{\partial \tilde{u}}{\partial \varepsilon_{ij}}\bigg|_0 \varepsilon_{ij} + \frac{1}{2}\frac{\partial^2 \tilde{u}}{\partial \varepsilon_{ij}\partial \varepsilon_{kl}}\bigg|_0 \varepsilon_{ij}\varepsilon_{kl} + \cdots \tag{1.57}$$

对 ε_{ij} 求偏导,得

$$\frac{\partial \tilde{u}}{\partial \varepsilon_{ij}} = 0 + \beta_{ij} + \frac{1}{2}(\alpha_{ijkl}\varepsilon_{kl} + \alpha_{klij}\varepsilon_{ij}) \tag{1.58}$$

这里 β_{ij} 和 α_{ijkl} 是常数。那么,由此得到

$$\sigma_{ij} - \sigma_{ij}^0 = C_{ijkl}\varepsilon_{kl} \tag{1.59}$$

式中:$\sigma_{ij}^0 = \beta_{ij}$为残余应力;$\alpha_{ijkl} = 1/2(C_{ijkl} + C_{klij}) = C_{ijkl}$,为对称刚度张量(见式(1.56))。式(1.59)为式(1.55)包含残余应力项的一般形式。

指标缩写形式下,广义 Hooke 定律变为

$$\begin{Bmatrix} \sigma_1 \\ \sigma_2 \\ \sigma_3 \\ \sigma_4 \\ \sigma_5 \\ \sigma_6 \end{Bmatrix} = \begin{bmatrix} C_{11} & C_{12} & C_{13} & C_{14} & C_{15} & C_{16} \\ C_{12} & C_{22} & C_{23} & C_{24} & C_{25} & C_{26} \\ C_{13} & C_{23} & C_{33} & C_{34} & C_{35} & C_{36} \\ C_{14} & C_{24} & C_{34} & C_{44} & C_{45} & C_{46} \\ C_{15} & C_{25} & C_{35} & C_{45} & C_{55} & C_{56} \\ C_{16} & C_{26} & C_{36} & C_{46} & C_{56} & C_{66} \end{bmatrix} \begin{Bmatrix} \epsilon_1 \\ \epsilon_2 \\ \epsilon_3 \\ \gamma_4 \\ \gamma_5 \\ \gamma_6 \end{Bmatrix} \quad (1.60)$$

再次说明,当 $\sigma_\alpha = 0(\alpha \neq 1)$ 时,退化为一维情况。此时 $\sigma_1 = \sigma, \varepsilon_1 = \epsilon, C_{11} = E$。

1.12.1 各向异性材料

式(1.60)代表了一个完全各向异性材料的应力 - 应变关系。这种材料的性能与方向有关。例如,即使载荷在各个方向上相同,图1.9所示的各向异性材料体在 P、T、Q 方向上的变形也不同。描述各向异性材料的独立常数为21个。

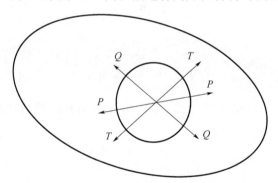

图1.9 各向异性材料

刚度矩阵的逆矩阵为柔度阵,$[S] = [C]^{-1}$。本构方程(3D Hooke 定律)可写成柔度阵形式,即

$$\begin{Bmatrix} \epsilon_1 \\ \epsilon_2 \\ \epsilon_3 \\ \gamma_4 \\ \gamma_5 \\ \gamma_6 \end{Bmatrix} = \begin{bmatrix} S_{11} & S_{12} & S_{13} & S_{14} & S_{15} & S_{16} \\ S_{12} & S_{22} & S_{23} & S_{24} & S_{25} & S_{26} \\ S_{13} & S_{23} & S_{33} & S_{34} & S_{35} & S_{36} \\ S_{14} & S_{24} & S_{34} & S_{44} & S_{45} & S_{46} \\ S_{15} & S_{25} & S_{35} & S_{45} & S_{55} & S_{56} \\ S_{16} & S_{26} & S_{36} & S_{46} & S_{56} & S_{66} \end{bmatrix} \begin{Bmatrix} \sigma_1 \\ \sigma_2 \\ \sigma_3 \\ \sigma_4 \\ \sigma_5 \\ \sigma_6 \end{Bmatrix} \quad (1.61)$$

柔度阵$[S]$也是对称矩阵,拥有 21 个独立常数。对于一维情况,如果 $p \neq 1$,则 $\sigma = 0$。那么 $\sigma_1 = \sigma, \epsilon_1 = \epsilon, S_{11} = 1/E$。

1.12.2 单对称材料

若材料只有一个对称面,如图 1.10 所示,这种材料称为单对称材料。单对称材料有 13 个独立常数。单对称材料意味着关于对称面对称的点的性能一致(如图 1.10 中的点 Z 和 $-Z$)。

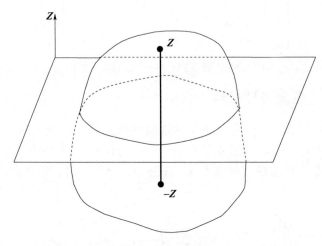

图 1.10 单对称材料

假设材料关于 1-2 平面对称,与 1-2 平面对称位置的性能一致。在这种情况下,式(1.21)中 a-矩阵为

$$\begin{array}{cccc} & x_1 & x_2 & x_3 \\ e_1'' & 1 & 0 & 0 \\ e_2'' & 0 & 1 & 0 \\ e_3'' & 0 & 0 & 1 \end{array} \qquad (1.62)$$

式中"符号用来避免与单层板坐标系统混淆。本章节中不使用( )′表达单层板坐标系统,但其他章节使用( )′表示单层板坐标系统。从式(1.40)中,得

$$[\overline{T}] = \begin{bmatrix} 1 & 0 & 0 & 0 & 0 & 0 \\ 0 & 1 & 0 & 0 & 0 & 0 \\ 0 & 0 & 1 & 0 & 0 & 0 \\ 0 & 0 & 0 & -1 & 0 & 0 \\ 0 & 0 & 0 & 0 & -1 & 0 \\ 0 & 0 & 0 & 0 & 0 & 1 \end{bmatrix} \qquad (1.63)$$

[$\overline{T}$]的作用是将[C]矩阵的第4、5行和第4、5列分量乘以 -1。C_{44} 和 C_{55} 保持正值是因为它们被乘了两次。因此,$C''_{i4} = -C_{i4}(i \neq 4,5)$,$C''_{i5} = -C_{i5}(i \neq 4,5)$。除此之外,其他项无任何变化。由于在单对称材料中的材料性能关于平面对称,所以必然满足:$C_{4i} = C_{i4} = 0, i \neq 4,5$;$C_{5i} = C_{i5} = 0, i \neq 4,5$。由此 3D Hooke 定律表达式可以退化为

$$\begin{Bmatrix} \sigma_1 \\ \sigma_2 \\ \sigma_3 \\ \sigma_4 \\ \sigma_5 \\ \sigma_6 \end{Bmatrix} = \begin{bmatrix} C_{11} & C_{12} & C_{13} & 0 & 0 & C_{16} \\ C_{12} & C_{22} & C_{23} & 0 & 0 & C_{26} \\ C_{13} & C_{23} & C_{33} & 0 & 0 & C_{36} \\ 0 & 0 & 0 & C_{44} & C_{45} & 0 \\ 0 & 0 & 0 & C_{45} & C_{55} & 0 \\ C_{16} & C_{26} & C_{36} & 0 & 0 & C_{66} \end{bmatrix} \begin{Bmatrix} \epsilon_1 \\ \epsilon_2 \\ \epsilon_3 \\ \gamma_4 \\ \gamma_5 \\ \gamma_6 \end{Bmatrix} \quad (1.64)$$

对应的柔度矩阵为

$$\begin{Bmatrix} \epsilon_1 \\ \epsilon_2 \\ \epsilon_3 \\ \gamma_4 \\ \gamma_5 \\ \gamma_6 \end{Bmatrix} = \begin{bmatrix} S_{11} & S_{12} & S_{13} & 0 & 0 & S_{16} \\ S_{12} & S_{22} & S_{23} & 0 & 0 & S_{26} \\ S_{13} & S_{23} & S_{33} & 0 & 0 & S_{36} \\ 0 & 0 & 0 & S_{44} & S_{45} & 0 \\ 0 & 0 & 0 & S_{45} & S_{55} & 0 \\ S_{16} & S_{26} & S_{36} & 0 & 0 & S_{66} \end{bmatrix} \begin{Bmatrix} \sigma_1 \\ \sigma_2 \\ \sigma_3 \\ \sigma_4 \\ \sigma_5 \\ \sigma_6 \end{Bmatrix} \quad (1.65)$$

1.12.3 正交各向异性材料

正交各向异性材料有 3 个和坐标平面一致的对称面。可以证明,如果材料存在两个正交的性能对称,则一定存在第三个正交的性能对称面。这种材料只有 9 个独立常数。

对称面可以是笛卡儿坐标平面,如图 1.11 所示,也可以对应其他任何坐标系统(如柱坐标系、球坐标系等)。例如,树的树干是关于柱坐标系的正交各向

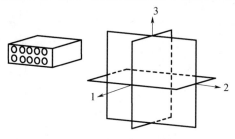

图 1.11 正交各向异性材料

异性材料,因为树干有年轮特性。但是大部分材料还是表现为笛卡儿坐标系统下的正交各向异性特性。单向纤维增强复合材料可以考虑为正交各向异性材料,其中一个对称面与纤维方向垂直,另一个对称面与纤维方向平行,第三个对称面与前两个对称面相互垂直。

在1.12.2节中讨论的材料关于1-2平面对称,若1-3平面为第二个材料对称面,这种情况下,a-矩阵为

$$[a] = \begin{bmatrix} 1 & 0 & 0 \\ 0 & -1 & 0 \\ 0 & 0 & 1 \end{bmatrix} \tag{1.66}$$

式(1.40)中$\overline{T}$-矩阵变为

$$[\overline{T}] = \begin{bmatrix} 1 & 0 & 0 & 0 & 0 & 0 \\ 0 & 1 & 0 & 0 & 0 & 0 \\ 0 & 0 & 1 & 0 & 0 & 0 \\ 0 & 0 & 0 & -1 & 0 & 0 \\ 0 & 0 & 0 & 0 & 1 & 0 \\ 0 & 0 & 0 & 0 & 0 & -1 \end{bmatrix} \tag{1.67}$$

该矩阵使得:$C_{i6} = -C_{i6}(i \neq 4,6), C_{i4} = -C_{i4}(i \neq 4,6)$。因为材料关于1-3平面对称,这就意味着$C_{i6} = C_{6i} = 0, i \neq 6$。这种情况下,3D胡克定律退化为

$$\begin{Bmatrix} \sigma_1 \\ \sigma_2 \\ \sigma_3 \\ \sigma_4 \\ \sigma_5 \\ \sigma_6 \end{Bmatrix} = \begin{bmatrix} C_{11} & C_{12} & C_{13} & 0 & 0 & 0 \\ C_{12} & C_{22} & C_{23} & 0 & 0 & 0 \\ C_{13} & C_{23} & C_{33} & 0 & 0 & 0 \\ 0 & 0 & 0 & C_{44} & 0 & 0 \\ 0 & 0 & 0 & 0 & C_{55} & 0 \\ 0 & 0 & 0 & 0 & 0 & C_{66} \end{bmatrix} \begin{Bmatrix} \epsilon_1 \\ \epsilon_2 \\ \epsilon_3 \\ \gamma_4 \\ \gamma_5 \\ \gamma_6 \end{Bmatrix} \tag{1.68}$$

使用柔度矩阵表达如下

$$\begin{Bmatrix} \epsilon_1 \\ \epsilon_2 \\ \epsilon_3 \\ \gamma_4 \\ \gamma_5 \\ \gamma_6 \end{Bmatrix} = \begin{bmatrix} S_{11} & S_{12} & S_{13} & 0 & 0 & 0 \\ S_{12} & S_{22} & S_{23} & 0 & 0 & 0 \\ S_{13} & S_{23} & S_{33} & 0 & 0 & 0 \\ 0 & 0 & 0 & S_{44} & 0 & 0 \\ 0 & 0 & 0 & 0 & S_{55} & 0 \\ 0 & 0 & 0 & 0 & 0 & S_{66} \end{bmatrix} \begin{Bmatrix} \sigma_1 \\ \sigma_2 \\ \sigma_3 \\ \sigma_4 \\ \sigma_5 \\ \sigma_6 \end{Bmatrix} \tag{1.69}$$

注意到,如果材料有两个对称平面,那么它自动会有第三个对称面。该例子中若再增加2-3平面作为对称面进行分析的话得到的结果(式(1.68)、

式(1.69))一致。

1.12.4 横观各向同性材料

横观各向同性材料有一个对称轴。例如,单向纤维增强复合材料的纤维任意分布在横截面中,其纤维方向可以作为一个对称轴(图 1.12)。这种情况下,任何包含纤维方向的平面均为该材料的对称面。横观各向同性材料有 5 个独立常数。若纤维方向(1 - 方向)作为对称轴,3D Hooke 定律退化为

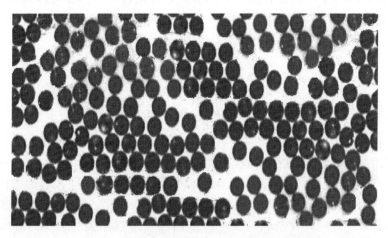

图 1.12 随机分布的 E - 玻璃纤维,放大 200 倍

$$\begin{Bmatrix} \sigma_1 \\ \sigma_2 \\ \sigma_3 \\ \sigma_4 \\ \sigma_5 \\ \sigma_6 \end{Bmatrix} = \begin{bmatrix} C_{11} & C_{12} & C_{13} & 0 & 0 & 0 \\ C_{12} & C_{22} & C_{23} & 0 & 0 & 0 \\ C_{13} & C_{23} & C_{33} & 0 & 0 & 0 \\ 0 & 0 & 0 & (C_{22}-C_{23})/2 & 0 & 0 \\ 0 & 0 & 0 & 0 & C_{66} & 0 \\ 0 & 0 & 0 & 0 & 0 & C_{66} \end{bmatrix} \begin{Bmatrix} \epsilon_1 \\ \epsilon_2 \\ \epsilon_3 \\ \gamma_4 \\ \gamma_5 \\ \gamma_6 \end{Bmatrix} \quad (1.70)$$

使用柔度矩阵为

$$\begin{Bmatrix} \epsilon_1 \\ \epsilon_2 \\ \epsilon_3 \\ \gamma_4 \\ \gamma_5 \\ \gamma_6 \end{Bmatrix} = \begin{bmatrix} S_{11} & S_{12} & S_{13} & 0 & 0 & 0 \\ S_{12} & S_{22} & S_{23} & 0 & 0 & 0 \\ S_{13} & S_{23} & S_{33} & 0 & 0 & 0 \\ 0 & 0 & 0 & 2(S_{22}-S_{23}) & 0 & 0 \\ 0 & 0 & 0 & 0 & S_{66} & 0 \\ 0 & 0 & 0 & 0 & 0 & S_{66} \end{bmatrix} \begin{Bmatrix} \sigma_1 \\ \sigma_2 \\ \sigma_3 \\ \sigma_4 \\ \sigma_5 \\ \sigma_6 \end{Bmatrix} \quad (1.71)$$

注意,如果对称轴不是1-方向上述等式将会不同。考虑到2-方向和3-方向互换性,以下工程常数(1.13节)之间的关系式适用于横观各向同性材料:

$$\begin{cases} E_2 = E_3 \\ \nu_{12} = \nu_{13} \\ G_{12} = G_{13} \end{cases} \tag{1.72}$$

另外,在2-3平面上任何两个相互垂直方向均可作为轴,换句话说,2-3平面内是各向同性的。因此,式(1.73)在2-3平面内成立是仅对各向同性材料而言的(见习题1.14)。

$$G_{23} = \frac{E_2}{2(1+\nu_{23})} \tag{1.73}$$

1.12.5 各向同性材料

工业应用中最普通的材料为各向同性材料,像铝合金、钢等。各向同性材料有无数的性能对称面,性能参数对于方向来讲是独立的。对于弹性材料,只需要两个独立常数。这两个独立常数可以是弹性模量 E 和泊松比 ν,当然也可以为了表达方便选取其他一对独立常数。但是,选取的任何一对独立常数都必须与各向同性材料相关的任一对独立常数相关。例如,可以采用 E 和 G 表达各向同性材料,这里各向同性材料的剪切模量 G 与 E 和 ν 的关系如下

$$G = \frac{E}{2(1+\nu)} \tag{1.74}$$

另外,为了方便,有时候也采用拉梅常数描述各向同性材料,这时两个常数分别为

$$\lambda = \frac{\nu E}{(1+\nu)(1-2\nu)} \tag{1.75}$$
$$\mu = G$$

为了形成另一对,可以将上述任何性能通过体积模量 k 来替代,k 表达如下

$$k = \frac{E}{3(1-2\nu)} \tag{1.76}$$

其将流体静压 p 和体积应变联系起来,有

$$p = k(\epsilon_1 + \epsilon_2 + \epsilon_3) \tag{1.77}$$

对于各向同性材料,3D Hooke定律可以写成只有两个常数 C_{11} 和 C_{12} 的形式,即

$$\begin{Bmatrix} \sigma_1 \\ \sigma_2 \\ \sigma_3 \\ \sigma_4 \\ \sigma_5 \\ \sigma_6 \end{Bmatrix} = \begin{bmatrix} C_{11} & C_{12} & C_{12} & 0 & 0 & 0 \\ C_{12} & C_{11} & C_{12} & 0 & 0 & 0 \\ C_{12} & C_{12} & C_{11} & 0 & 0 & 0 \\ 0 & 0 & 0 & (C_{11}-C_{12})/2 & 0 & 0 \\ 0 & 0 & 0 & 0 & (C_{11}-C_{12})/2 & 0 \\ 0 & 0 & 0 & 0 & 0 & (C_{11}-C_{12})/2 \end{bmatrix} \begin{Bmatrix} \epsilon_1 \\ \epsilon_2 \\ \epsilon_3 \\ \gamma_4 \\ \gamma_5 \\ \gamma_6 \end{Bmatrix}$$
(1.78)

柔度矩阵也可以使用两个常数 S_{11} 和 S_{12} 表达,即

$$\begin{Bmatrix} \epsilon_1 \\ \epsilon_2 \\ \epsilon_3 \\ \gamma_4 \\ \gamma_5 \\ \gamma_6 \end{Bmatrix} = \begin{bmatrix} S_{11} & S_{12} & S_{12} & 0 & 0 & 0 \\ S_{12} & S_{11} & S_{12} & 0 & 0 & 0 \\ S_{12} & S_{12} & S_{11} & 0 & 0 & 0 \\ 0 & 0 & 0 & 2s & 0 & 0 \\ 0 & 0 & 0 & 0 & 2s & 0 \\ 0 & 0 & 0 & 0 & 0 & 2s \end{bmatrix} \begin{Bmatrix} \sigma_1 \\ \sigma_2 \\ \sigma_3 \\ \sigma_4 \\ \sigma_5 \\ \sigma_6 \end{Bmatrix}$$
(1.79)

$$s = S_{11} - S_{12}$$

各向同性材料的各种独立常数不仅成对相关,而且对于实际材料,这些常数的取值也有一定的限制。因为弹性模量和剪切模量必须是正值,所以泊松比必须满足 $\nu > -1$。另外,因为体积模量必须是正值,还可以得到 $\nu < \frac{1}{2}$。最后,得到各向同性材料的泊松比的取值范围为 $-1 < \nu < \frac{1}{2}$。

1.13 工程常数

请注意,从此处开始( )′符号表示单层板坐标系。下一步工作是把正交各向异性材料的刚度矩阵和柔度矩阵分量采用工程常数表达出来。而使用工程常数表达柔度矩阵(刚度矩阵的逆矩阵)更容易实现。在单层板坐标系中,$[S'] = [C']^{-1}$。柔度矩阵用来建立应变-应力关系,正交各向异性材料的应变-应力关系见式(1.69)。重写式(1.69)中第一项,该项对应 1-方向(纤维方向)的应变,有

$$\epsilon'_1 = S'_{11}\sigma'_1 + S'_{12}\sigma'_2 + S'_{13}\sigma'_3$$
(1.80)

让我们做一个思维实验。注意到 $[S']$ 提示我们处在单层板坐标系中。首先沿着 1 方向(纤维方向)施加拉伸应力,如图 1.13 所示,其他应力均为 0。计算产生的 1 方向应变为

$$\epsilon_1' = \frac{\sigma_1'}{E_1} \tag{1.81}$$

然后,仅在 2 方向施加一个应力,利用适当的泊松比计算 1 方向的应变[1],有

$$\epsilon_1' = -\nu_{21}\frac{\sigma_2'}{E_2} \tag{1.82}$$

现在,再仅在 3 方向施加一个应力,利用适当的泊松比计算 1 方向的应变,有

$$\epsilon_1' = -\nu_{31}\frac{\sigma_3'}{E_3} \tag{1.83}$$

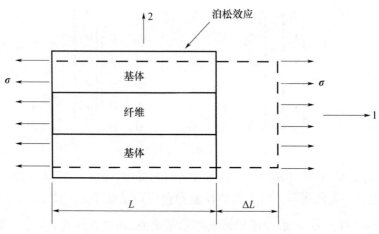

图 1.13 纵向加载

总的 1 - 方向应变为式(1.81)、式(1.82)和式(1.83)的和,即

$$\epsilon_1' = \frac{1}{E_1}\sigma_1' - \frac{\nu_{21}}{E_2}\sigma_2' - \frac{\nu_{31}}{E_3}\sigma_3' \tag{1.84}$$

对比式(1.84)和式(1.80),得

$$S_{11}' = \frac{1}{E_1}; S_{12}' = -\frac{\nu_{21}}{E_2}; S_{13}' = -\frac{\nu_{31}}{E_3} \tag{1.85}$$

对 ϵ_2' 和 ϵ_3' 重复上述等式建立过程,获得式(1.69)中柔度矩阵的第 2 和第 3 行系数。

对于剪切项,使用式(1.69)中柔度矩阵的第 4、5 和 6 行建立关系。例如,从图 1.14 中,得

$$\sigma_6' = \epsilon_6'G_{12} = 2\varepsilon_6'G_{12} \tag{1.86}$$

与式(1.69)中柔度矩阵的第 6 行对比可以得到 $S_{66}' = 1/G_{12}$。

对于正交各向异性材料,单层板坐标系中的柔度矩阵 $[S']$ 定义为

$$[S'] = \begin{bmatrix} \dfrac{1}{E_1} & -\dfrac{\nu_{21}}{E_2} & -\dfrac{\nu_{31}}{E_3} & 0 & 0 & 0 \\ -\dfrac{\nu_{12}}{E_1} & \dfrac{1}{E_2} & -\dfrac{\nu_{32}}{E_3} & 0 & 0 & 0 \\ -\dfrac{\nu_{13}}{E_1} & -\dfrac{\nu_{23}}{E_2} & \dfrac{1}{E_3} & 0 & 0 & 0 \\ 0 & 0 & 0 & \dfrac{1}{G_{23}} & 0 & 0 \\ 0 & 0 & 0 & 0 & \dfrac{1}{G_{13}} & 0 \\ 0 & 0 & 0 & 0 & 0 & \dfrac{1}{G_{12}} \end{bmatrix} \quad (1.87)$$

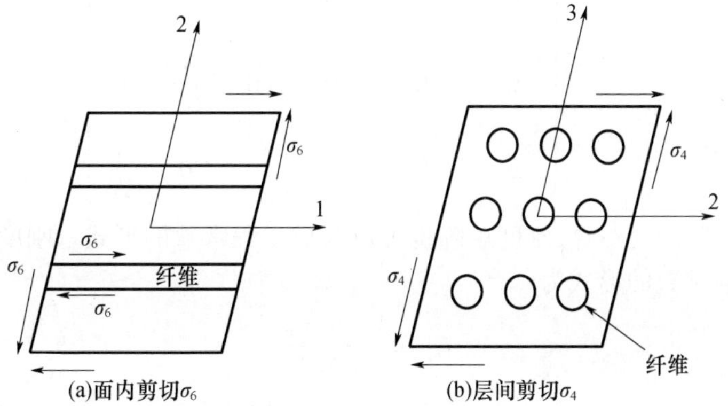

(a)面内剪切σ_6 (b)层间剪切σ_4

图1.14 剪切加载

式中:E_i,G_{ij},ν_{ij}分别为弹性模量、剪切模量和泊松比。另外,这里的下标代表的是单层板坐标系统,即

$$\nu_{ij} = \nu_{x'_i y'_j}, E_{ii} = E_{x'_i} \quad (1.88)$$

由于$[S']$的对称性,下式必然成立

$$\frac{\nu_{ij}}{E_{ii}} = \frac{\nu_{ji}}{E_{jj}} (i,j = 1,\cdots,3) \quad (1.89)$$

另外,根据泊松比的定义,横向应变为

$$\epsilon_j = -\nu_{ij}\epsilon_i \quad (1.90)$$

在 ANSYS 软件中,泊松比的定义与本书不同。ANSYS 软件中采用 PRXY,PRXZ 和 PRYZ 分别表示 ν_{xy},ν_{xz},ν_{yz};采用 NUXY,NUXZ 和 NUYZ 分别表示 ν_{yx},ν_{zx},ν_{zy}。而 Abaqus 软件和本书采用相同的标准符号记法。也就是说,符号 NU12、NU13、NU23 遵循式(1.90)描述的约定。

计算 S_{ij} 之后,应力分量可利用式(1.46)或者式(1.49)计算获得。只要应变足够小,该公式就可以预测有限位移和转动条件下的实际力学行为。但该公式应用起来较为"昂贵",因为需要 18 个状态变量:在初始构型中计算的 12 个应变位移矩阵分量($u_{i,j}$ 和 $u_{r,i} u_{r,j}$)和 6 个方向余弦[a]代表有限旋转。

然而,因为[S']必然对称(见式(1.93)),式(1.87)中只有 9 个独立常数,所以

$$[S'] = \begin{bmatrix} \dfrac{1}{E_1} & -\dfrac{\nu_{12}}{E_1} & -\dfrac{\nu_{13}}{E_1} & 0 & 0 & 0 \\ -\dfrac{\nu_{12}}{E_1} & \dfrac{1}{E_2} & -\dfrac{\nu_{23}}{E_2} & 0 & 0 & 0 \\ -\dfrac{\nu_{13}}{E_1} & -\dfrac{\nu_{23}}{E_2} & \dfrac{1}{E_3} & 0 & 0 & 0 \\ 0 & 0 & 0 & \dfrac{1}{G_{23}} & 0 & 0 \\ 0 & 0 & 0 & 0 & \dfrac{1}{G_{13}} & 0 \\ 0 & 0 & 0 & 0 & 0 & \dfrac{1}{G_{12}} \end{bmatrix} \quad (1.91)$$

通过矩阵的逆运算,刚度矩阵也可以写成工程常数的形式。刚度矩阵各分量写成工程常数的形式为

$$\begin{aligned} C'_{11} &= \frac{1 - \nu_{23}\nu_{32}}{E_2 E_3 \Delta} \\ C'_{12} &= \frac{\nu_{21} + \nu_{31}\nu_{23}}{E_2 E_3 \Delta} = \frac{\nu_{12} + \nu_{32}\nu_{13}}{E_1 E_3 \Delta} \\ C'_{13} &= \frac{\nu_{31} + \nu_{21}\nu_{32}}{E_2 E_3 \Delta} = \frac{\nu_{13} + \nu_{12}\nu_{23}}{E_1 E_2 \Delta} \\ C'_{22} &= \frac{1 - \nu_{13}\nu_{31}}{E_1 E_3 \Delta} \\ C'_{23} &= \frac{\nu_{32} + \nu_{12}\nu_{31}}{E_1 E_3 \Delta} = \frac{\nu_{23} + \nu_{21}\nu_{13}}{E_1 E_2 \Delta} \\ C'_{33} &= \frac{1 - \nu_{12}\nu_{21}}{E_1 E_2 \Delta} \\ C'_{44} &= G_{23} \\ C'_{55} &= G_{13} \\ C'_{66} &= G_{12} \\ \Delta &= \frac{1 - \nu_{12}\nu_{21} - \nu_{23}\nu_{32} - \nu_{31}\nu_{13} - 2\nu_{21}\nu_{32}\nu_{13}}{E_1 E_2 E_3} \end{aligned} \quad (1.92)$$

至此,对于正交各向异性材料,得到了$[S']$和$[C']$矩阵,它们均为有9个独立常数的6×6矩阵。如果材料是横观各向同性的,那么$G_{13}=G_{12}$,$\nu_{13}=\nu_{12}$,$E_3=E_2$。

注意到由于柔度矩阵(式(1.91))的对称性,工程常数需要满足如下条件

$$\frac{\nu_{ij}}{E_i}=\frac{\nu_{ji}}{E_j}(i,j=1,\cdots,3;i\neq j) \tag{1.93}$$

其他弹性常数值的限制可以从柔度矩阵和刚度矩阵的所有对角线元素必须是正值推导出。因为所有工程弹性常数必须为正值($E_1,E_2,E_3,G_{12},G_{23},G_{31}>0$),所以需要满足下面两个条件时,刚度矩阵的所有对角线元素就将会是正值。第一个条件为$(1-\nu_{ij}\nu_{ji})>0;i,j=1,\cdots,3;i\neq j$。该条件推导出第一个工程常数值的限制如下:

$$0<\nu_{ij}<\sqrt{\frac{E_i}{E_j}}(i,j=1,\cdots,3;i\neq j) \tag{1.94}$$

第二个条件为

$$\Delta=1-\nu_{12}\nu_{21}-\nu_{23}\nu_{32}-\nu_{31}\nu_{13}-2\nu_{21}\nu_{32}\nu_{13}>0 \tag{1.95}$$

这些工程常数的限制条件可以用来检验试验数据的正确性。例如,考虑一个测试方案:E_1和ν_{12}参数通过纵向(纤维方向加载)拉伸试验进行测量,拉伸试片上粘贴两个应变片,分别测量纵向和横向;E_2和ν_{21}参数通过横向(垂直纤维方向加载)拉伸试验进行测量。这个测试项目若有效,测得的4个数据值E_1,E_2,ν_{12}和ν_{21}在误差允许范围内必须满足式(1.93)~式(1.95)。

例1.4 Sonti等[6]对拉挤玻璃纤维增强复合材料进行了一系列试验。纵向拉伸的8次试验平均值为$E_1=19.981\text{GPa}$,$\nu_{12}=0.274$。横向拉伸的8次试验平均值为$E_2=11.389\text{GPa}$,$\nu_{21}=0.192$,试问这些数据是否在弹性常数的限制条件范围内?

解 首先对于$i,j=1,2$,分别计算式(1.93)的两边项,有

$$\begin{cases}\dfrac{E_1}{\nu_{12}}=\dfrac{19.981}{0.274}=72.9\text{GPa}\\[2mm]\dfrac{E_2}{\nu_{21}}=\dfrac{11.381}{0.192}=59.3\text{GPa}\end{cases}$$

横向结果比预期低23%,可以推测要么是测得E_2的值太低,要么测得的ν_{21}值比理论值高23%。在任何情况下,23%的误差量都值得仔细审查。

下一步检验式(1.94),有

$$\text{abs}(\nu_{12}) < \sqrt{\frac{E_1}{E_2}}$$

$$0.274 < 1.32$$

$$\text{abs}(\nu_{21}) < \sqrt{\frac{E_2}{E_1}}$$

$$0.192 < 0.75$$

最后，因为没有足够的数据，无法检验式(1.95)的弹性常数限制条件。

1.14 从3D退化为平面应力状态

针对正交各向异性材料，令式(1.69)中 $\sigma_3 = 0$，意味着可以不使用柔度矩阵中的第3行和第3列。

$$\begin{Bmatrix} \epsilon_1' \\ \epsilon_2' \\ \epsilon_3' \\ \gamma_4' \\ \gamma_5' \\ \gamma_6' \end{Bmatrix} = \begin{bmatrix} S_{11}' & S_{12}' & S_{13}' & 0 & 0 & 0 \\ S_{12}' & S_{22}' & S_{23}' & 0 & 0 & 0 \\ S_{13}' & S_{23}' & S_{33}' & 0 & 0 & 0 \\ 0 & 0 & 0 & S_{44}' & 0 & 0 \\ 0 & 0 & 0 & 0 & S_{55}' & 0 \\ 0 & 0 & 0 & 0 & 0 & S_{66}' \end{bmatrix} \begin{Bmatrix} \sigma_1' \\ \sigma_2' \\ \sigma_3' = 0 \\ \sigma_4' \\ \sigma_5' \\ \sigma_6' \end{Bmatrix} \quad (1.96)$$

所以，前两个方程加上最后一个方程组成一个退化的 3×3 柔度矩阵 $[S]$

$$\begin{Bmatrix} \epsilon_1' \\ \epsilon_2' \\ \gamma_6' \end{Bmatrix} = \begin{bmatrix} S_{11}' & S_{12}' & 0 \\ S_{12}' & S_{22}' & 0 \\ 0 & 0 & S_{66}' \end{bmatrix} \begin{Bmatrix} \sigma_1' \\ \sigma_2' \\ \sigma_6' \end{Bmatrix} \quad (1.97)$$

第三个等式很少用到

$$\epsilon_3' = S_{13}' \sigma_1' + S_{23}' \sigma_2' \quad (1.98)$$

剩下的两个等式单独写出来，如下：

$$\begin{Bmatrix} \gamma_4' \\ \gamma_5' \end{Bmatrix} = \begin{bmatrix} S_{44}' & 0 \\ 0 & S_{55}' \end{bmatrix} \begin{Bmatrix} \sigma_4' \\ \sigma_5' \end{Bmatrix} \quad (1.99)$$

为了根据应变计算应力分量，式(1.97)可以反向书写成 $\{\sigma\} = \{Q\}\{\epsilon\}$ 或者表述如下：

$$\begin{Bmatrix} \sigma_1' \\ \sigma_2' \\ \sigma_6' \end{Bmatrix} = \begin{bmatrix} Q_{11}' & Q_{12}' & 0 \\ Q_{12}' & Q_{22}' & 0 \\ 0 & 0 & Q_{66}' \end{bmatrix} \begin{Bmatrix} \epsilon_1' \\ \epsilon_2' \\ \gamma_6' \end{Bmatrix} \quad (1.100)$$

式中:矩阵$[Q'] = [S'_{3\times3}]^{-1}$,是退化后的平面应力状态的刚度矩阵。注意到,简化后的柔度矩阵$[S'_{3\times3}]$各分量与6×6柔度矩阵各对应分量的数值相同。但是简化后刚度矩阵$[Q']$各分量的值与6×6的刚度矩阵$[C']$不是对应相同的,因此刚度矩阵符号有所变化。这是因为3×3矩阵的逆矩阵与6×6矩阵的逆矩阵对应分量的值不同。剩下的两组方程写为

$$\begin{Bmatrix} \sigma'_4 \\ \sigma'_5 \end{Bmatrix} = \begin{bmatrix} C'_{44} & 0 \\ 0 & C'_{55} \end{bmatrix} \begin{Bmatrix} \gamma'_4 \\ \gamma'_5 \end{Bmatrix} \qquad (1.101)$$

式中:C'_{44},C'_{55}与6×6刚度矩阵里对应分量的值相同,因为式(1.101)中的2×2矩阵为对角阵。

例 1.5 证明与面内的伸长量$a\epsilon_1$和$b\epsilon_2$相比,一个平板厚度方向上的变化$t\epsilon_3$可以忽略不计。复合材料平板参数如下:厚度$t = 0.635\text{mm}$,边长$a = 279\text{mm}$,$b = 203\text{mm}$;$E_1 = 19.981\text{GPa}$,$E_2 = 11.389\text{GPa}$,$\nu_{12} = 0.274$。

解 假设0.635mm厚度的玻璃纤维增强聚酯基复合材料板为横观各向同性材料,取$E_3 = E_2 = 11.389\text{GPa}$,$\nu_{12} = \nu_{21} = 0.274$,$G_{31} = G_{12}$。Sonti等[6]通过8次扭转试验得到面内剪切模量平均值$G_{12} = 3.789\text{GPa}$。由于缺乏实验数据,假设$\nu_{23} \approx \nu_m = 0.3$,$G_{23} \approx G_m = 0.385\text{GPa}$,聚酯基体的性能取自文献[1]。式(1.91)中剩余参数的性能可利用式(1.93)计算得到,即

$$\begin{cases} \nu_{21} = \nu_{12}\dfrac{E_2}{E_1} = 0.274\left(\dfrac{11.389}{19.981}\right) = 0.156 \\ \nu_{31} = \nu_{13}\dfrac{E_3}{E_1} = 0.274\left(\dfrac{11.389}{19.981}\right) = 0.156 \\ \nu_{32} = \nu_{23}\dfrac{E_3}{E_2} = 0.3\left(\dfrac{11.389}{11.389}\right) = 0.3 \end{cases}$$

因为横观各向同性,$G_{13} = G_{12} = 3.789\text{GPa}$。现在,假设一个平面应力状态$\sigma'_1 = \sigma'_2 = 0.1\text{GPa}$,$\sigma'_4 = \sigma'_5 = \sigma'_6 = 0$,$\sigma'_3 = 0$。利用式(1.97),得

$$\begin{cases} \epsilon'_1 = S'_{11}\sigma'_1 + S'_{12}\sigma'_2 = \dfrac{0.1}{19.981} - \dfrac{0.1(0.156)}{11.389} = 3.635\times 10^{-3} \\ \epsilon'_2 = S'_{12}\sigma'_1 + S'_{22}\sigma'_2 = -\dfrac{0.1(0.156)}{11.389} + \dfrac{0.1}{11.389} = 7.411\times 10^{-3} \\ \epsilon'_3 = S'_{13}\sigma'_1 + S'_{23}\sigma'_2 = -\dfrac{0.274}{19.981} - \dfrac{0.3(0.1)}{11.389} = -4.005\times 10^{-3} \end{cases}$$

最后得到

$$t\epsilon'_3 = -0.635(4.005\times 10^{-3}) = -2.543\times 10^{-3}\text{mm}$$
$$a\epsilon'_1 = 279(3.635\times 10^{-3}) = 1.014\text{mm}$$
$$b\epsilon'_2 = 203(7.411\times 10^{-3}) = 1.504\text{mm}$$

由于法向延伸量很小,在文献[1]中推导该平板方程时忽略了法向变形。

1.15 层合板表观性能

对于一个拥有 N 层平衡、对称铺层的层合板,其刚度矩阵 $[C]$ 由在层合板坐标系中叠加每一层单层板的刚度矩阵得到。叠加时,每一层单层板刚度矩阵需要乘以该单层的厚度比 t_k/t。这里 t 为层合板的厚度,t_k 为第 k 层单层板的厚度。

$$[C] = \sum_{k=1}^{N} \frac{t_k}{t}[C_k] \tag{1.102}$$

注意到,柔度既不是求和也不是求平均。层合板的柔度阵由 6×6 刚度矩阵求逆后得到,即

$$[S] = [C]^{-1} \tag{1.103}$$

如果层合板的 θ 和 $-\theta$ 方向的单层总厚度相同,则该层合板为平衡的。这种层合板拥有正交各向异性的刚度矩阵 $[C]$ 和柔度矩阵 $[S]$。该柔度矩阵写成表观工程参数的形式为

$$[S] = \begin{bmatrix} \dfrac{1}{E_x} & \dfrac{-\nu_{yx}}{E_y} & \dfrac{-\nu_{zx}}{E_z} & 0 & 0 & 0 \\ \dfrac{-\nu_{xy}}{E_x} & \dfrac{1}{E_y} & \dfrac{-\nu_{zy}}{E_z} & 0 & 0 & 0 \\ \dfrac{-\nu_{xz}}{E_x} & \dfrac{-\nu_{yz}}{E_y} & \dfrac{1}{E_z} & 0 & 0 & 0 \\ 0 & 0 & 0 & \dfrac{1}{G_{yz}} & 0 & 0 \\ 0 & 0 & 0 & 0 & \dfrac{1}{G_{xz}} & 0 \\ 0 & 0 & 0 & 0 & 0 & \dfrac{1}{G_{xy}} \end{bmatrix} \tag{1.104}$$

因为柔度矩阵对称,所以必然满足式(1.93),其中 $i,j = x,y,z$。因此,层合板表观力学性能可以由层合板柔度矩阵分量计算,即

$$\begin{aligned} E_x &= 1/S_{11} & \nu_{xy} &= -S_{21}/S_{11} \\ E_y &= 1/S_{22} & \nu_{xz} &= -S_{31}/S_{11} \\ E_z &= 1/S_{33} & \nu_{yz} &= -S_{32}/S_{22} \\ G_{yz} &= 1/S_{44} \\ G_{xz} &= 1/S_{55} \\ G_{xy} &= 1/S_{66} \end{aligned} \tag{1.105}$$

例 1.6 计算层合板力学性能。层合板铺层顺序为 $[0/90/\pm 30]_s$,每一层厚度 $t_k = 1.5\text{mm}$,$E_f = 241\text{GPa}$,$\nu_f = 0.2$,$E_m = 3.12\text{GPa}$,$\nu_m = 0.38$,纤维体积含量 $V_f = 0.6$。这里的 f,m 分别代表纤维和基体。

解 首先使用式(6.8)介绍的周期性微观结构计算方法获得单层板力学性能(单位:MPa)。

$$E_1 = 145880 \quad G_{12} = 4386 \quad \nu_{12} = \nu_{13} = 0.263$$
$$E_2 = 13312 \quad G_{23} = 4528 \quad \nu_{23} = 0.470$$

然后,使用式(1.91)计算单层板柔度矩阵 $[S']$、使用式(1.34)计算坐标转换矩阵 $[T]$、使用式(1.53)计算层合板坐标系下的单层板柔度 $[S]$,每单层板在层合板坐标系下的刚度由 $[C] = [S]^{-1}$ 计算得到。然后,使用式(1.102)对每单层刚度进行叠加得到层合板总刚度阵并求逆,最后利用式(1.105)得

$$E_x = 78901 \quad G_{xy} = 17114 \quad \nu_{xy} = 0.320$$
$$E_y = 47604 \quad G_{yz} = 4475 \quad \nu_{yz} = 0.364$$
$$E_z = 16023 \quad G_{xz} = 4439 \quad \nu_{xz} = 0.280$$

习题

习题 1.1 一个长度为 L 的锥形细长杆,其原点固定并在自由端施加轴向拉伸载荷。利用虚功原理(PVW),求出在 $0 < x < L$ 范围内的二次位移函数 $u(x)$。杆件的截面面积呈线性变化,固定端和自由端的面积分别为 A_1、A_2,且 $A_1 > A_2$。杆件采用均匀各向同性材料,其模量为 E。

习题 1.2 一个截面为圆形,长度为 L 的锥形细长杆,其原点固定并在自由端施加扭矩 T。利用虚功原理,求出在 $0 < x < L$ 范围内的二次转角函数 $\theta(x)$。杆件的截面面积呈线性变化,固定端和自由端的面积分别为 A_1,A_2,且 $A_1 > A_2$。杆件采用均匀各向同性材料,其剪切模量为 G。

习题 1.3 构造依次由(a) $x-y$ 平面,(b) $x-z$ 平面,(c) $y-z$ 平面对称产生的旋转矩阵 $[a]$。得到的系统不遵循右手规则。

习题 1.4 关于(a) x 轴,(b) y 轴,(c) z 轴旋转 $\theta = \pi$,分别构造 3 个旋转矩阵 $[a]$。

习题 1.5 利用

$$\boldsymbol{\sigma} = \begin{bmatrix} 10 & 2 & 1 \\ 2 & 5 & 1 \\ 1 & 1 & 3 \end{bmatrix}$$

和例 1.2 中的 $[a]$,验证式(1.29)与式(1.26)得到的结果相同。

习题1.6 编写一个计算程序,使用工程常数计算柔度和刚度矩阵。从一个文件中获取输入并将结果输出到另一个文件。使用读者自己的例子验证这个程序。读者可以使用文献[1]中表1.3、表1.4提供的材料属性,并假设材料如1.12.4节所述是横观各向同性的。

习题1.7 编写一个计算程序,通过绕 z-轴旋转 θ 角(见图1.7),将单层板坐标系统下的刚度和柔度矩阵 C', S' 转换到另一个坐标系下的 C, S。建立一个输入文件用于读取数据 C', S', θ。输出的 C, S 写入另一个文件。读者可以使用自己的例子验证该程序。可以使用文献[1]中表1.3、表1.4提供的材料属性,并假设材料如1.12.4节所述是横观各向同性的。

习题1.8 根据读者选择的材料属性,对照 $[S]^{-1}$ 计算并验证式(1.92)。读者可以使用文献[1]中表1.3、表1.4提供的材料属性,并如1.12.4节所述,假设材料是横观各向同性的。

习题1.9 以下数据是单向碳纤维–环氧预浸料(体积含量为63%的MR50碳纤维、LTM25环氧树脂基体)的实验数据。确定是否满足弹性常数的限制。

$$E_1 = 156.403 \text{GPa}, \quad E_2 = 7.786 \text{GPa}$$

$$\nu_{12} = 0.352, \quad \nu_{21} = 0.016$$

$$G_{12} = 4386 \text{GPa}$$

$$\sigma_{1t}^u = 1.826 \text{GPa}, \quad \sigma_{1c}^u = 1.134 \text{GPa}$$

$$\sigma_{2t}^u = 19 \text{MPa}, \quad \sigma_{2c}^u = 131 \text{MPa}$$

$$\sigma_6^u = 75 \text{MPa}$$

$$\epsilon_{1t}^u = 11900 \times 10^{-6}, \quad \epsilon_{1c}^u = 8180 \times 10^{-6}$$

$$\epsilon_{2t}^u = 2480 \times 10^{-6}, \quad \epsilon_{2c}^u = 22100 \times 10^{-6}$$

$$\gamma_{12}^u = 20000 \times 10^{-6}$$

习题1.10 解释应力和应变的指标符号缩写规则。

习题1.11 什么是正交各向异性材料?描述它需要多少个常数?

习题1.12 什么是横观各向同性材料?描述它需要多少个常数?

习题1.13 使用习题1.4中的3个旋转矩阵计算验证式(1.48)。

习题1.14 利用式(1.71)和式(1.91)证明式(1.73)。

习题1.15 证明具有两个正交对称面的材料也同时具有第三个对称面。使用1.12.3节中介绍的推导过程,将关于2–3平面的对称应用到式(1.68)中。

习题1.16 什么是平面应力假设?

习题 1.17 编写一个计算程序计算对称平衡层合板的表观力学性能。所有铺层的材料都是相同的。输入数据包括横观各向同性材料的所有工程常数、层数 N、所有层的厚度和角度 t_k、θ_k($k=1,\cdots,N$)。使用 1.15 节、1.12.4 节和 1.13 节中的相关知识。

第2章 有限元分析介绍

在本书中,有限元法(FEM)是作为一种解决实际工程问题的工具。本书中的大部分例题均利用 Abaqus 商业软件进行分析。计算编程仅限于实现材料模型和后处理算法。当商业程序中缺乏所需功能时,使用了文献[5]中提供的其他程序。任何有限元软件的有效使用都需要对有限元法有一个基本的认识。因此,本章包含了对有限元法的简短介绍,旨在为那些没有系统学习过有限元法的读者提供参考。此外,还介绍了 Abaqus/CAE 图形用户界面(GUI),使读者熟悉适用于任何商业软件的有限元建模典型步骤。

2.1 基本的 FEM 步骤

考虑一个杆件的轴向变形问题。使用常微分方程(ODE)表示杆件的变形如下:

$$-\frac{\mathrm{d}}{\mathrm{d}x}\left(EA\frac{\mathrm{d}u}{\mathrm{d}x}\right) - f = 0 \quad (0 \leq x \leq L) \tag{2.1}$$

式中:E,A 分别为杆件的弹性模量和横截面积;f 为作用在杆件上的分布力。如图 2.1 所示,其边界条件为

$$u(0) = 0$$
$$\left[\left(EA\frac{\mathrm{d}u}{\mathrm{d}x}\right)\right]_{x=L} = P \tag{2.2}$$

图 2.1(a)所示的实际几何模型可以简化为图 2.1(b)所示的线状数学模型,这是材料力学教科书中的惯用做法。杆件在 x 轴长度方向上的范围是 $[0,L]$。

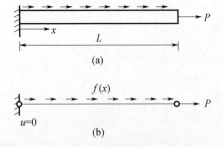

图 2.1 物理和数学(理想化)模型

2.1.1 离散

下一步把杆件在长度范围内分割成有限个单元,如图2.2所示。

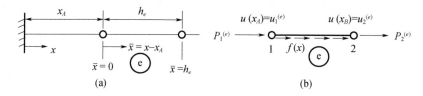

图2.2 单元离散

2.1.2 单元方程

利用常微分方程(ODE)的积分形式导出单元方程,通过对ODE乘以权函数 v 后进行积分得到该积分形式,即

$$0 = \int_{x_A}^{x_B} v \left[-\frac{\mathrm{d}}{\mathrm{d}x}\left(EA\frac{\mathrm{d}u}{\mathrm{d}x}\right) - f \right] \mathrm{d}x \tag{2.3}$$

上述方程称为弱形式,因为其解 $u(x)$ 在 $[0,L]$ 内不一定要满足ODE式(2.1)。相反,解 $u(x)$ 只需要在加权平均意义上满足ODE式(2.3)。因此,找到一个弱形式解比找到一个强形式解容易得多。尽管上述杆件求解问题的强形式解(精确解)是已知的,但是大多数的复合材料力学问题均没有精确解。对式(2.3)进行分部积分,得到控制方程为

$$0 = \int_{x_A}^{x_B} EA \frac{\mathrm{d}v}{\mathrm{d}x}\frac{\mathrm{d}u}{\mathrm{d}x} \mathrm{d}x - \int_{x_A}^{x_B} v f \mathrm{d}x - \left[v\left(EA\frac{\mathrm{d}u}{\mathrm{d}x}\right) \right]_{x_A}^{x_B} \tag{2.4}$$

式中:$v(x)$ 为权函数,通常将其设置为主变量 $u(x)$。从边界项可以看出:

(1)在 x_A 或 x_B 点处指定 $v(x)$ 为本质边界条件。

(2)在杆件的两端指定 $\left(EA\frac{\mathrm{d}u}{\mathrm{d}x}\right)$ 为自然边界条件。

$u(x)$ 为主变量,那么 $\left(EA\frac{\mathrm{d}u}{\mathrm{d}x}\right) = EA\epsilon_x = A\sigma_x$ 为次变量。令

$$\begin{aligned} u(x_A) &= u_1^e \\ u(x_B) &= u_2^e \\ -\left[\left(EA\frac{\mathrm{d}u}{\mathrm{d}x}\right)\right]_{x_A} &= P_1^e \\ \left[\left(EA\frac{\mathrm{d}u}{\mathrm{d}x}\right)\right]_{x_B} &= P_2^e \end{aligned} \tag{2.5}$$

那么控制方程变为

$$0 = \int_{x_A}^{x_B} \left(EA \frac{dv}{dx} \frac{du}{dx} - vf \right) dx - P_1^e v(x_A) - P_2^e v(x_B) = B(v,u) - l(v) \tag{2.6}$$

式中

$$B(v,u) = \int_{x_A}^{x_B} EA \frac{dv}{dx} \frac{du}{dx} dx$$

$$l(v) = \int_{x_A}^{x_B} vf dx + P_1^e v(x_A) + P_2^e v(x_B) \tag{2.7}$$

2.1.3 一个单元上的近似解

现在,未知函数 $u(x)$ 可以近似看作为已知函数 $N_i^e(x)$ 和未知系数 a_j^e 的线性组合(级数展开),即

$$u_e(x) = \sum_{j=1}^{n} a_j^e N_j^e(x)$$

式中:a_j^e 为需要求解的系数;$N_j^e(x)$ 为插值函数。

对于权函数 $v(x)$,可使用 Ritz 方法[4],令 $v(x) = N_j^e(x)$。将其代入控制方程式(2.6),得

$$\sum_{j=1}^{n} \left(\int_{x_A}^{x_B} EA \frac{dN_i^e}{dx} \frac{dN_j^e}{dx} dx \right) a_j^e = \int_{x_A}^{x_B} N_i^e f dx + P_1^e N_i^e(x_A) + P_2^e N_i^e(x_B) \tag{2.8}$$

上式可以简化写为

$$\sum_{j=1}^{n} K_{ij}^e a_j^e = F_i^e \tag{2.9}$$

或者写成矩阵形式

$$[K^e]\{a^e\} = \{F^e\} \tag{2.10}$$

式中:$[K^e]$ 为单元刚度矩阵;$\{F^e\}$ 为单元等效力列阵;$\{a^e\}$ 为单元未知参数。

2.1.4 插值函数

虽然任何一组完整的线性无关函数均可以作为插值函数,但为了方便起见,设置插值函数时可使得未知系数表示节点位移,即 $a_i = u_i$。对于在 $x_e \leq x \leq x_{e+1}$ 区间上的双节点单元,可以使用下面的线性插值函数(图2.3):

$$\begin{cases} N_1^e = \dfrac{x_{e+1} - x}{h_e} \\ N_2^e = \dfrac{x - x_e}{h_e} \end{cases} \tag{2.11}$$

式中：$h_e = x_{e+1} - x_e$ 为单元的长度。上式的插值函数满足以下条件：

$$N_i^e(x_j) = \begin{cases} 0 & (i \neq j) \\ 1 & (i = j) \end{cases} \quad (2.12)$$

$$\sum_{i=1}^{2} N_i^e(x) = 1 \quad (2.13)$$

这样就能保证未知系数代表的是节点位移，即 $a_i = u_i$。

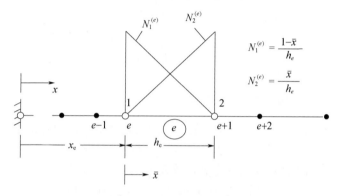

图 2.3　双节点杆单元的线性插值函数

当然还可以使用许多其他的插值函数形式，每种函数均存在优缺点。插值函数与单元的节点数密切相关。图 2.4 说明了 8 节点壳单元中的插值函数 N_1 和 N_5（对应于节点 1 和节点 5）的形状。

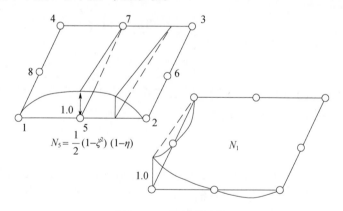

图 2.4　二维插值函数

一般说来，单元节点越多，意味着更高的求解精度且不必细化网格，但同时也意味着计算时间成本的提高。图 2.5 说明了当单元数从 2 增加到 4，或者单元节点数为 2 的线性单元增加到节点数为 3 的二次单元时，近似解如何收敛到精确解。

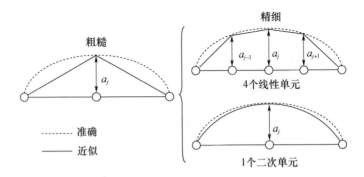

图 2.5　离散化误差

2.1.5　一个特定问题的单元方程

对于满足式(2.12)、式(2.13)条件的插值函数,可以重写式(2.10)为

$$[K^e]\{a^e\} = \{F^e\} \tag{2.14}$$

式中:$\{u^e\}$为节点位移;$[K^e]$为单元刚度矩阵。

$$[K^e] = \begin{bmatrix} \int_{x_A}^{x_B} EA \dfrac{dN_1^e}{dx}\dfrac{dN_1^e}{dx}dx & \int_{x_A}^{x_B} EA \dfrac{dN_1^e}{dx}\dfrac{dN_2^e}{dx}dx \\ \int_{x_A}^{x_B} EA \dfrac{dN_2^e}{dx}\dfrac{dN_1^e}{dx}dx & \int_{x_A}^{x_B} EA \dfrac{dN_2^e}{dx}\dfrac{dN_2^e}{dx}dx \end{bmatrix} \tag{2.15}$$

式中:$\{F^e\}$为单元等效力矩阵,有

$$\{F^e\} = \begin{Bmatrix} \int_{x_A}^{x_B} N_1^e f dx + P_1^e \\ \int_{x_A}^{x_B} N_2^e f dx + P_2^e \end{Bmatrix} \tag{2.16}$$

对于双节点的单元号为 e 的杆单元,其横截面积 A_e,单元长度 h_e,模量 E 这些参数都是固定的。这些参数定义的拉-压单元刚度为

$$k^e = \frac{EA_e}{h_e} \tag{2.17}$$

单元受到的外载为分布载荷 f_e,在节点 1 处载荷值为 P_1^e,在节点 2 处载荷值为 P_2^e。利用这些已知参数、线性插值函数式(2.11)以及式(2.15)、式(2.16),单元的刚度矩阵和等效节点力变为

$$[K^e] = \begin{bmatrix} k^e & -k^e \\ -k^e & k^e \end{bmatrix} = \frac{EA_e}{h_e}\begin{bmatrix} 1 & -1 \\ -1 & 1 \end{bmatrix} \tag{2.18}$$

$$\{F^e\} = \frac{f_e h_e}{2}\begin{Bmatrix} 1 \\ 1 \end{Bmatrix} + \begin{Bmatrix} P_1^e \\ P_2^e \end{Bmatrix} \tag{2.19}$$

2.1.6 单元方程的组装

单元未知参数对应的是单元节点的位移。两个相邻单元在共用节点处具有相同的位移且位移值唯一。例如,考虑到图 2.6 所示单元之间的连接特性,每个单元的节点位移采用唯一的标识符号进行表示,上标表示单元号,下标表示节点号,有

$$\begin{cases} u_1^1 = U_1 \\ u_2^1 = U_2 = u_1^2 \\ u_2^2 = U_3 = u_1^3 \\ u_2^3 = U_4 \end{cases} \quad (2.20)$$

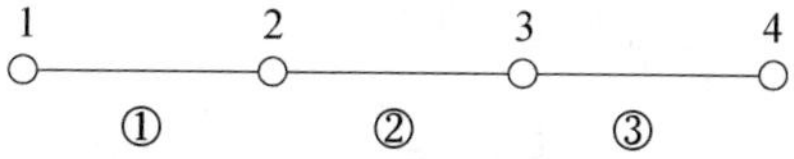

图 2.6　3 个双节点单元之间的连接

现在,所有的单元方程可以组装到整体系统中。首先,1#单元组装后如下:

$$\begin{bmatrix} k^1 & -k^1 & 0 & 0 \\ -k^1 & k^1 & 0 & 0 \\ 0 & 0 & 0 & 0 \\ 0 & 0 & 0 & 0 \end{bmatrix} \begin{Bmatrix} U_1 \\ U_2 \\ U_3 \\ U_4 \end{Bmatrix} = \begin{Bmatrix} f_1 h_1/2 \\ f_1 h_1/2 \\ 0 \\ 0 \end{Bmatrix} + \begin{Bmatrix} P_1^1 \\ P_2^1 \\ 0 \\ 0 \end{Bmatrix} \quad (2.21)$$

组装 2#单元后如下:

$$\begin{bmatrix} k^1 & -k^1 & 0 & 0 \\ -k^1 & k^1+k^2 & -k^2 & 0 \\ 0 & -k^2 & 0 & 0 \\ 0 & 0 & 0 & 0 \end{bmatrix} \begin{Bmatrix} U_1 \\ U_2 \\ U_3 \\ U_4 \end{Bmatrix} = \begin{Bmatrix} f_1 h_1/2 \\ f_1 h_1/2 + f_2 h_2/2 \\ f_2 h_2/2 \\ 0 \end{Bmatrix} + \begin{Bmatrix} P_1^1 \\ P_2^1 + P_1^2 \\ P_2^2 \\ 0 \end{Bmatrix} \quad (2.22)$$

最后,组装 3#单元获得完整的组装后的单元方程

$$\begin{bmatrix} k^1 & -k^1 & 0 & 0 \\ -k^1 & k^1+k^2 & -k^2 & 0 \\ 0 & -k^2 & k^2+k^3 & -k^3 \\ 0 & 0 & -k^3 & k^3 \end{bmatrix} \begin{Bmatrix} U_1 \\ U_2 \\ U_3 \\ U_4 \end{Bmatrix} = \frac{1}{2} \begin{Bmatrix} f_1 h_1 \\ f_1 h_1 + f_2 h_2 \\ f_2 h_2 + f_3 h_3 \\ f_3 h_3 \end{Bmatrix} + \begin{Bmatrix} P_1^1 \\ P_2^1 + P_1^2 \\ P_2^2 + P_1^3 \\ P_2^3 \end{Bmatrix} \quad (2.23)$$

2.1.7 边界条件

由于平衡条件(图 2.2),两个单元在共用节点处的内力相互抵消,即

$$\begin{cases} P_2^1 + P_1^2 = 0 \\ P_2^2 + P_1^3 = 0 \end{cases} \tag{2.24}$$

剩下的 P_1^1 和 P_2^3 为杆件首尾处的载荷。如果杆件的任一端固定，那么固定处的位移必然是 0。即若杆件 $x=0$ 处固定，那么其位移 $U_1=0$。若杆件的另一端 $x=L$ 处自由，那么 P_2^3 必须明确规定，因为 $U_4 \neq 0$。如果没有指定，则假设载荷为 0。

2.1.8 方程的解

因为 $U_1=0$，消除刚度矩阵的第一行和第一列，得到一个只有 3 个未知量的 3×3 代数方程组。3 个未知量分别为 U_2, U_3, U_4。一旦 U_2 求解得到，其反力 P_1^1 可通过式(2.23)中第一个方程得到，即

$$-k^1 U_2 = \frac{f_1 h_1}{2} + P_1^1 \tag{2.25}$$

2.1.9 单元内部的解

现在，沿着杆件的 4 个节点处的解 U_i 已经得到。下一步通过插值函数计算在杆件上任意位置 x 处的位移解，有

$$U^e(x) = \sum_{j=1}^{2} U_j^e N_j^e(x) \tag{2.26}$$

或者

$$u(x) = \begin{cases} U_1 N_1^1(x) + U_2 N_2^1(x) & (0 \leq x \leq h_1) \\ U_2 N_1^2(x) + U_3 N_2^2(x) & (h_1 \leq x \leq h_1 + h_2) \\ U_3 N_1^3(x) + U_4 N_2^3(x) & (h_1 + h_2 \leq x \leq h_1 + h_2 + h_3) \end{cases} \tag{2.27}$$

2.1.10 派生结果

1. 应变

通过式(1.5)直接根据单元内已知的位移值计算得到应变，例如

$$\epsilon_x = \frac{\mathrm{d}u}{\mathrm{d}x} = \sum_{j=1}^{2} \frac{\mathrm{d}N_j^e}{\mathrm{d}x} \tag{2.28}$$

注意到，如果 $N_j^e(x)$ 是线性函数，则该单元上的应变即为常数。一般来说，应变的精度比主变量(位移)低一阶。

2. 应力

应力值通常根据本构方程计算得到，该例子中，一维应力－应变关系为

$$\sigma_x = E\epsilon_x \tag{2.29}$$

注意到,应力的精度和应变相同。

2.2 通用的有限元法过程

对于任意单元类型的单元方程推导、组装和求解均与 2.1 节所述的一维杆单元类似,除了采用虚功原理(PVW,式(1.16))代替控制方程式(2.1)。PVW 提供了一个类似式(2.4)的弱解形式。针对三维变形状态,展开式(1.16),其内部虚功为

$$\delta W_I = \int (\sigma_{xx}\delta\epsilon_{xx} + \sigma_{yy}\delta\epsilon_{yy} + \sigma_{zz}\delta\epsilon_{zz} + \sigma_{yz}\delta\gamma_{yz} + \sigma_{xz}\delta\gamma_{xz} + \sigma_{xy}\delta\gamma_{xy})\mathrm{d}V$$
$$= \int_V \underline{\sigma}^T \delta\underline{\epsilon}\,\mathrm{d}V$$

$$\tag{2.30}$$

上式中

$$\underline{\sigma}^T = \{\sigma_{xx},\sigma_{yy},\sigma_{zz},\sigma_{yz},\sigma_{xz},\sigma_{xy}\}$$
$$\delta\underline{\epsilon}^T = \{\delta\epsilon_{xx},\delta\epsilon_{yy},\delta\epsilon_{zz},\delta\gamma_{yz},\delta\gamma_{xz},\delta\gamma_{xy}\} \tag{2.31}$$

下一步,外部功为

$$\delta W_E = \int_V \underline{f}^T \delta \underline{u}\,\mathrm{d}V + \int_S \underline{t}^T \delta \underline{u}\,\mathrm{d}S \tag{2.32}$$

上式中单位体积内的体力和单位面积上的表面力为

$$\underline{f}^T = \{f_x, f_y, f_z\}$$
$$\underline{t}^T = \{t_x, t_y, t_z\} \tag{2.33}$$

式中的下划线(_)表示一维数组,而不一定是一个向量。例如,$\underline{u}$ 是一个矢量,但 $\underline{\sigma}$ 是六元数组中的 6 个应力分量。虚应变是由虚位移 $\delta u(x)$ 产生的应变。因此,利用应变 – 位移方程式(1.5),由虚位移计算得到虚应变,矩阵形式表示如下

$$\underline{\epsilon} = \underline{\underline{\partial}}\,\underline{u}$$
$$\delta\underline{\epsilon} = \underline{\underline{\partial}}\,\delta\underline{u} \tag{2.34}$$

式中

$$\underline{\underline{\partial}} = \begin{bmatrix} \frac{\partial}{\partial x} & 0 & 0 & \frac{\partial}{\partial y} & 0 & \frac{\partial}{\partial z} \\ 0 & \frac{\partial}{\partial y} & 0 & \frac{\partial}{\partial x} & \frac{\partial}{\partial z} & 0 \\ 0 & 0 & \frac{\partial}{\partial z} & 0 & \frac{\partial}{\partial y} & \frac{\partial}{\partial x} \end{bmatrix}^T \tag{2.35}$$

那么,虚功原理写成矩阵形式如下

$$\int_V \underline{\sigma}^T \underline{\partial} \delta \underline{u} dV = \int_V \underline{f}^T \delta \underline{u} dV + \int_S \underline{t}^T \delta \underline{u} dS \tag{2.36}$$

实体在体积 V 和表面 S 上的积分可以在 m 个单元上逐个分解,有

$$\sum_{e=1}^m \left[\int_{V_e} \underline{\sigma}^T \underline{\partial} \delta \underline{u} dV \right] = \sum_{e=1}^m \left[\int_{V_e} \underline{f}^T \delta \underline{u} dV + \int_{S_e} \underline{t}^T \delta \underline{u} dS \right] \tag{2.37}$$

只要两个单元共面时,第二个积分项的贡献就会取消,就像 2.1.7 节中消除内部载荷一样。利用本构方程给出应力分量。对于线性材料

$$\underline{\sigma} = \underline{\underline{C}} \underline{\epsilon} \tag{2.38}$$

其中 $\underline{\underline{C}}$ 由式(1.68)给出。每个单元的内部虚功变为

$$\delta W_I^e = \int_{V_e} \underline{\sigma}^T \delta \underline{\epsilon} dV = \int_{V_e} \underline{\epsilon}^T \underline{\underline{C}} \delta \underline{\epsilon} dV \tag{2.39}$$

位移的展开式可以用矩阵形式写为

$$\underline{u} = \underline{\underline{N}} \underline{a} \tag{2.40}$$

其中: $\underline{\underline{N}}$ 包含了单元的插值函数; $\underline{a}$ 为单元的节点位移,如第 2.1.4 节所述。因此,应变为

$$\underline{\epsilon} = \underline{\partial} \underline{u} = \underline{\partial} \underline{\underline{N}} \underline{a} = \underline{\underline{B}} \underline{a} \tag{2.41}$$

式中: $\underline{\underline{B}} = \underline{\partial} \underline{\underline{N}}$ 是应变 – 位移矩阵。现在每个单元内部虚功的离散形式为

$$\delta W_I^e = \int_{V_e} \underline{a}^T \underline{\underline{B}}^T \underline{\underline{C}} \underline{\underline{B}} \delta \underline{a} dV = \underline{a}^T \int_{V_e} \underline{\underline{B}}^T \underline{\underline{C}} \underline{\underline{B}} dV \delta \underline{a} = \underline{a}^T \underline{\underline{K}}^e \delta \underline{a} \tag{2.42}$$

式中单元刚度阵 $\underline{\underline{K}}^e$ 为

$$\underline{\underline{K}}^e = \int_{V_e} \underline{\underline{B}}^T \underline{\underline{C}} \underline{\underline{B}} dV \tag{2.43}$$

外部虚功变为

$$\delta W_E^e = \int_{V_e} \underline{f}^T \delta \underline{u} dV + \int_{S_e} \underline{t}^T \delta \underline{u} dS$$

$$= \left(\int_{V_e} \underline{f}^T \underline{\underline{N}} dV + \int_{S_e} \underline{t}^T \underline{\underline{N}} dS \right) \delta \underline{a} = (\underline{P}^e)^T \delta \underline{a} \tag{2.44}$$

式中单元等效力为

$$\underline{P}^e = \int_{V_e} \underline{\underline{N}}^T \underline{f} dV + \int_{S_e} \underline{\underline{N}}^T \underline{t} dS \tag{2.45}$$

在单元体积 V_e 和单元表面 S_e 上的积分通常采用高斯积分法进行数值计算。对于体积积分,需要在体积内的几个点上对被积函数进行计算。这样的点,称为高斯点。高斯点非常重要:首先,本构矩阵 $\underline{\underline{C}}$ 在这些位置上进行计算。其次,应变(和应力)的最精确的计算值也是在这些位置上计算得到的。

将单个单元方程的 δW_I^e 和 δW_E^e 组装到整体 PVW 中的做法与 2.1.6 节中介

绍的过程类似。显然，该过程比杆件单元的组装要复杂得多。这种组装过程的细节及其计算程序是有限元技术的一部分，超出了本书的介绍范围。最后，所有单元的刚度矩阵 $\underline{\underline{K}}^e$ 和单元等效力 $\underline{P}^e$ 组装进整体系统中。

$$\underline{\underline{K}}\,\underline{a} = \underline{P} \tag{2.46}$$

下一步，与 2.1.6 节中的求解过程类似，对式(2.46)应用边界条件。然后，求解代数方程(2.46)，得到整体系统的节点位移列阵 $\underline{a}$。由于每个单元的节点位移结果在 $\underline{a}$ 中都有对应值，所以可以通过式(2.34)和式(2.38)来计算单元内任何位置的应变和应力。

例 2.1 针对使用一个单元进行离散的杆，计算其单元刚度矩阵式(2.43)和单元等效力式(2.45)。使用式(2.11)中的线性插值函数。将结果与式(2.18)、式(2.19)进行比较。

解 令 A_e 和 h_e 分别代表杆件的横截面积和单元长度，$x_e = 0$，$x_{e+1} = h_e$。将上述这些值代入式(2.11)中的线性插值函数，得到如下插值函数列阵：

$$\underline{N}^{\mathrm{T}} = \begin{bmatrix} N_1^e \\ N_2^e \end{bmatrix} = \begin{bmatrix} \dfrac{x_{e+1} - x}{h_e} \\ \dfrac{x - x_e}{h_e} \end{bmatrix} = \begin{bmatrix} 1 - x/h_e \\ x/h_e \end{bmatrix}$$

应变-位移列阵的计算如下

$$\underline{\underline{B}}^{\mathrm{T}} = \underline{\underline{\partial}}\,\underline{N}^{\mathrm{T}} = \begin{bmatrix} \partial N_1^e/\partial x \\ \partial N_2^e/\partial x \end{bmatrix} = \begin{bmatrix} -1/h_e \\ 1/h_e \end{bmatrix}$$

该杆单元具有线弹性行为，且处于一维应变-应力状态。因此

$$\underline{\underline{C}} = E$$

利用式(2.43)，得

$$\underline{\underline{K}}^e = \int_{V_e} \underline{\underline{B}}^{\mathrm{T}}\,\underline{\underline{C}}\,\underline{\underline{B}}\,\mathrm{d}V = \int_0^{h_e} \begin{bmatrix} -1/h_e & 1/h_e \end{bmatrix} E \begin{bmatrix} -1/h_e \\ 1/h_e \end{bmatrix} A_e \mathrm{d}x$$

通过积分计算得到单元的刚度阵，即

$$[K^e] = \dfrac{EA_e}{h_e} \begin{bmatrix} 1 & -1 \\ -1 & 1 \end{bmatrix}$$

为了计算单元等效力，定义 f_e 为作用在单元上的分布力，$x=0$ 处的载荷为 P_1^e，$x = h_e$ 处的载荷为 P_2^e。代入式(2.45)，得

$$\underline{P}^e = \int_{V_e} \underline{N}^{\mathrm{T}} f \mathrm{d}V + \int_{S_e} \underline{N}^{\mathrm{T}} \underline{t}\,\mathrm{d}S = \int_0^{h_e} \begin{bmatrix} 1 - x/h_e \\ x/h_e \end{bmatrix} f_e \mathrm{d}x + \begin{bmatrix} P_1^e \\ P_2^e \end{bmatrix}$$

通过积分，得到单元等效力为

$$\underline{P}^e = \frac{f_e h_e}{2}\begin{bmatrix}1\\1\end{bmatrix} + \begin{bmatrix}P_1^e\\P_2^e\end{bmatrix}$$

2.3 实体建模、分析和可视化

许多商业软件对不同的工程学科问题均有较强的有限元分析能力。它们有助于解决从简单的线性静态分析到非线性瞬态分析的各种问题。少数软件(如 ANSYS 和 Abaqus)具有分析复合材料问题的能力,这两种软件均接受常规界面输入、用户自定义本构方程和单元类型。由于这些软件包不仅提供分析工具、几何建模和结果可视化,而且还可以集成到更大的设计、生产和产品生命周期的过程中,所以它们通常称为完备的分析平台或计算机辅助工程(CAE)系统。

FEA 软件通常分为 3 个模块,即前处理器、求解器和后处理器。在前处理器中,通过定义几何形状、材料特性和单元类型等建立模型。此外,载荷和边界条件也在前处理器中进行输入,当然它们也可以在求解阶段中进行输入。利用这些输入信息,求解器可以计算刚度矩阵和单元等效力。下一步,对代数式(2.46)进行求解,得到位移解。最后在后处理器中计算派生的结果,如应力、应变和失效指数等。计算结果可以使用图形工具进行查看。

本章接下来以 Abaqus 软件[7]为例,对基本的有限元分析(FEA)过程和具体步骤进行介绍。本书中例子均使用 Abaqus 软件进行求解,使用 ANSYS 软件[9]求解类似的例子可以参考文献[8]。

有限元模型建模步骤:①建立几何模型;②赋予构成几何模型的各部分的材料性能;③将荷载和边界条件施加到几何模型上;④将几何体离散为有限个单元,其由节点和单元连接关系定义。总而言之,需针对要解决的问题类型选择合适的单元类型,然后对模型进行求解并计算派生结果及可视化。

2.3.1 模型几何体

通过指定所有节点及位置信息和单元连接关系来建立模型几何体。连接关系信息使程序能够组装单元刚度矩阵和单元等效力,以获得 2.1.6 节所示的整体平衡方程。

有两种方法可以生成模型:第一种方法是手动直接创建网格;第二种是采用实体几何建模,然后对几何实体进行网格化以得到节点和单元分布。

1. 手动创建网格模型

在商业软件包中的实体建模模块广泛应用之前,手动创建网格模型是唯一可用方法。在一些早期的软件版本和定制软件中这种方法仍然是唯一选择。手动生成网格时,用户首先创建节点,然后连接节点形成单元。接下来,用户可以直接在节点和(或)元素上施加边界条件和载荷。例 2.2 中给出了使用手动创建网格模型的过程。

例 2.2 采用手动创建网格模型的方法建立平面应力状态下的二维正方形板有限元模型,板的尺寸为 20mm×20mm×4mm。在 Abaqus 中划分 2×2 个正方形的连续单元,采用的单元类型为 CPS4R。

左侧边缘(图 2.7 中的节点 1、4、7)施加水平方向约束。此外,节点 1 施加垂直方向约束。在右侧边缘(节点 3、6、9)施加 $-9.5\text{MN}/\text{mm}^2$ 的均匀面载荷。板的材料为钢,$E = 210000\text{MPa}$,$\nu = 0.3$。在 DOS shell 中运行模型。然后,使用 Abaqus/CAE 可视化结果。

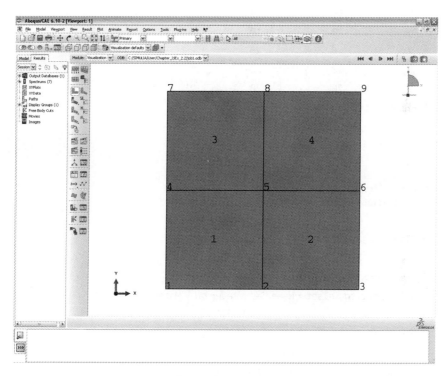

图 2.7 例 2.2 中的正方形网格模型

解 下面列出的命令行可以通过手动划分网格创建几何模型,并将其写入输入文件中(Ex_2.2.inp)。该文件可在网站[5,文件名:Ex_2.2.inp]上找到。

*Node 命令行后面是节点信息列表,每行包含节点号和每个节点的坐标 x、y 和可能的 z 坐标。

*Element 命令行后面是单元信息列表,每行包含单元号和围绕单元边界逆

时针(ccw)方向的节点编号序列(图 2.7)。

其余命令行的含义要么很明显,要么可以在 Abaqus 关键词参考手册 (Abaqus Keywords Reference Manual)[10]中找到。

```
* Heading
Example 2.2 FEA of Composite Materials: using Abaqus
* Part, name = Part1
* Node
    1,        0.,        0.
    2,       10.,        0.
    3,       20.,        0.
    4,        0.,       10.
    5,       10.,       10.
    6,       20.,       10.
    7,        0.,       20.
    8,       10.,       20.
    9,       20.,       20.
* Element, type = CPS4R
    1,  1,  2,  5,  4
    2,  2,  3,  6,  5
    3,  4,  5,  8,  7
    4,  5,  6,  9,  8
* Nset, nset = NamedSet, generate
    1,  9,  1
* Elset, elset = NamedSet, generate
    1,  4,  1
* * Section: Section - 1
* Solid Section, elset = NamedSet, material = Material-1
4.,
* End Part
* Assembly, name = Assembly
* Instance, name = Part1-1, part = Part1
* End Instance
* Nset, nset = Nodeset1, internal, instance = Part1-1
1
* Nset, nset = Nodeset2, internal, instance = Part1-1
4, 7
* Elset, elset = Elset1, internal, instance = Part1-1
    2, 4
```

```
* Surface, type = ELEMENT, name = Surface1, internal
Elset1, S2
* End Assembly
* Material, name = Material-1
* Elastic
210000. , 0.3
* Step, name = Step-1
* Static
1. , 1. , 1e-05, 1.
* Boundary
Nodeset1, 1, 2
Nodeset2, 1, 1
* Dsload
Surface1, P,-9.5
* Restart, write, frequency = 0
* Output, field, variable = PRESELECT
* Output, history, variable = PRESELECT
* End Step
```

要使用命令运行 Abaqus 软件,首先必须进入 DOS shell,如下:在 Windows 7 系统中单击 Start,然后在 Search Programs and Files 窗口中键入 cmd,然后按回车键。然后,在 DOS shell 中,执行以下操作:

```
> cd c:\SIMULIA
> mkdir User
> cd User
```

这里 cd 代表变更目录,mkd 为创建新文件夹(mkd 为"make directory"的缩写)[1]。

接下来,将文件 Ex_2.2.inp 从网站[5]中复制到文件夹 c:\Simula\User,或者使用记事本或其他文本编辑器[2]键入上面列出的命令行。运行 Abaqus 的命令行是:

```
> abaqus job = job1 input = Ex_2.2
```

现在,单击 Start 在 Windows 7 中启动 CAE,然后单击 All Programs,再单击

---

[1] 如果系统管理员限制了对 C:\SIMULIA 的写访问,请将用户目录设置在其他不受限制的位置,如 C:\TEMP,或参阅附录 C 以获取其他详细信息。

[2] 我们建议使用在 http://notepad-plus-plus.org 上免费提供的 Notepad++。

Abaqus 6.xx,再单击 Abaqus/CAE。一旦软件启动,关闭弹出的窗口。然后,在屏幕顶部①的菜单栏上依次单击 File,Open,导航到 c:\Simulia\User\文件夹,将文件类型(File Filter)更改为 *.odb,然后选择 Job1.odb。此时,应该看到如图 2.7 所示的网格模型。

要显示单元和节点编号,可以在可视化模块中执行以下操作:在顶部菜单栏上,依次单击 Options,Common,Labels(图 2.8),选中 Show element labels 和 Show node labels。在同一窗口中,要更改标签的字体,可以执行以下操作:Set Font。为便于后续参考,本段落中给出的操作说明总结如下(见表 2.1 中的注释)。

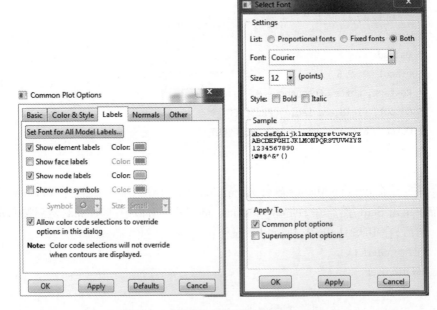

图 2.8　以更大的字体显示单元和节点号

```
Menu: File,
    Open[c:\Simulia\User\], File Filter[*.odb], File Name[Job1.odb]
Menu: Options,
Common, Labels,
    # checkmark[Show element labels]
    # checkmark[Show node labels]
    Set Font for All Model Labels, Size[24], OK
OK
```

---

① 也称为显示器、监视器或窗口。

2. 实体建模

用户可以使用体、面、线和点等特征创建实体模型。边界条件、载荷和材料属性可在划分网格前分配给实体模型的相应部分。这样，若重新划分网格就不会丢失载荷和边界条件。模型的网格划分在求解之前进行即可。在本书的示例 2.3 中展示了实体建模步骤。

Abaqus/CAE 具有在同一 CAE 环境中构造实体模型、划分网格、求解和可视化结果的所有功能。也可使用更先进的实体建模软件，如 SolidWorks[11]，生成实体模型，然后再导入 CAE 中定义边界条件和载荷，调用求解器计算，最后对计算结果进行可视化。下面简要描述 Abaqus 的 CAE 图形用户界面(GUI)。表 2.1 列出了用于描述在 CAE 中构造模型所需步骤的部分符号。

表 2.1 描述用户与 Abaqus/CAE 交互的首字母缩略词和命令等伪代码符号①

伪代码符号	含义
=	等于
= =	恒等
(+)	展开左侧模型树上的项目
(−)	折叠左侧模型树上的项目
[ ]	用户输入，如文件名、值等
OK	用于完成一系列命令的按钮名称
×	用于取消一系列命令的按钮名称
Cont	Continue，是表示 OK 的按钮名称
Dis	Dismiss，是表示取消的按钮名称
Enter	键盘上的 Enter 键
#	注释行
# pick	根据 WS 底部窗口中的请求选择对象
>	DOS 提示符
> > >	Python 提示符
Esc	退出键
ctrl-z	取消步骤
ctrl-y	重做步骤
F6	根据工作区自动调整
crtl-click	从所选内容中取消选择

---

① 命令是操作、指令、步骤、鼠标单击、选择、输入数据以及用户在图形界面(GUI)上执行的所有其他操作。

(续)

伪代码符号	含义
shift-click	添加到所选内容
mouse wheel	放大/缩小
CS	坐标系统
BC	边界条件
WD	工作目录
WS	工作区

3. CAE 操作窗口

CAE 操作窗口①(图 2.7)分为以下几个区域:

(1)菜单栏(Menu):顶部菜单栏(在窗口顶部的水平方向上)是动态的,也就是说,它根据正在处理的问题而变化。具体而言,它会根据不同的模块(Module)进行更改,下面会有详细解释。

(2)工具栏(Toolbar):在菜单栏下方是动态工具栏。可以依次把鼠标放置在每个工具图标上来了解工具的名称。它们是动态的,取决于所选择的模块(Module)。

(3)模型树(Trees):在窗口左边,有两个选项卡:模型(Model)和结果(Results);每个选项卡都由模型树组成。可以右击模型树中所有项目以修改每个项目中存储的信息。这些模型树对应的数据库文件为 .mdb 和 .odb,如下所示:

①模型(Model):.mdb;

②结果(Results):.odb。

(4)下拉菜单:在 Model 和 Results 选项卡的右边,有几个动态的下拉菜单,例如 Module、Model 和 Part。② Module:中包含一些选项,如下:

①部件(Part);

②特性(Property);

③装配(Assembly);

④分析步(Step);

⑤相互作用(Interaction);

⑥载荷(Load);

⑦网格(Mesh);

⑧作业(Job);

---

① 此时,用户应已进入 Abaqus/CAE 窗口。
② 由于下拉菜单是动态的,在图 2.7 中只显示了 Module 和 ODB。

⑨可视化(Visualization);
⑩草图(Sketch)。

(5)工作区:工作区占据了大部分屏幕。它位于下拉菜单下面,在Model/Results选项卡的右边。

(6)在Model/Results数据树和工作区之间有一系列图标,与上面的下拉菜单动态链接。如果将光标悬停在图标上,图标的名称会显示出来。图标是菜单栏上相关命令的快捷方式。

(7)在工作区下面有一个信息窗口,通过单击左边的选项卡(图2.7)切换到Python shell①。Python shell通过＞＞＞提示符识别。

例2.3 使用Abaqus/CAE中的实体建模技术,创建图2.9所示的几何模型并施加边界条件和载荷。对问题进行求解并可视化结果。部件的厚度为4.0mm。其材料性能为$E=195000\text{MPa}, \nu=0.3$。

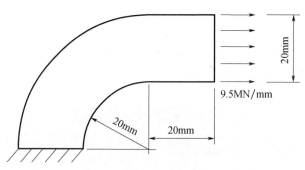

图2.9 例2.3中的曲梁

解 本例的解决步骤分散在以下各个小节中,每个小节描述了建模、求解和可视化所涉及的各个步骤。读者最好在计算机上运行Abaqus/CAE,并在阅读本章节时执行所有指令进行实操学习。

要在Windows 7系统中启动Abaqus/CAE,请依次单击Start,All Programs,Abaqus 6.xx,Abaqus/CAE②。一旦软件启动,关闭弹出的窗口。除了没有划分网格,屏幕内容应该基本如图2.7所示。

要做的第一件事是确保这个示例的文件保存到一个单独的文件夹中,这样它们就不会与其他示例文件混淆。最简单的方法是立即将这个示例保存到一个新文件夹中,即使它是空的。在屏幕顶部的菜单栏上,依次单击File,Save As,导

---

① Shell为一种程序窗口,用来接收文字(打字输入)命令。Python shell接收Python代码命令,DOS shell接收DOS命令,如cd等。

② 快速操作提示:可以把Abaqus/CAE快捷方式拖拽到Windows Start按钮上方。这样就可以把Abaqus/CAE快捷方式与Windows直接连接起来。

航到选择的新文件夹,或者根据需要创建一个文件夹;然后在 File name 中输入 Ex_2.3,单击 OK。这些步骤总结如下。

  Menu: File, Save as, New directory[name of new directory],
    Select new directory,[name of file], OK

  首先需要创建一个或多个部件,这些部件表示模型的几何。由于这个例子非常简单,一个部件就足够了,而更复杂的模型需要多个部件。有许多方法可以定义部件的几何。本例中,使用弧线(Arcs)连接的点(Points)和直线(Lines)来定义。要创建部件,必须将 Module 设置为 Part。记住

Module:指 Model/result 选项卡右侧的第一个下拉框(图 2.7)。

Menu:指位于 CAE 窗口顶部的菜单栏。

这样,下面操作步骤将打开如图 2.10 所示的对话框:

 Module: Part
  Menu: Part, Create,
  Name[Part-1], 2D, Deformable, Shell, Approx size[200], Cont
  # sketch mode is now active

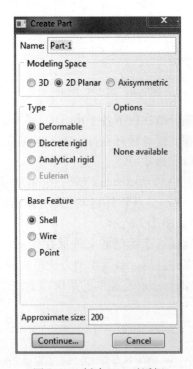

图 2.10 创建 Part 对话框

接下来,定义点并用弧线连接它们,步骤如下。参见图 2.11 和下面的分步说明。
Menu: Add, Point,
　　[0,0] # type in the input window below the WS, then[Enter]
　　[-20,0]
　　[-40,0]
　　[0,20]
　　[0,40]
　　[20,20]
　　[20,40]
　　F6　# zoom to fit
Menu: Add, Arc, Center/Endpoints,
　　# pick center point (see Fig. 2.11)
　　# pick point left of center point
　　# pick point above of center point, the arc is done
　　# again pick center point (see Fig. 2.11)
　　# pick point left - most of center point
　　# pick point above - most of center point, the outer arc is done
　　X # to finalize the Add feature (do not click Done)

此时,工作区应如图 2.11 所示。

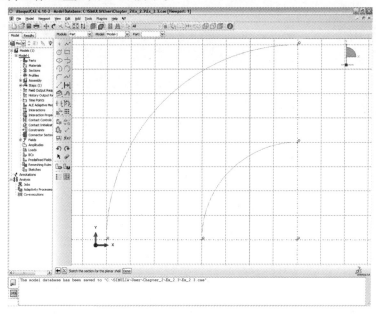

图 2.11　定义两个弧线

草图模式(Sketch mode)与SolidWorks[11]类似。弧线、点和其他几何构型称为实体。添加/编辑实体时,必须遵循WS下面显示的指令。位于指令行左侧的×用于取消当前操作。

仍参考图2.11,添加直线条以闭合图形边界,一旦图形闭合,则成功创建部件。

Menu: Add, Lines, Connected,
    # pick two nodes to close bottom left of the part
    X　# to finalize the Line drawing
Menu: Add, Lines, Connected,
    # pick 4 nodes in sequence to close rectangle at top right
    X　# to finalize the Line drawing
Done　　# to finish the sketch, thus creating the part

此时,工作区应如图2.12所示。

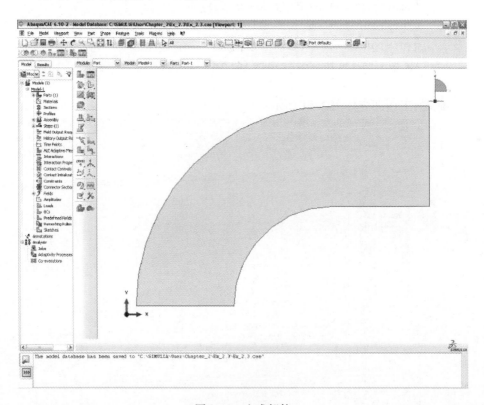

图2.12　生成部件

2.3.2　材料及截面属性

部件必须被赋予材料性能。依据分析类型,材料的性能可以是线性的(线弹性分析)或非线性的(如损伤力学分析),各向同性的或正交各向异性的,常数的或随温度变化的。正确的材料性能输入是成功分析复合材料的重要方面之一。本书对材料属性给予了极大的关注。现在,只要使用一种线弹性的各向同性材料来说明这个过程就足够了。对于结构分析,弹性特性必须根据 1.12 节相关内容来定义。其他力学性能,如强度、密度和热膨胀系数是可选的,它们定义与否取决于分析的目的。

在 Abaqus/CAE 中,材料输入需要进入 Module：Property, Menu：Material。弹出的窗口如图 2.13 所示,在其中输入材料属性。步骤如下：

```
Module: Property
    Menu: Material, Create,
        Name[Material-1], Mechanical, Elasticity, Elastic,
        Type, Isotropic[195000, 0.3], OK
```

所有单元均需要被赋予材料属性,而结构性(structural)单元还需要额外的参数,这些参数随单元类型的不同而不同。这些参数是在形成单元方程时,由 3D 控制方程进行积分得到。例如,式(2.1)中出现的截面面积 A,是由 3D 偏微分方程在杆的横截面上积分得到,从而形成常微分方程式(2.1)。

梁(beam)单元需要截面面积和惯性矩。层合板壳(laminated shell)单元需要层合板的堆叠顺序(LSS,见图 3.4)。连续体(continuum)单元,如 3D 实体单元(表 2.2),不需要额外的参数,只需要材料属性,因为网格可以完全描述几何特性。然而,代表层合板复合材料的连续体单元仍然需要 LSS。

常规(conventional)壳和梁是典型的结构性(structural)单元。Abaqus 也用 conventional 这个词来强调壳单元不是连续体。另一方面,连续体壳(continuum shell)为连续体单元,其引入了运动约束用于模拟壳的行为。

因此,除了材料(material)属性外,还需要创建截面(section)属性来提供结构性单元所需的附加参数。在 Abaqus 中,附加参数称为截面参数(section parameters);在 ANSYS 中,它们被称为截面常数(section constants)[①]。连续体单元不需要截面参数,但截面属性必须建立。

在这个简单的例子中,截面参数用于定义图 2.9 所示的处于平面应力状态下的部件。复杂的模型可能有几个不同的截面属性。截面也在 Module：Property

---

① 在旧版的 ANSYS 软件中,截面常数被称为"真实常数(real constants)"。

中定义,如下所示:

Menu: Section, Create,
 # plane stress/strain is considered solid even though
 # pstres needs a thickness
 # pstran does not
 # shells are not solids, they need a thickness
 # beams are not solids, they need area and moments of inertia
 Name[Section-1], Solid, Homogeneous, Cont
 Material: Material-1,
 Plane stress/strain thickness[4.0], OK

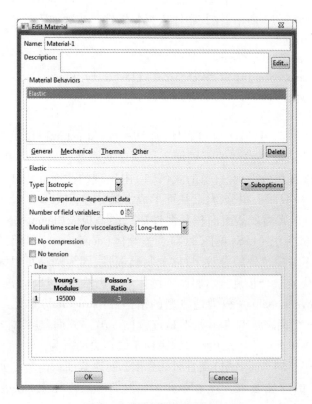

图 2.13　材料属性定义窗口

表 2.2　Abaqus 和 ANSYS 软件中的一些单元类型

Abaqus	ANSYS	节点	DOF	单元描述
结构性单元				
T2D2	LINK180	2	u_X, u_Y	线性杆/桁架,2D 空间

(续)

Abaqus	ANSYS	节点	DOF	单元描述
T3D2	LINK180	2	u_X, u_Y, u_Z	线性杆/桁架,3D 空间
—	COMBIN14	2	u_X, u_Y, u_Z	弹簧/阻尼,3D 空间
B21	BEAM188	2	u_X, u_Y θ_X, θ_Y	2D 空间线性梁
B31	BEAM188	2	u_X, u_Y, u_Z $\theta_X, \theta_Y, \theta_Z$	3D 空间线性梁
CPE4R	PLANE182	4	u_X, u_Y	2D 空间中的四边形实体
CPE8R	PLANE183	8	u_X, u_Y	2D 空间中的四边形实体
S4R	SHELL181	4	u_X, u_Y, u_Z $\theta_X, \theta_Y, \theta_Z$	3D 空间中的四边形壳(常规)
S8R	SHELL281	8	u_X, u_Y, u_Z $\theta_X, \theta_Y, \theta_Z$	3D 空间中的四边形壳(常规)
S8R5	—	4	u_X, u_Y, u_Z θ_X, θ_Y	3D 空间中的四边形薄壳(常规)
连续体单元					
C3D8	SOLID185	8	u_X, u_Y, u_Z	3D 空间中的六面体实体
C3D20	SOLID186	20	u_X, u_Y, u_Z	3D 空间中的六面体实体
SC8R	SOLID190	8	u_X, u_Y, u_Z	3D 空间中的六面体壳(连续体)

创建和编辑截面属性窗口,依次如图 2.14 所示。它们用于定义截面属性。

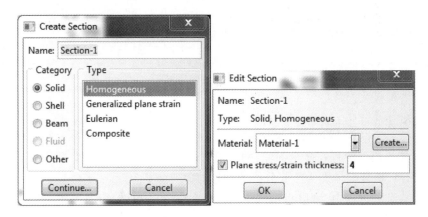

图 2.14　创建和编辑截面属性窗口

最后,我们需要告诉软件每个部件的截面属性是什么,这需要将截面属性赋予相应部件来完成。在这个简单的模型中,只有一个部件和一个截面属性,所以赋予很简单。请注意,截面赋予是在同一个模块中完成的,即 Property 模块。操作过程如图 2.15 所示,总结如下:

Menu: Assign, Section
  # pick all (click on part), Done
  OK

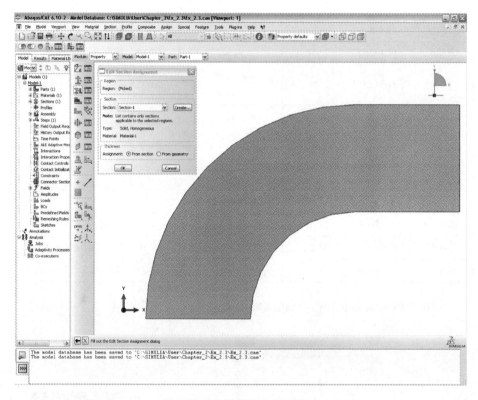

图 2.15 为部件赋予截面属性

2.3.3 装配

如果模型中存在多个部件,则必须将这些部件组合到一起,即为装配体(assembly),该装配体即为需要分析的物理对象。

在装配过程中,可以指定网格与部件的关系。也就是说,网格可以依赖(dependent)也可以独立(independent)于部件。如果部件是网格依赖的,在一个装配体中多次使用该部件(实例)时,该部件的所有实例都将被完全相同地网格化。而网格独立意味着部件的每个实例都必须独立地进行网格划分。后者为部件的某些实例提供了改进网格的灵活性,但也增加了更多的网格划分带来的时间。由于本示例只有一个部件,所以装配过程非常简单,如下所示(另见图 2.16):

Module: Assembly
Menu: Instance, Create,
  Independent, OK

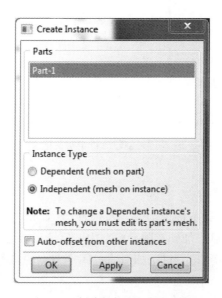

图 2.16 创建部件实例的对话框

2.3.4 分析步

接下来,分析过程通常分为多个分析步,每个分析步代表不同的加载和约束条件。分析步最少有两个,即初始分析步和至少一个附加分析步。初始分析步不能施加载荷条件,只能施加边界条件。

当前示例仅使用两个分析步,如下所示:

Module: Step
Menu: Step, Create,
  Name[Step-1], General, Static/General, Cont
  OK

创建分析步的对话框后面跟着编辑分析步的对话框以完成该分析步的数据输入,如图 2.17 所示。

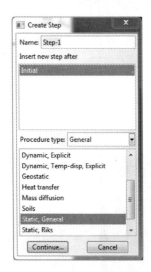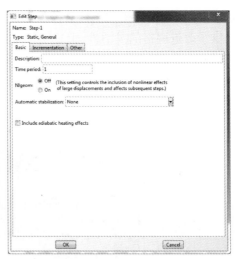

图 2.17　创建分析步对话框

2.3.5　载荷

在结构分析中,荷载条件由力、压力、惯性力(如重力)和指定位移来定义,所有这些载荷形式都可加到有限元模型中。后续章节会详述如何对有限元模型施加不同类型的载荷,也会讨论如何通过固定节点的自由度(位移和旋转)获得反作用力。

荷载可以通过集中力和力矩的形式施加在节点上,如例 2.4。此外,载荷也可以分布在各个单元上,如面载荷、体载荷、惯性载荷或其他耦合场载荷(如热应变)。在例 2.5 中使用的是面载荷形式。

面载荷是施加在表面上的分布载荷,如压力。体载荷是一种体积载荷,例如在结构分析中,由于温度的增加,材料的膨胀产生的载荷。惯性载荷是指由于物体的惯性产生的载荷,如重力加速度、角速度和角加速度。

施加在节点上的集中载荷通过直接添加力矢量实现。然而,分布载荷情况下,则使用单元插值函数来计算节点的等效力矢量。

在 Abaqus/CAE 中,结构的载荷和边界(支撑)条件在载荷模块中施加,具体如下:

Module: Load
Menu: Load, Create,
　　Name[load-1], Mechanical, Pressure, Cont
　　# pick the vertical edge on the right, Done
　　Magnitude[-9.5], OK

创建载荷的对话框(图 2.18 左上角)之后是选择用于施加荷载(或边界条件)的表面(图 2.18 中 $x-y-z$ 坐标轴下方的注释对话框)。然后,在编辑载荷对话框中输入载荷值(图 2.18 中右上角)。

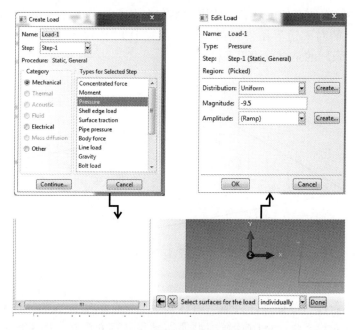

图 2.18　创建载荷对话框

2.3.6　边界条件

边界条件是边界上自由度(DOF)的已知值。在结构分析中,DOF 为位移和旋转。有了自由度信息,分析软件就知道式(2.46)中的 a 哪些值已知哪些值未知。

1. 位移和旋转自由度约束

通常,一个节点拥有多于一个的 DOF。例如,如果有限元模型使用 2D 空间中的梁单元,则该梁单元节点有 3 个 DOF,即水平位移自由度、垂直位移自由度和围绕垂直于该 2D 平面的轴的旋转自由度。施加不同的边界条件需要准确约束 DOF。在 2D 梁单元中,仅对水平位移自由度和垂直位移自由度进行约束,可以得到简支边界条件,但若约束所有自由度则会得到固支边界条件。

2. 对称条件

施加对称条件可以在不损失计算精度的情况下减小模型的计算规模。对称条件必须同时满足 4 种对称类型:几何对称、边界条件对称、材料对称和载荷对称。在这些条件下,解也是对称的。例如,关于 $y-z$ 平面对称意味着对称平面上的节点具有以下约束

$$u_x = 0; \theta_y = 0; \theta_z = 0 \qquad (2.47)$$

式中:u_x 为沿 x 方向的位移自由度;θ_y,θ_z 分别为 y 和 z 轴的旋转自由度(图2.19)。请注意,壳理论中旋转的定义(ϕ_i,见 3.1 节)与遵循右手规则的旋转 θ_i 定义不同。Abaqus① 中所有的旋转都是使用满足右手规则的旋转角 θ_i 来描述的。对称平面上节点的对称边界条件包括面外的位移 DOF 约束和面内的旋转 DOF 约束。例 2.5 采用的是对称边界条件。

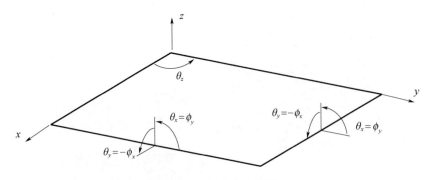

图 2.19　板或壳的旋转

3. 反对称条件

反对称条件与对称条件类似。当计算模型仅荷载条件表现为反对称性时,可以应用反对称条件。此时计算模型还需满足几何对称性、边界条件对称性和材料对称性。反对称平面上节点的反对称边界条件包括面内的位移 DOF 约束和面外旋转 DOF 的约束。

4. 周期性条件

当材料、载荷、边界条件和几何特征表现为周期性$(x,y,z) = (2a_i, 2b_i, 2c_i)$时,只需对尺寸为$(2a_i, 2b_i, 2c_i)$的结构部分进行建模。结构的周期性重复意味着解也是周期性的。周期性条件可以通过不同的方式施加,可以利用在 DOF 中的约束方程(CE)(见 6.2 节)也可以利用拉格朗日乘子实现。

5. 在 Abaqus/CAE 中施加边界条件

在 Abaqus/CAE 中,创建边界条件(Create Boundary Condition)对话框后跟着编辑边界(Edit Boundary Condition)条件对话框,如图 2.20 所示。进入该对话框步骤如下:

---

① 引用"除了轴对称的单元,自由度号对应关系如下:1:x - 位移,2:y - 位移,3:z - 位移,4:绕 x 轴旋转,弧度表示,5:绕 y 轴旋转,弧度表示,6:绕 z 轴旋转,弧度表示,此处,x、y 和 z 方向分别与全局 X、Y 和 Z 方向一致;但是,如果在节点上定义了局部坐标转换(参见转换坐标系[7,2.1.5 节]),则它们与定义的转换后的局部方向一致。"结束引用文献[7]。

Module: Load
Menu: BC, Create,
    Name[BC-1], Step: Initial, Mechanical, Displacement/Rot, Cont
    # pick the horizontal edge on bottom, Done
    # checkmark U1, U2, UR3, OK
Step: Step-1 # to see the load and BC together

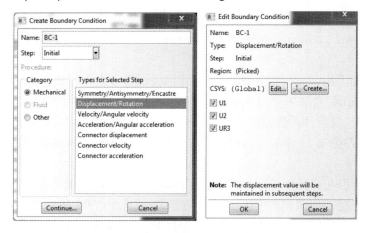

图 2.20　创建边界条件对话框

若要一起查看施加的 BC 和载荷，必须将分析步切换至 Step-1，因为在 Step：Initial 时，只有 BC 存在。请在 Module 右侧的 Step 下拉框中切换分析步。切换后，窗口应该如图 2.21 所示。

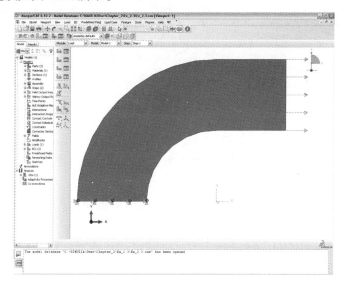

图 2.21　同时显示载荷和边界条件

2.3.7 划分网格和单元类型

接下来,装配体每个实例均需要进行网格化分。网格化分有许多种方式。对于这个例子,可以使用一种非常简单的方法,称为全局种子法,如下:

Module: Mesh
Menu: Seed, Instance,
　　　Approx global size[5.0], Apply, OK
Menu: Mesh, Controls, Quad, Structured, OK

全局种子法,如图2.22中的左边对话框所示,是指定网格密度的最简单方法。最好在OK之前先单击Apply。这样的话可以先查看种子的分布情况,视情况再修改它,确定种子分布形式后再提交。当然,所有参数都可以再编辑。

然后,在图2.22中的右边对话框中,网格控制(Mesh Control)的选项用于指定单元特征。

有限元分析(FEA)程序均有单元库(Element Library),其中包含许多不同的单元类型。使用的单元类型决定了单元形式,如自由度集、插值函数、单元是否为2D单元或3D单元等。单元类型分为拉伸-压缩杆、弯曲梁、实体、壳、层合板壳等,每种商业软件均使用不同的标签标识不同的单元类型。

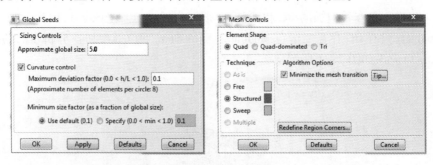

图2.22　全局种子法(左对话框)和网格控制(右对话框)

一部分单元类型的标识和基本特征见表2.2。此外,每种单元类型均有不同的可选项。例如,在平面实体单元中,允许在平面应变和平面应力单元之间进行选择。

在单元类型(Element Type)对话框中选择所需的单元类型如图2.23所示。最后一步是完成(实例)网格划分。该步骤如下:

Menu: Mesh, Element type, Standard, Linear, Plane stress, OK
Menu: Mesh, Instance, Yes

最后,生成的模型网格应该与图 2.24 中显示的网格一致。

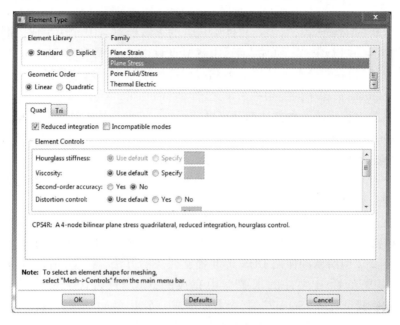

图 2.23　单元类型

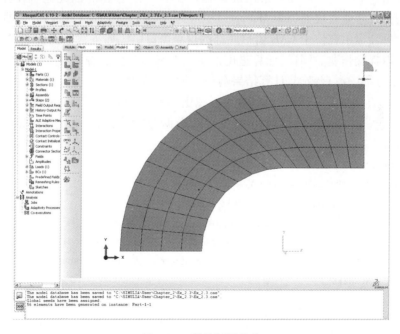

图 2.24　部件划网格分

2.3.8 求解

在分析的求解阶段，有限元软件中包含的求解器对有限元法生成的联合方程组式(2.46)进行求解。通常，通过求解节点自由度值，即位移和转角，得到基础结果。然后，在积分点上计算应力和应变等派生结果。基础结果称为节点解(nodal solutions)，派生的结果称为单元解(element solutions)。

解联立方程组的几种方法都是可以使用的。有些方法更适合于规模较大的模型，有些方法求解非线性问题速度更快，另一些方法则允许通过并行计算方式分块求解。商用有限元程序以批量的方式求解这些方程组。直接求解法是目前常用的一种方法，因为它对有限元分析效率高。当非线性分析时，必须进行多次求解，这样会大大增加计算时间。

要提交求解一个模型，必须先创建一个任务(Job)。然后，该求解任务从CAE中提交给Abaqus求解器执行。在CAE中提交任务步骤较为简单，所以应避免如例2.2中所示的使用DOS shell提交任务的方法。但是，请注意，任务的求解过程是在CAE之外的求解器中完成的，CAE只包含了模型数据文件(.mdb)，在提交任务(Submit)时，会自动生成一个由Abaqus求解器读取的输入文件(.inp)。虽然求解器会对输入文件进行错误检查，但它无法找到所有可能的模型错误。由于求解器的错误检查能力远小于前处理器，因此通常在提交任务之前先在作业管理器(Job Manager)中进行数据检查(Data Check)。数据检查相关信息会显示在位于CAE窗口底部的信息窗口中。如果数据检查通过，下一步就可以提交作业。同样，计算进度信息也会显示在信息窗口中。一旦求解完成，求解过程中没有错误，计算结果会写入输出数据文件(.odb)中。此时，单击Results后直接从任务(Job)窗口进入可视化(Visualization)模块。最后一步按需求改变可视化显示模式，读取.odb文件中数据，并可视化显示计算结果。该过程步骤如下：

```
Module: Job
Menu: Job, Manager, # opens the Job Manager window
    Create, Name[Job-1], Model:[Model-1],
        Description:[User's heading goes here], Cont, OK
    Data check       # watch the execution window below the WS
    Submit           # watch the execution window below the WS
    Results
```

图2.25所示为作业管理器(Job Manager)和创建作业(Create Job)对话框。在作业管理器中，有写入输入文件(Write Input)、数据检查(Data Check)、提交

（Submit）、继续（Continue）、监控（Monitor）、结果（Results）和终止（Kill）的按钮。这些功能允许用户控制 Abaqus 求解器的执行状态，并在任务完成后直接跳转至可视化模式。

在描述（Description）中允许用户输入计算任务的说明标题，该标题将显示在可视化模块上，这需要事先在 Viewport Annotation Options，General，Show title box 选项前勾选。在输入文件 .inp 中，该标题出现在关键字 *Heading 之后，如例 2.2 所示。

图 2.25　任务管理和创建任务对话框

2.3.9　后处理和可视化

一旦求解完毕，就可以使用后处理模块来查看和分析结果。结果可以通过图形或列出具体数值的方式进行查看。但求解的模型通常包含大量的结果信息，所以最好使用图形工具查看计算结果。商业软件的后处理程序可以生成应力和应变云图、变形云图等。这些软件通常还包括额外的计算能力，如误差估计、加载工况组合或路径操作等。

可视化模块可以从作业管理器中直接进入（图 2.25 中的 Results 按钮），也可以通过 Module 的下拉菜单切换至 Visualization 模块。但后者要求用户事先通过 CAE 执行 File，Open，File Filter，[.odb]，File Name，[job1.odb]步骤将结果文件 .odb 导入。若已通过单击作业管理器中的 Results 按钮进入可视化模块，则基本的可视化操作如下所示：

```
Module: Visualization
Menu: Plot, Contours, On Deformed Shape
```

如图 2.26 所示，在模型变形基础上显示 von Mises 应力云图。应力最大值为 65.78MPa。最后，在退出程序之前，保存包含 Abaqus/CAE 当前会话状态的

各种文件,如下所示:
    Menu: File, Save   # save the .cae and .mdb files
    Menu: File, Exit   # exit CAE

.mdb 文件包含所有模型信息。.cae 文件允许用户可以稍后恢复会话状态。Ex_2.3.cae 文件的副本可以在文献[5]上获得。

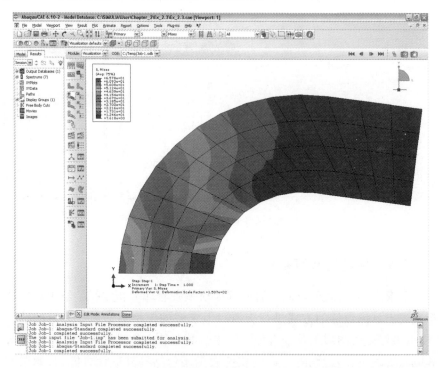

图 2.26　在变形模式下显示 von Mises 应力云图

例 2.4 和例 2.5 分别包含了通过列表和图形输出形式查看结果的命令。

例 2.4　文献[8,例 2.5]杆件一端固支,另一端有轴向载荷,使用商业有限元软件计算杆件的轴向位移。该杆件材料为钢,$E = 200000$MPa,直径 $d = 9$mm,长度 $L = 750$mm,载荷 $P = 100000$N。另外计算应力和应变。使用 3 个双节点(线性)桁架(truss)单元(在 ANSYS 中称为连接(link)单元)。

解　Abaqus/CAE 操作步骤如下所示。CAE 文件可在网站[5,Ex_2.4.cae]中获得。对杆件进行一维分析时不需要泊松比参数,但在 Abaqus/CAE 中定义各向同性材料时必须输入泊松比,因此该例子使用的泊松比为 $\nu = 0.3$。

ⅰ. 创建几何模型。
Module: Part

```
Menu: Create, Name:[Part-1], 2D, Deformable, Wire,
    Approx size:[2000], Cont
Menu: Add, Line, Connected Lines,
    # enter coordinates in the dialog box below the WS
    [0,0], Enter,
    [750,0], Enter, X, Done
```

ⅱ．输入材料和截面属性。
```
Module: Property
Menu: Material, Create, Name[Material-1],
    Mechanical, Elasticity, Elastic, Isotropic,[200E3, 0.3], OK
Menu: Section, Create, Name:[Section-1], Beam, Truss, Cont
    Material-1, Cross - sectional area:[63.617], OK
Menu: Assign, Section, # pick the truss, Done, OK
```

ⅲ．创建装配体，本例子的创建过程很简单。
```
Module: Assembly
Menu: Instance, Create,[Part-1], Independent, OK
```

ⅳ．按照求解策略创建分析步。
```
Module: Step
Menu: Step, Create, Name[Step-1], Static, General, Cont, OK
```

ⅴ．指定载荷和边界条件并显示。
```
Module: Load
Menu: BC, Create, Name:[BC-1],
    Step: Initial, Category: Mechanical, Types: Symm, Cont
    # pick the origin, Done,
    Encastre, OK
Menu: Load, Create, Name:[Load-1],
    Step: Step-1, Category: Mechanical, Types: Concentrated, Cont
    # pick the end @ x=750, Done,
    CF1[100E3], OK
Menu: View, Assembly display options,
    Attribute, Symbol, Size:[24], Arrows:[24], Apply, OK
    F6 # roll the mouse wheel forward to see the part smaller
```

ⅵ. 划分网格。
Module: Mesh
Menu: Seed, Instance, Apply, # to see the proposed mesh seeds
　　# change it to get 3 elements
　　Approx global size[250], Apply, OK
Menu: Mesh, Element Type, Family: Truss, Linear, OK
Menu: Mesh, Instance, Yes
Menu: View, Assembly display options,
　　Tab: Mesh, # checkmark Show node labels, Apply
　　# checkmark Show element labels, Apply, OK
　　# to save these options as default for the future
Menu: File, Save Options, Current, OK

ⅶ. 创建任务并提交 Abaqus 求解器求解。
Module: Job
Menu: Job, Manager,
　　Create, Name[Job-1], Cont, OK
　　Submit, # monitor the progress window below the WS
　　Results, # switches to Module: Visualization

ⅷ. 通过点击作业管理器的 Results 按钮，可以进入可视化模块。
Module: Visualization
　　# to get arrows with magnitude proportional to the displacement
　　# max. value read on the color scale is 5.895 mm
Menu: Result, Field Output,
　　Tab: Symbol Variable, Name: U, Vector: Resultant, Apply, OK
　　# to get the stress
　　# max. value read on the color scale is 1572 MPa
Menu: Result, Field Output, Tab: Primary Variable,
　　Name: S, Component: S11, Apply, Contour, OK
　　# to get the strain
　　# max. value read on the color scale 7860E-3 (micro-strain)
Menu: Result, Field Output, Tab: Primary Variable,
　　Name: E, Component: E11, Apply, Contour, OK
　　# further modify the way things are seen on the screen
　　# and save the settings
Menu: View, ODB Display Options, Tab: Entity Display,

```
Symbol size[24]
Menu: File, Save Options, Current, OK
```

本例中,方便起见,单位统一为 N、mm 和 MPa = N/mm²。通过材料力学计算可以很容易地验证分析结果,如下所示:

$$U_x = \frac{PL}{AE} = \frac{(750)(100000)}{(63.617)(200000)} = 5.894 \text{mm}$$

$$\sigma = \frac{P}{A} = \frac{100000}{63.617} = 1571.9 \text{MPa}$$

$$\epsilon = \frac{\sigma}{E} = 7.859 \times 10^{-3}$$

例 2.5 文献[8,例 2.6]使用商用有限元程序求矩形含缺口矩形条的应力集中系数。尺寸和载荷施加状态如图 2.27 所示。使用 8 节点(二次)四边形平面应力单元进行计算。

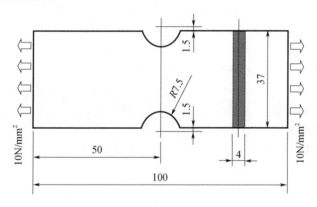

图 2.27 含缺口的矩形条尺寸和载荷施加状态

解 Abaqus/CAE 操作步骤如下所示。CAE 文件可在网站[5,Ex 2.5.cae]中获得。

i. 创建几何模型。

```
Module: Part,
Menu: Part, Create, Name:[Part-1],
    2D, Deformable, Shell, Approx size[200], Cont
    # a quarter model of the Figure in the textbook
Menu: Add, Point,
    [0,0]
    [50,18.5]
    X,
```

Menu: Add, Line, Rectangle, #pick diagonal points, X
　　Add, Point,
　　　　[0,20] # center
　　　　[7.5,20] # radius
　　X,
Menu: Add, Circle # select center and edge point
　　Edit, Auto trim # the unneeded lines/arches
　　X,
　　Done # with the sketch. finalize the part.
Menu: File, Save

ⅱ．输入材料和截面属性。
Module: Property
Menu: Material, Create, Name:[Material-1],
　　　Mechanical, Elasticity, Elastic,[190E3, 0.3], OK
Menu: Section, Create, Name:[Section-1], Solid, Homogeneous, Cont
　　　# checkmark Plane stress/strain thickness:[4], OK
Menu: Assign, Section, # pick the part, Done, OK

ⅲ．创建装配体。
Module: Assembly
Menu: Instance, Create, Name:[Part-1], Independent, OK

ⅳ．按照求解策略创建分析步。
Module: Step
Menu: Step, Create, Name:[Step-1], Static, General, Cont, OK

ⅴ．指定载荷和边界条件并显示。
Module: Load
Menu: BC, Create, Name:[BC-1], Step: Initial, Mechanical, Symm, Cont
　　# pick the horiz edge on bottom, Done, YSYMM, OK
Menu: BC, Create, Name:[BC-2], Step: Initial, Mechanical, Symm, Cont
　　# pick the vert edge on left, Done, XSYMM, OK
Menu: Load, Create, Name:[Load-1],
　　Step: Step - 1, Mechanical, Types: Pressure, Cont
　　# pick the vertical edge on the right, Done,
　　Distribution: Uniform, Magnitude[-10], OK

mouse wheel fwd to zoom out on the WS

ⅵ. 划分网格。
Module: Mesh
Above the WS, select Object: Assembly
Menu: Seed, Instance, Apply, #if you like it, OK,
    # otherwise Approx global size:[1.0], Apply, OK
Menu: Mesh, Controls, Quadratic, Structured, OK
Menu: Mesh, Element type, Standard, Quadratic,
      Family: Plane stress, Reduced Integration, OK
Menu: Mesh, Instance, Yes, # OK to mesh the part instance

ⅶ. 创建任务并提交 Abaqus 求解器求解。
Module: Job
Menu: Job, Manager
    Create, Name: Job-1, Cont, OK
    Submit # the job for execution by the Abaqus processor
    Results # switches to Module:Visualization

ⅷ. 可视化计算结果。
Module: Visualization
to get a report file
Menu: Report, Field output,
    Tab: Variable, Position: Integ. point, expand S, # checkmark S11
    Tab: Setup, Name:[abaqus.rpt],
    OK, # immediately the.rpt file is written in the current WD
    # min S11 = 0 at intersection of hole and horz symm line
    # max S11 ~ 28.1 at intersection of hole and vert symm line
    # or
    # integration point location varies upon mesh refinement
    # the node with max stress is always the same
    # unique nodal
    # displays average stress from neighbor elements at node
    # element nodal also, but displays the element number as well
    Tab: Variable, Position: Unique nodal, expand S, check S11
    Tab: Setup, Name:[abaqus.rpt], OK
    # open abaqus.rpt with notepad + +

```
# min S11 = 0 @  node 5
# max S11 ~ 27.9 @  node 7
# to see the node #'s
```
Module: Mesh, Object: Assembly
Menu: View, Assembly display options,
　　Tab: Mesh, # checkmark Show node labels, Apply

非应力集中区域应力为

$$\sigma_0 = \frac{P}{A} = \frac{10 \cdot 37 \cdot 4}{25 \cdot 4} = 14.8 \text{MPa}$$

由有限元计算结果,靠近缺口处的最大水平方向应力为27.9MPa。因此,应力集中系数为

$$k = \frac{\sigma_{\max}}{\sigma_0} = 1.885$$

例2.6　文献[8,例2.4]使用 Abaqus/CAE,采用不同的截面属性和两种材料,为具有不同单元类型(壳和梁单元)的穹顶结构创建有限元模型。通过这个例子,用户将熟悉使用 Abaqus/CAE 创建三维部件。

解　为便于在建模过程中操作部件,请按如下方式激活视图工具栏(Views toolbar):

Menu: View, Toolbars, Views.
Integrate it next to the other toolbars above the WS area.

创建三维圆顶结构。

ⅰ. 创建一个半球实体。

Module: Part
Menu: Part, Create, Name[Part-1], 3D, Deformable, Solid,
　　Revolution, Approximate size[2000], Cont,
Menu: Add, Arc, Center/Endpoint,
　　# in the dialog box below the WS enter the coordinates
　　[0,0],[0,500],[500,0], X # to end the Arc command, F6 # zoom/fit
Menu: Add, Line, Connected Lines,
　　# enter the coordinates
　　[0,500],[0,0],[500,0], X # to end the Line command, Done,
　　Angle[360], OK

3D 的半球体创建完成。

ⅱ. 创建侧壁。
Rotate the part so you can see the base of the semi-sphere,
Menu: View, Rotate, X # to end the Rotate command,
　　# roll the mouse-wheel backward to zoom in,
　　# save this view-point for use in future procedures,
　　# click the 'Save View' icon from the 'Views toolbar',
　　# in the pop-up window select[User 1], Save Current, OK,
Menu: Shape, Cut, Extrude, # pick the base of the semi - sphere,
　　# pick the edge of the base of the semi-sphere,
　　# Sketch mode is now active,
Menu: Add, Line, Rectangle, # enter the coordinates
　　[300,300],[-300,-300],
　　# create a new rectangle entering the coordinates
　　[600,600],[-600,-600], X # to end the Line command, Done,
　　Type:[Through All], OK

半球体已切割出 4 个相同的侧壁,如图 2.28 所示。

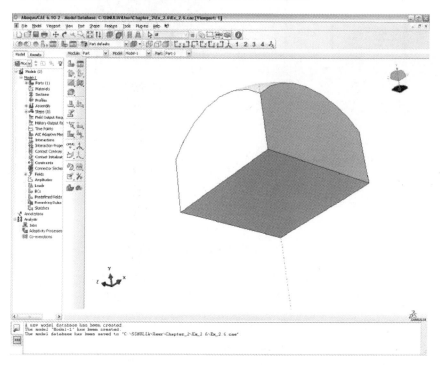

图 2.28　穹顶的侧壁

iii. 降低侧壁高度。

Menu: Shape, Cut, Extrude, # pick one of the side-walls,
 # in the dialog box below the WS select[horizontal on bottom],
 # pick the longest straight edge of the side-wall selected,
 # Sketch mode is now active,
 # roll the mouse-wheel forward to see the figure smaller,
 # Draw a rectangle in an area away from the figure,
Menu: Add, Line, Rectangle, # pick any two point on the screen,
 X # to end Line command

几何模型如图 2.29 所示。

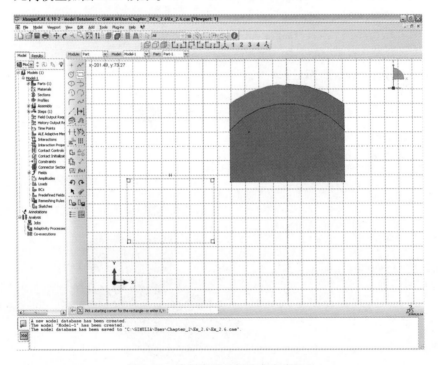

图 2.29 降低穹顶的侧壁高度步骤 1

Menu: Add, Dimension, # pick a vertical line of the rectangle
 # may need to move pointer and right-click to create dimension
 # enter the height of the rectangle[200],
 # pick a horizontal line of the rectangle,
 # enter the width of the rectangle[600],
 X # to end the Dimension command

几何模型如图 2.30 所示。

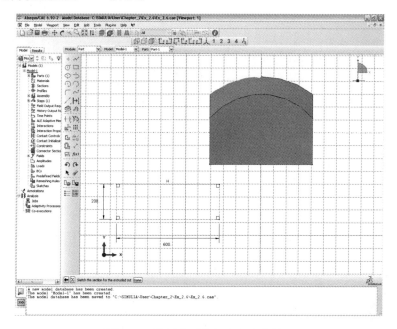

图 2.30　降低穹顶的侧壁高度步骤 2

Menu: Edit, Transform, Translate,
　　# in the notification area below the WS select[Move],
　　# pick the four lines forming the rectangle
　　# hold the 'shift-key' to individually select the lines, Done,
　　# pick the lower-left point of the rectangle,
　　# pick the lower-left point of the figure,
　　X # to end the Translate command, Done

矩形被移动到如图 2.31 所示位置。
　　Type:[Through All], OK

零件高度已降低,如图 2.32 所示。
ⅳ. 实体部件转化为壳部件。
Menu: Shape, Shell, From Solid,
　　# pick the part, Done, X # to end the shell command,
Menu: Tools, Geometry Edit, Face, Remove,
　　# pick the surface at the base of the part,
　　Done, X # to end the command, # close the pop-up window

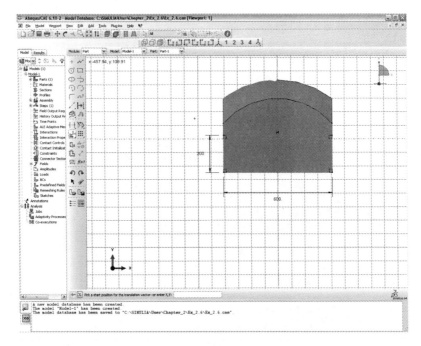

图 2.31　降低穹顶的侧壁高度步骤 3

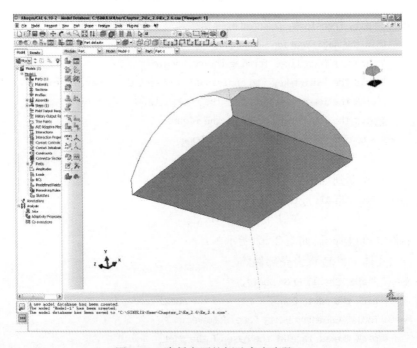

图 2.32　降低穹顶的侧壁高度步骤 4

生成的部件如图 2.33 所示。

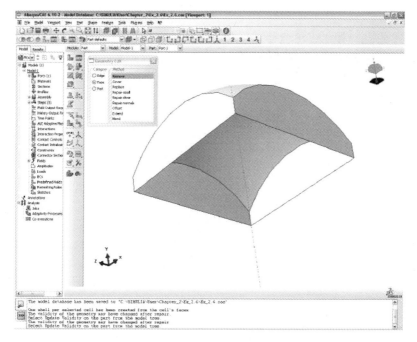

图 2.33 生成壳结构

Ⅴ. 创建支撑柱。

Menu: Shape, Wire, Sketch, # pick one of the side-walls,
    # in the dialog box below the WS select[horizontal on bottom],
    # pick the longest straight edge of the side-wall selected,
    # Sketch mode is now active,
    # roll the mouse-wheel forward to see the figure smaller,

Menu: Add, Line, Connected Lines,
    # pick the lower-left corner of the part,
    # pick any point directly below (it must create a vertical line),
    X # to end the Line command,

Menu: Add, Dimension, # pick the line,
    # enter the length of the line[200],
    X # to end the Dimension command,
    # Repeat the procedure to create a vertical line of length 200
    # starting at the lower-right point of the part,
    Done # to end the sketch

在对面的侧壁上创建另一个草绘,然后重复上面的步骤以创建另外两个垂直的支撑柱。现在生成的部件如图 2.34 所示。

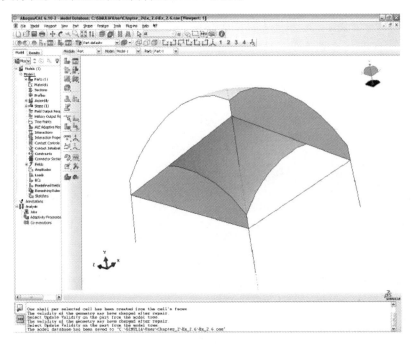

图 2.34 穹顶结构模型完成

材料和截面属性步骤如下。
ⅰ. 定义材料。
Menu: Material, Manager,
 Create, Name[Dome], Mechanical, Elasticity, Elastic, Isotropic,
 Young's Modulus[190e3], Poisson's ratio[0.27], OK,
 Create, Name[Columns], Mechanical, Elasticity, Elastic, Isotropic,
 Young's Modulus[200e3], Poisson's ratio[0.29], OK,
 # close the Material Manager pop-up window

ⅱ. 定义截面属性。
Menu: Section, Manager,
 Create, Name[Dome-Roof], Shell, Homogeneous, Cont,
 Shell thickness[6.0], Material[Dome], OK,
 Create, Name[Side-Walls], Shell, Homogeneous, Cont,
 Shell thickness[4.0], Material[Dome], OK,
 Create, Name[Columns], Beam, Truss, Cont, Material[Columns],

Cross-sectional area[100], OK,
close the Material Manager pop-up window

ⅲ. 分配截面属性。
Menu: Assign, Section,
    # pick the roof of the dome, Done, Section[Dome-Roof], OK,
    # pick the side-walls,
    # hold the 'shift-key' to individually select the surfaces,
    Done, Section[Side-Walls], OK,
    # pick the columns,
    # hold the 'shift-key' to individually select the lines,
    Done, Section[Columns], OK,
    X # to end the command

部件装配步骤如下。
Module: Assembly
Menu: Instance, Create, Parts[Part-1], Independent, OK

载荷和边界条件步骤如下。
ⅰ. 创建分析步。
Module: Step
Menu: Step, Create, Name[Step-1], General, Static/Gen., Cont, OK,
    # Return to the saved view,
    # click the 'Apply User 1 View' icon from the 'Views toolbar'

ⅱ. 定义边界条件和载荷。
Module: Load
Menu: BC, Manager,
    Create, Name[BC-1], Step: Initial, Mechanical, Sym/Ant/Enc, Cont,
    # pick the lower-end of the four columns,
    # hold the 'shift-key' to individually select the points, Done,
    Encastre, OK,
    Create, Name[BC-2], Step: Initial, Mechanical, Disp/Rota, Cont,
    # pick the vertical edges of the side-walls,
    # hold the 'shift-key' to individually select the sides, Done,
    # checkmark U1, and U3, OK,
    # Close the Boundary Condition Manager pop-up window,

Menu: Load, Create, Name[Load-1], Step: Step-1,
　　Mechanical, Pressure, Cont, # pick the dome-roof, Done,
　　# in the notification area below the WS select[Brown],
　　Distribution: Uniform, Magnitude[-100], OK,
　　# click the 'Apply Iso View' icon from the 'Views toolbar',
　　# roll the mouse-wheel backward to zoom in,
　　# increase the size of the BC and Load markers,
Menu: View, Assembly Display Options,
　　Tab: Attribute, Symbol, Size[18], Arrows[18], Face Density[10],
　　Apply, OK

现在生成的模型如图 2.35 所示。

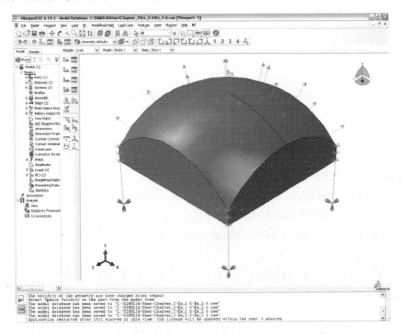

图 2.35　加载后的穹顶结构

部件划分网格步骤如下。
Return to the saved view,
click the 'Apply User 1 View' icon from the 'Views toolbar',
Module: Mesh
Menu: Seed, Instance, Approximate Global Size[30], Apply, OK,
Menu: Mesh, Element Type,
　　# pick the roof of the dome, Done, Standard, Linear, Family[Shell], OK,

pick the four side-walls, Done, Standard, Linear, Family[Shell], OK,
pick the four columns, Done, Standard, Linear, Family[Truss], OK,
X # to end the command,
Menu: Mesh, Instance,
in the notification area below the WS select[Yes],
click the 'Apply Iso View' icon from the 'Views toolbar',
roll the mouse-wheel backward to zoom in

现在生成的模型如图 2.36 所示。

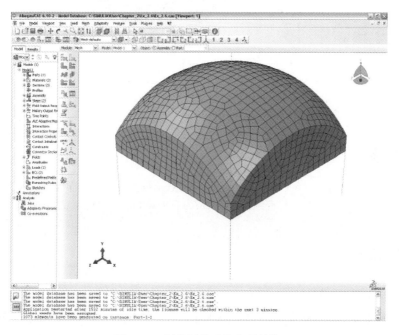

图 2.36 划分网格后的穹顶结构

求解并可视化结果步骤如下。
Module: Job
Menu: Job, Manager,
    Create, Name[Job-1], Cont, OK
    Submit, # when the Status message indicates 'Completed'
    Results
Module: Visualization
Menu: Plot, Contours, On Deformed Shape

计算结果云图如图 2.37 所示。该 CAE 文件可在文献[5,Ex 2.6.cae]中获得。

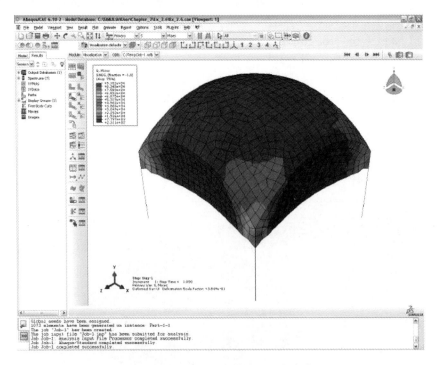

图 2.37　von Mises 应力云图(支撑柱上最大应力 91020MPa,
壳结构上最大应力 66922MPa)

习题

习题 2.1　按照 2.1 节的求解步骤,仅使用两个单元求解例 2.4。

习题 2.2　在习题 2.1 的求解过程中,计算在(a)$x=500\text{mm}$,(b)$x=700\text{mm}$点处的轴向位移。

习题 2.3　按照例 2.1 中相同的求解过程,采用二次插值函数计算三节点杆单元的单元刚度矩阵和等效力矢量。其插值函数为

$$N_1^e = \frac{x-x_2}{x_1-x_2}\frac{x-x_3}{x_1-x_3}, N_2^e = \frac{x-x_3}{x_2-x_3}\frac{x-x_1}{x_2-x_1}, N_3^e = \frac{x-x_1}{x_3-x_1}\frac{x-x_2}{x_3-x_2}$$

式中 x_1, x_2, x_3 分别为节点 1,2 和 3 的坐标位置。其中 $x_1=0, x_2=h/2$ 和 $x_3=h$,h 为单元长度。

习题 2.4　利用例 2.1 中获得的单元方程和 2.1.6 节中所示的单元方程组装过程编制 FE 程序。使用这段程序,求解例 2.4。

习题 2.5　利用习题 2.3 中获得的单元方程和 2.1.6 节中所示的单元方程组装过程编制 FE 程序。使用这段程序,求解例 2.4。

第3章 层合板的刚度和强度

大部分复合材料结构一般采用板和壳形式,因为这种结构形式承受面内载荷的效率更高,另一方面,厚度较大的复合材料层合结构制造比较困难。

例如:考虑一根各向同性材料梁,受到弯矩 M 载荷作用,其材料的拉伸和压缩强度为 σ_u。假如该梁为边长 $2c$ 的正方形实心截面形式,其截面面积 A、惯性矩 I 和截面模量 S 为

$$\begin{cases} A = 4c^2 \\ I = \dfrac{4}{3}c^4 \\ S = \dfrac{I}{c} = \dfrac{4}{3}c^3 \end{cases} \tag{3.1}$$

梁表面应力达到失效应力 σ_u 时,此时梁受到的单位面积弯矩为

$$m_u = \frac{M_u}{A} = \frac{S\sigma_u}{A} = \frac{1}{3}c\sigma_u \tag{3.2}$$

若该梁截面为方管形式(图 3.1),其边长为 $2c \times 2c$,壁厚为 $t, 2c \gg t$。那么,梁的截面属性为

$$\begin{cases} A = 4(2c)t = 8ct \\ I = 2\left[\dfrac{t(2c)^3}{12} + c^2(2ct)\right] = \dfrac{16}{3}tc^3 \\ S = \dfrac{I}{c} = \dfrac{16}{3}tc^2 \end{cases} \tag{3.3}$$

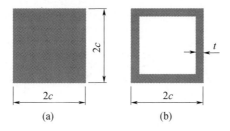

图 3.1 实体梁(a)和方管梁(b)

此时

$$m_u = \frac{M_u}{A} = \frac{S\sigma_u}{A} = \frac{\frac{16}{3}tc^2\sigma_u}{8ct} = \frac{2}{3}c\sigma_u \tag{3.4}$$

对于这两种截面形式的梁,方管形式下单位面积可承受的弯矩 m_u 是实体形式的两倍。

当然,对于该薄壁结构梁而言,失效弯矩主要由屈曲决定(见第4章)。这也是为什么对于复合材料结构而言,结构屈曲的研究是非常重要的原因。由于复合材料结构的厚度小、强度大,大部分复合材料结构的设计约束条件往往是结构屈曲。因此,复合材料结构很难像金属结构那样(如屈服应力)出现材料强度破坏,而是以屈曲破坏为主。

平板是一类没有初始曲率的特殊壳,因此本章后续主要以壳作为研究对象。壳的长和宽两个尺寸远大于壳的厚度,所以一般采用二维结构进行模拟。壳的控制方程中去除了厚度坐标,这将3D问题简化为了2D问题。但在壳的建模过程中,需要提供厚度参量。

复合材料层合板的建模与常规材料结构的建模不同,主要体现在3个方面:首先,每单一铺层的本构方程均为正交各向异性(见1.12.3节);其次,单元的本构方程取决于所采用板壳理论的运动学假设及具体实现形式;最后,当需要使用对称条件时,材料对称性与几何对称性和载荷对称性同样重要。

3.1 壳的运动学假设

壳单元基于多种壳理论,而这些理论均建立在不同的壳运动学假设基础上。而这些关于基本的材料变形假设正是将控制方程从3D简化为2D所需要的,这些假设条件适用于多种情形,将在下面讨论。

3.1.1 一阶剪切变形理论

一阶剪切变形理论(FSDT)是最普遍采用的复合材料板壳理论,它基于如下假设:

i. 直法线假设。当板壳发生变形时,厚度方向上的法线会旋转且始终保持直线不变。相对变形后中面法向形成的角度定义为 ϕ_x 和 ϕ_y,分别对应于 $x-z$ 面和 $y-z$ 面(见图2.19和图3.2)。

ii. 当板壳发生变形时,忽略厚度的变化。

对于大部分层合板壳结构,上述的假设条件已通过试验得到证明,但需要满足以下条件:

纵横比 $r = a/t$。纵横比定义为板壳的短边 a 与厚度 t 之比,需要大于10。

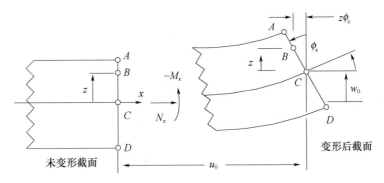

图 3.2 FSDT 变形假设[①]

在壳坐标系(x,y,z)中各单层板的刚度相差不大于两个数量级。该限制条件将夹层板壳排除,其夹芯比面板刚度弱得多。

基于上述假设,对于壳中任意一点B,它的位移可以通过中面上C点的位移和转角进行描述,如下:

$$\begin{cases} u(x,y,z) = u_0(x,y) - z\phi_x(x,y) \\ v(x,y,z) = v_0(x,y) - z\phi_y(x,y) \\ w(x,y,z) = w_0(x,y) \end{cases} \quad (3.5)$$

式(3.5)中右侧中面的变量仅为 2 个坐标(x 和 y)的函数,因此该板壳理论是 2D 的。而等式左侧,位移为 3 个坐标的函数,因此是一种 3D 材料的表述方式。在 3D 形式下,一般采用 3D 本构方程式(1.68)和 3D 应变-位移方程(1.5)进行描述。现在也可以通过 2D 的参数量进行描述,即

$$\begin{cases} \epsilon_x(x,y,z) = \dfrac{\partial u_0}{\partial x} - z\dfrac{\partial \phi_x}{\partial x} = \epsilon_x^0 + zk_x \\ \epsilon_y(x,y,z) = \dfrac{\partial v_0}{\partial y} - z\dfrac{\partial \phi_y}{\partial y} = \epsilon_y^0 + zk_y \\ \gamma_{xy}(x,y,z) = \dfrac{\partial u_0}{\partial y} + \dfrac{\partial v_0}{\partial x} - z\left(\dfrac{\partial \phi_x}{\partial y} + \dfrac{\partial \phi_y}{\partial x}\right) = \gamma_{xy}^0 + zk_{xy} \\ \gamma_{yz}(x,y) = -\phi_y + \dfrac{\partial w_0}{\partial y} \\ \gamma_{xz}(x,y) = -\phi_x + \dfrac{\partial w_0}{\partial x} \\ \epsilon_z = 0 \end{cases} \quad (3.6)$$

---

① 转载自 E. J. Barbero,*Introduction to Composite Materials Design*,图 6.2,版权所有(1999),经 Taylor 和 Francis 许可。

式中

中面应变 $\epsilon_x^0, \epsilon_y^0, \gamma_{xy}^0$，也称膜应变，代表中面面内的拉伸和剪切效应；

曲率 k_x, k_y, k_{xy}，其近似等于中面的几何曲率，而对于 3.1.2 节讨论的 Kirchhoff 理论，它们是准确的；

层间剪切应变 γ_{xz}、γ_{yz}，其为厚度截面上的剪切变形。对于复合材料层合板来说，它们虽然较小但不能忽略，因为与面内弹性模量 E_1 相比，层间模量 G_{23} 和 G_{13} 小得多。另外，层合板的层间剪切强度 F_4、F_5 相对面内强度 F_{1t}、F_{1c} 较小，因此对复合材料层合板进行层间剪切强度评价很有必要。对于金属材料而言，剪切模量（$G = E/2(1+\nu)$）一般较大，所以面外剪切应变往往可以忽略。而金属材料的剪切强度与拉伸强度相差不大，面外剪切应力又小于面内应力，因此对于各向同性金属壳，可以不必校核面外剪切强度。但对于多层粘接的金属壳，有必要通过层间剪切应力校核各层壳之间的粘接强度。

虽然 3D 本构方程描述的是应变和应力之间的关系，但是层合板本构方程描述的是中面应变与曲率之间的关系，其可以利用应力合力的定义获得。对于 3D 弹性体，每个材料点均在应力的作用下；但是壳受载的形式是应力合力（图 3.3），其为在壳厚度方向上的应力分量的简单积分，即

$$\begin{Bmatrix} N_x \\ N_y \\ N_{xy} \end{Bmatrix} = \sum_{k=1}^{N} \int_{z_{k-1}}^{z_k} \begin{Bmatrix} \sigma_x \\ \sigma_y \\ \sigma_{xy} \end{Bmatrix}^k dz$$

$$\begin{Bmatrix} V_y \\ V_x \end{Bmatrix} = \sum_{k=1}^{N} \int_{z_{k-1}}^{z_k} \begin{Bmatrix} \sigma_{yz} \\ \sigma_{xz} \end{Bmatrix}^k dz \quad (3.7)$$

$$\begin{Bmatrix} M_x \\ M_y \\ M_{xy} \end{Bmatrix} = \sum_{k=1}^{N} \int_{z_{k-1}}^{z_k} \begin{Bmatrix} \sigma_x \\ \sigma_y \\ \sigma_{xy} \end{Bmatrix}^k z dz$$

式中：N 为层合板的铺层总数；z_{k-1}，z_k 分别为第 k 层铺层底面和顶面的坐标（图 3.4）。在壳局部坐标系下，将每层铺层的 3D 本构方程（式（1.100）、式（1.101））中的平面应力部分代入上式积分并叠加，得

$$\begin{Bmatrix} N_x \\ N_y \\ N_{xy} \\ M_x \\ M_y \\ M_{xy} \end{Bmatrix} = \begin{bmatrix} A_{11} & A_{12} & A_{16} & B_{11} & B_{12} & B_{16} \\ A_{12} & A_{22} & A_{26} & B_{12} & B_{22} & B_{26} \\ A_{16} & A_{26} & A_{66} & B_{16} & B_{26} & B_{66} \\ B_{11} & B_{12} & B_{16} & D_{11} & D_{12} & D_{16} \\ B_{12} & B_{22} & B_{26} & D_{12} & D_{22} & D_{26} \\ B_{16} & B_{26} & B_{66} & D_{16} & D_{26} & D_{66} \end{bmatrix} \begin{Bmatrix} \epsilon_x^0 \\ \epsilon_y^0 \\ \gamma_{xy}^0 \\ k_x \\ k_y \\ k_{xy} \end{Bmatrix} \quad (3.8)$$

$$\begin{Bmatrix} V_x \\ V_y \end{Bmatrix} = \begin{bmatrix} H_{44} & H_{45} \\ H_{45} & H_{55} \end{bmatrix} \begin{Bmatrix} \gamma_{yz} \\ \gamma_{xz} \end{Bmatrix}$$

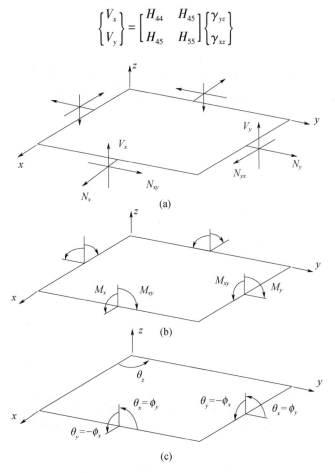

图 3.3 板或壳单元上的应力合力
(a) 单位长度上的作用力；(b) 单位长度上的弯矩；
(c) 板壳理论中转角 ϕ 与遵循右手准则的转角 θ 定义的比较。

式中

$$A_{ij} = \sum_{k=1}^{N} (\overline{Q}_{ij})_k t_k \quad (i,j = 1,2,6)$$

$$B_{ij} = \sum_{k=1}^{N} (\overline{Q}_{ij})_k t_k \bar{z}_k \quad (i,j = 1,2,6)$$

$$D_{ij} = \sum_{k=1}^{N} (\overline{Q}_{ij})_k \left(t_k \bar{z}_k^2 + \frac{t_k^3}{12} \right) \quad (i,j = 1,2,6)$$

$$H_{ij} = \frac{5}{4} \sum_{k=1}^{N} (\overline{Q}_{ij}^*)_k \left[t_k - \frac{4}{t^2} \left(t_k \bar{z}_k^2 + \frac{t_k^3}{12} \right) \right] \quad (i,j = 4,5) \tag{3.9}$$

其中：$(\bar{Q}_{ij})_k$ 为层合板坐标中的第 k 层铺层的平面应力刚度矩阵系数；t_k 为第 k 层铺层的厚度；$\bar{z}_k$ 为第 k 层铺层的中面坐标。

对其他项的进一步详细讨论见文献［1］。总之，A_{ij} 为层合板面内刚度系数；D_{ij} 为层合板弯曲刚度系数；B_{ij} 为层合板拉－弯耦合刚度系数；H_{ij} 为层合板层间剪切刚度系数。所有系数均可由式(3.9)计算，且已被各种商业软件广泛采用，例如 CADEC[12]。

当面内变形和弯曲变形不耦合时（如对称层合板），FSDT 的控制方程采用 3 个变量(w^0, ϕ_x, ϕ_y)描述弯曲问题，采用两个变量(u^0, v^0)描述面内变形问题。对于拉－弯耦合问题，则必须同时采用上述 5 个变量进行描述。而对于有限元软件，不论是否耦合，均采用 5 个变量同时进行描述。

板的平衡方程可利用 PVW(1.16 节)推导得到。此外，将本构方程式(3.8)代入平衡方程，即可得到控制方程。

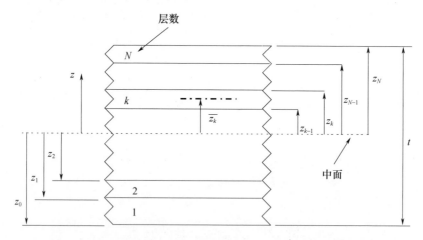

图 3.4　层合板中铺层界面(z_k)和中面($\bar{z}_k$)在 z－坐标轴上的定义

3.1.2　Kirchhoff 理论

历史上，Kirchhoff 理论更常被使用，因为控制方程可以只需 1 个壳的横向变形 w_0 变量即可描述。在计算机时代之前，Kirchhoff 理论相对于需要 3 个变量的 FSDT 理论更容易获得解析解。因此，基于 Kirchhoff 理论，产生了包含大量的闭环设计方程和近似解的工程设计手册[13]。这些简易的设计计算方法仍可用于复合材料层合板壳结构的初步设计中，当然这需要我们详细了解这些方法的局限性。金属材料壳一直都是采用 Kirchhoff 理论进行分析。FSDT 控制方程可以退化为 Kirchhoff 控制方程，其封闭解的推导见文献[14]。

在 Kirchhoff 理论中，假设层间（面外）剪切应变为 0，从式(3.6)中最后两个

等式,可得

$$\begin{cases} \phi_x = \dfrac{\partial w_0}{\partial x} \\ \phi_y = \dfrac{\partial w_0}{\partial y} \end{cases} \quad (3.10)$$

将式(3.10)引入式(3.6)中的前 3 个方程,得

$$\begin{cases} \epsilon_x(x,y,z) = \dfrac{\partial u_0}{\partial x} - z\dfrac{\partial^2 w_0}{\partial x^2} = \epsilon_x^0 + zk_x \\ \epsilon_y(x,y,z) = \dfrac{\partial v_0}{\partial y} - z\dfrac{\partial^2 w_0}{\partial y^2} = \epsilon_y^0 + zk_y \\ \gamma_{xy}(x,y,z) = \dfrac{\partial u_0}{\partial y} + \dfrac{\partial v_0}{\partial x} - 2z\dfrac{\partial^2 w_0}{\partial x \partial y} = \gamma_{xy}^0 + zk_{xy} \end{cases} \quad (3.11)$$

式(3.11)中,变量 ϕ_x,ϕ_y 已经被消除,Kirchhoff 理论仅使用 $u_0(x,y),v_0(x,y)$ 和 $w_0(x,y)$ 3 个变量描述。因此 Kirchhoff 理论的解析解容易获得,但 Kirchhoff 理论的数值计算比较复杂。应变的计算需要 w_0 的二阶偏导数,而通过弱解形式(式(2.30))可以得到 w_0 的二阶导数。由于需要用到二阶导数,因此需要插值函数(见 2.1.4 节)一阶可导(C^1 连续)。也就是说,插值函数不仅要位移连续,而且位移的一阶导数也要连续。这就要求在任意相邻单元的边界处,其位移 w_0 和偏导数 $\partial w_0/\partial x,\partial w_0/\partial y$ 都必须相同,要满足上述条件比较困难。

以梁的弯曲为例,受分布载荷 $\hat{q}(x)$ 作用下变形的常微分方程为

$$EI\dfrac{\mathrm{d}^4 w_0}{\mathrm{d}x^4} = \hat{q}(x) \quad (3.12)$$

可参照式(2.3)得到其弱解形式,即

$$0 = \int_{x_A}^{x_B} v\left[-EI\dfrac{\mathrm{d}^4 w_0}{\mathrm{d}x^4} + \hat{q}(x)\right]\mathrm{d}x \quad (3.13)$$

对上式进行两次分部积分,得

$$\begin{aligned} 0 &= \int_{x_A}^{x_B} \dfrac{\mathrm{d}^2 v}{\mathrm{d}x^2} EI \dfrac{\mathrm{d}^2 w_0}{\mathrm{d}x^2}\mathrm{d}x + vEI\dfrac{\mathrm{d}^3 w_0}{\mathrm{d}x^3} - \dfrac{\mathrm{d}v}{\mathrm{d}x}EI\dfrac{\mathrm{d}^2 w_0}{\mathrm{d}x^2} - \int_{x_A}^{x_B} v\hat{q}(x)\mathrm{d}x \\ 0 &= B(v,w_0) + [vQ_x]_{x_A}^{x_B} - \left[\dfrac{\mathrm{d}v}{\mathrm{d}x}M_x\right]_{x_A}^{x_B} - \int_{x_A}^{x_B} v\hat{M}(x)\mathrm{d}x \\ 0 &= B(v,w_0) + L(v) \end{aligned} \quad (3.14)$$

如 2.1.6 节所述组装单元后,相邻单元 i 和 $i+1$ 之间共用节点处的变形一致,但剪力 Q_x 和弯矩 M_x 相反,即

$$\begin{cases} w^i = w^{i+1} \\ Q^i = -Q^{i+1} \\ M^i = -M^{i+1} \end{cases} \quad (3.15)$$

对于剪力可通过式(2.24)中的关系进行消除,那么现在求解只需要满足 $v^i = v^{i+1}$。而上式因为单元的 C^0 连续性在共用节点处满足 $w^i = w^{i+1}$ 而自然满足。对于弯矩也可通过式(2.24)中的关系进行消除,这时求解需要满足 $\mathrm{d}w^i/\mathrm{d}x = \mathrm{d}w^{i+1}/\mathrm{d}x$。而上式需要单元满足一阶导数连续($C^1$ 连续),即在单元共用节点处满足 $\mathrm{d}w^i/\mathrm{d}x = \mathrm{d}w^{i+1}/\mathrm{d}x$。满足上述条件的单元使用起来较为困难(见文献[15],P276)。

对于 FSDT 理论,应变的计算式(3.6)中仅需要用到一阶导数。因此,弱解形式(2.30)中仅存在一阶导数,并且按照式(2.24),根据单元的 C^0 连续性,所有共用节点处的内力均可消除。

3.1.3 简支边界条件

具有耦合效应的复合材料层合板即使在纯弯曲、纯剪切或纯面内载荷作用下,也可能发生耦合的弯曲、剪切和面内变形(见文献[1],图6.7)。虽然简支支撑意味着对横向变形 $w(x,y)$ 限制,但是并没有唯一指定是限制面内变形的 u_n 还是 u_s。u_n 和 u_s 分别表示边界的法向和切向。本书在分析过程中,通常限制 u_n 或 u_s 中的一个。因此,边界条件存在以下可能,即

$$SS\text{-}1: w = u_s = \phi_s = 0; N_n = \hat{N}_n; M_n = \hat{M}_n$$

$$SS\text{-}2: w = u_n = \phi_s = 0; N_{ns} = \hat{N}_{ns}; M_n = \hat{M}_n$$

在 SS-1 条件下,指定了法向力和力矩。在 SS-2 条件下,指定了剪切力和力矩。这里对于旋转的命名规则与图3.3中对于力矩合力的命名规则相同,其中下标 $()_n$ 表示垂直于壳边缘的方向,下标 $()_s$ 表示与壳边缘相切的方向(见文献[14],图6.2.1)。注意旋转矢量 ϕ_s 垂直于方向 s,于是 $\phi_s \approx \dfrac{\partial w}{\partial s}$;因此,SS-1 和 SS-2 都将旋转定义为 0。$(\hat{\ })$ 表示已知的固定值,可以为 0 或不为 0。

对于仅承受弯矩的对称层合板几何线性分析,不会出现明显的 u_n、u_s 变形。因此 SS-1 和 SS-2 两种条件下,计算结果应该几乎相同。其他情况下,差异可能很明显。此外,$\phi_s = 0$ 的条件不应忽略。例如,对于使用 36 个 S4R 单元的夹层板进行计算[16],当取消 $\phi_s = 0$ 条件时,夹层板中心变形增加了 11.5%。

3.2 层合板的有限元分析

复合材料层合板的变形和应力可以在几个不同的层次中进行分析(图3.5)。

根据实际的后处理需求,采用不同精细层级的模型。

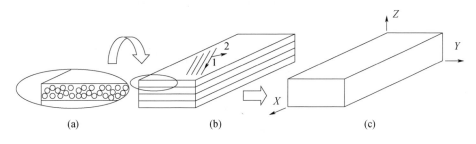

图 3.5　不同层次分析
(a)细观力学层次;(b)单层板层次;(c)层合板宏观层次。

当进行细观力学层次分析时(图 3.5(a)),应变和应力在组分层面上进行计算,即纤维和基体层面。这种情况下,需要细观结构的描述信息,包括纤维形状、几何分布、各组分的材料性能等。第 6 章中详细描述了采用细观力学模型获得纤维和基体不同组合形式下力学性能的方法。当复合材料为编织形式、铺层较厚或者研究如自由边效应的局部问题时,此时复合材料结构应该如第 5 章所述一样,采用实体进行建模并分析。但必须注意的是:大部分复合材料层合结构都可采用 3.1 节中介绍的简化板或壳进行分析。

另一个极端情况(图 3.5(c)),如果要进行宏观层次分析,复合材料可以被等效成均匀材料。这种情况下,如第 1 章所述,结构行为可采用正交各向异性材料性能进行分析。如果层合板在宏观上被当作均质材料进行分析(图 3.5(c)),那么将无法获得层合板的应力分布。但对于仅进行位移、屈曲载荷及模态或振动频率及模态等问题的分析,采用等效的均质材料性能已满足计算精度要求。上述分析问题中,仅需要层合板的刚度参数(式(3.8))(见 3.2.9 节)。甚至在某些情况下,可采用更简化的分析方法。例如,层合板为单向板时,或者层合板为平衡对称结构时(见文献[1],6.3 节),此时层合板可简化为一层正交各向异性材料的单层板进行分析(3.2.10 节)。

大部分情况下,层合板的每层铺层的应力和应变均需要计算。因此,在计算程序中需输入真实的铺层(LSS)信息(3.2.11 节),包括每层铺层的弹性参数、厚度、纤维方向等。这种方法即为中尺度分析方法(图 3.5(b))。

单向板可以认为是横观各向同性材料,因此单向板可以认为是满足 $E_3 = E_2$、$G_{23} = E_3/2(1+v_{23})$ 的正交各向异性材料。单向板的材料性能可以通过第 6 章的细观力学分析方法或单向板测试获得。表 3.1 所列为部分复合材料单向板性能。

表 3.1 碳纤维/环氧复合材料单向板材料性能[①]

性能	单位	AS4D/9310	T300/5208
E_1	GPa	133.86	136.00
$E_1 = E_3$	GPa	7.706	9.80
$G_{12} = G_{13}$	GPa	4.306	4.70
G_{23}	GPa	2.76	4.261
$v_{12} = v_{13}$		0.301	0.280
v_{23}		0.396	0.150
V_f		0.55	
ρ	[g/cm^3]	1.52	1.54
α_1	[10^{-6}/℃]	0.32	
α_2	[10^{-6}/℃]	25.89	
F_{1t}	MPa	1830	1550
F_{1c}	MPa	1096	1090
$F_{2t} = F_{3t}$	MPa	57	59
$F_{2c} = F_{3c}$	MPa	228	207
F_4	MPa	141	128
F_6	MPa	71	75

① F_4 为 $\alpha_0 = 54°$ 时的计算结果(见文献[1],式(4.109))。

对大多数复合材料进行结构分析时,通常避免使用细观力学分析方法,而是通过实验方法获得单向板性能或层合板性能。但是实验方法并不万能,因为在设计过程中材料组分或纤维体积分数的变化将导致材料性能变化,这需要重新对性能进行测试。所以最好的方法是利用软件,如文献[12](另见 6.1 节)所示,通过细观力学方法计算单层板的弹性性能。然而不幸的是,细观力学方法中的公式无法准确预测强度,因此计算不能完全替代实验。

总之,层合板性能可以通过以下两种方式指定:

(1)通过指定本构刚度矩阵 **A**,**B**,**D** 和 **H**。

(2)通过定义层合板的铺层顺序(LSS),以及每层铺层的材料性能。

当采用本构刚度矩阵 **A**,**B**,**D** 和 **H** 方式定义层合板时,壳单元无法区分不同的铺层。该方法只将广义力和力矩与相应的广义应变和曲率联系起来。另一种方法,即利用层合板铺层顺序(LSS)以及每层铺层的材料性能,层合壳单元可以计算得到层合板的性能。

3.2.1 单元类型和命名规则

壳单元适用于边厚比大于10的薄壳和中厚度壳。这些壳单元一般有3或4个节点,也可以有利用高阶插值函数描述的8个节点。壳单元定义在3D空间,每个节点有5或6个自由度(DOF),分别是$x-,y-,z-$三个方向的平动和绕$x-,y-,z-$轴的3个转动自由度。第6个自由度(绕z轴转动)可用于弯折壳的模拟,如果壳表面是平滑的,则不需要此自由度。

Abaqus中单元的分类很多,首次学习阅读会有困难,我们概括如下:

(1)常规壳单元。常规壳单元需要几何(网格)来表示3D空间中的2D平板和曲面,常规壳单元包括薄壳单元(如S8R5)和厚壳单元(如S8R)。

(2)连续体壳单元。连续体单元需要几何精确模拟壳的厚度,如同3D实体单元一样在厚度方向上划分网格。连续体壳单元通过特殊的插值函数施加了FSDT约束(见3.1.1节),其类型包括SC6R和SC8R。

(3)薄壳单元。薄壳单元无论基于薄板理论的(如STRI3)还是数值近似的(如S8R5),都会被施加Kirchhoff条件约束(见3.1.2节)。因此,薄壳单元的横向剪切变形假设为0或忽略。如果层合板较厚和(或)某一或几层铺层的横向剪切模量G_{23}较小,采用薄壳单元计算精度不够,其剪切变形会被低估。所有薄壳单元均属于常规壳单元,其包括STRI3、S4R5、STRI65、S8R5、S9R5、SAXA1N和SAXA2N。薄壳单元没有绕面法向轴的转动自由度。

(4)厚壳单元。厚壳单元仅施加FSDT条件约束(见3.1.1节),因此,横向剪切变形不为0。厚壳单元可以是常规壳单元(S3R,S4,S4R,S8R,SAX1,SAX2)或者是连续体壳单元(SC6R和SC8R)。

(5)通用壳单元。通用壳单元适用于模拟厚壳和薄壳,其包括厚的常规壳单元(如S3,S3R,S3RS,S4,S4R,S4RS,S8R,S4RSW,SAX1,SAX2,SAX2T)和厚的连续体壳单元,如SC6R和SC8R。

(6)3D实体单元。3D实体单元也称实体3D连续体单元,其对3D实体直接进行离散化,不需采用任何壳的运动学假设。它们不是壳单元,一般计算量较大。实体单元可用在应力、应变变化梯度较大的局部区域进行详细分析。实体单元需要3D网格,即对壳几何实体进行详细的3D网格划分,包括厚度方向。Abaqus中对于壳单元和实体单元的过渡连接可采用壳-实体耦合约束(shell-to-solid coupling constraints)方法实现。

在Abaqus中,2D和3D概念也易混淆。壳理论中,壳为一个2D表面,其上任意一点只需要2个曲线坐标即可定位。但在Abaqus中,壳被说成是3D的,因为壳本身处在3D空间中,这与2D空间平面不同。

另外，Abaqus 中使用下述命名规则来描述壳单元：

(1) SnRsW 常规壳。

S 壳

n 节点数

R 减缩积分

s 小量膜应变

W 考虑翘曲

(2) SCnRT 连续体壳。

S 壳

C 连续体

n 节点数

R 减缩积分

T 热 – 力耦合

(3) STRInm 三角形壳。

S 壳

TRI 三角形

n 节点数

m 自由度数

(4) SAXAxN 轴对称壳。

S 壳

AXA 轴对称

1N 经向线性插值，N Fourier 模式

2N 经向二次插值，N Fourier 模式

本节中提到的单元类型将在后文进一步解释。

3.2.2　薄壳单元

当壳的厚度较薄，材料的剪切模量较大，并且壳在变形过程中，中面的法线始终与中面保持垂直。上述就是 Kirchhoff 理论的假设，也称作 Kirchhoff 约束（见 3.1.2 节）

Kirchhoff 约束规定：垂直于参考面的直线，当壳发生变形时，该直线依然保持垂直于参考面。这意味着壳的横向剪切变形 γ_{xz}、γ_{yz} 为 0。另外的假设条件在 Kirchhoff 理论和一阶剪切变形理论（FSDT，3.2.4 节）中通用：法线保持直线状态且不可拉伸（$\varepsilon_{zz}=0$）。施加 Kirchhoff 约束的壳单元称作薄壳单元，而厚壳单元施加的是 FSDT 约束。所有的薄壳单元均为常规壳单元，其可以使用 2D 平板/

曲面描述 3D 空间中的壳几何。而连续体壳单元需要几何有精确的厚度信息，并且需要采用 3D 实体单元进行网格划分。

薄壳单元，例如 STRI3、S4R5、STRI65、S8R5、S9R5、SAXA1N 和 SAXA2N，都可用来描述薄壳几何。薄壳单元对薄壳的参考面进行离散，厚度通过截面参数施加。STRI3 单元采用解析方式施加 Kirchhoff 约束，因此它得到的是薄壁结构的理论解，甚至对于厚壳结构也一样。S4R5、STRI65、S8R5、S9R5、SAXA1N 和 SAXA2N 单元采用数值方法施加 Kirchhoff 约束。上述这些单元对于横向剪切变形很重要的情况是不适用的，即壳的厚度较大，剪切模量较小，或者两种都有的情况(参见例 3.4a)。

所有单元名字最后的数字为 5 的单元表示该单元节点有 5 个自由度，即 3 个平动自由度和 2 个面内转动自由度，没有沿平面法向轴转动的自由度(见 3.2.8 节)。5 自由度单元，如 S4R5、STRI65、S8R5、S9R5，比其他单元计算效率更高，但这些单元仅适用于模拟薄壳，不适用于模拟厚壳。

3.2.3 厚壳单元

复合材料层合板应被认为是厚壳，尽管其厚度较薄，因为其剪切模量相对较小，需要像厚壳一样考虑剪切变形。Abaqus 提供了 S3R、S3RS、S4、S4R、S4RS、S4RSW、S8R、SC6R 和 SC8R 单元用于描述层合板结构。另外，如果需要更精确的横向剪切变形分析，可以选择连续体壳单元 SC6R 和 SC8R。参见例 3.4b。

3.2.4 通用壳单元

一阶剪切变形理论(FSDT，见 3.1.1 节)表明：与壳参考面垂直的直线，在壳变形后依然保持直线并且不能拉伸变形，但没有要求该直线与变形后参考面垂直，这就允许了非 0 的横向剪切变形。

通用壳单元，如 S3、S3R、S3RS、S4、S4R、S4RS、S8R、S4RSW、SAX1、SAX2、SAX2T、SC6R 和 SC8R，这些单元包含横向剪切变形，但它们也可以用来分析薄壳。采用 S3、S3R、S3RS、S4、S4R、S4RS、S8R、S4RSW、SAX1、SAX2、SAX2T 等单元对壳参考面进行网格划分，并通过截面参数形式赋予厚度。S3、S3R、S3RS、S4、S4R、S4RS 和 S8R 单元节点有 6 个自由度，包括绕平面法向轴的转动自由度，因此这些单元可分析弯折壳。SC6R 和 SC8R 单元为连续体壳单元(见 3.2.5 节)，其也被认为是通用壳单元，因为它们均可以用于分析薄壳和厚壳。

3.2.5 连续体壳单元

连续体壳(CS)单元，如 SC4R，是在 3D 实体单元基础上通过特殊的插值函数施加 FSDT 约束的单元[17]。与常规壳单元相比,它们在模型的厚度方向上节点数更多。

连续体壳单元只有 3 个平动自由度,无转动自由度。因此,连续体壳单元可以堆叠并可以与实体单元连接。采用连续体壳单元对壳的 3D 几何进行网格离散的方式与 3D 实体单元的方式一样。

连续体壳单元可以通过在一个单元内设置铺层信息来描述层合板或层合板子铺层。而且,通过连续体壳单元的堆叠可以更精确地描述层合板厚度方向的横向剪切变形。在这种情况下,每一个堆叠单元均代表层合板的一个子铺层。如果只通过一层单元来描述层合板整个厚度方向的铺层,那么连续体壳与 FSDT 单元使用效果相同。不过,由于连续体壳单元可以进行多层单元堆叠,所以它比 FSDT 单元优势明显。通过在厚度方向上连续体壳单元的多层堆叠,法向变形可以精确地进行求解。由于厚度方向上连续体壳单元数量可以不受限制,因此对于平面问题可求得精确的弹性解。连续体壳单元相对于 3D 实体单元的优势是它不存在纵横比问题,即厚度相对于长和宽很小。这是由于连续体壳单元融合了法向不可压缩的假设($\epsilon_z = 0$,式 3.6),和 FSDT 单元一样。

因为壳几何在厚度方向也要进行网格划分,连续体壳单元的单元划分一般较困难。此外,边界条件也更难施加,因为连续体壳单元不能施加典型的壳边界,而是要在壳的顶面和底面施加位移约束来模拟边界条件。例如:如果壳的中面上没有连续体壳单元的网格节点,那么简支边界将很难施加。固支边界和对称边界相对容易施加(参见例 3.4c)。

3.2.6 夹层壳

对于夹层壳,夹芯比面板要软很多,因此不论夹层的厚度是多少,横向剪切变形都是很重要的。由于常规壳单元可能没有所需要的剪切弹性,因此可使用连续体壳单元代替。对于大部分精细的夹层壳模型,可以在厚度上采用三层连续体壳单元进行堆叠,一层模拟夹芯,另两层分别模拟面板(参见例 3.7)。

3.2.7 节点和曲率

3 节点单元 STRI3 和 4 节点单元 S4R5 都是平板单元。对于曲面壳,最好使用 6 节点、8 节点或 9 节点单元分析,例如 STRI65、S8R5 和 S9R5。SR85 单元有一个内部隐藏的第 9 节点,对于双曲面壳,其可能位于参考面之外。此时若采用该单元进行屈曲载荷分析将不准确,而选用 S9R5 单元会更合适。

3.2.8 面外转动

常规壳单元均基于壳理论,其通过满足一些基本的运动学假设,对壳的三维连续体变形进行约束,这些假设包括 Kirchhoff 假设、FSDT 假设或者其他一些假设。在连续介质理论中,三维实体中一点的变形可以通过两个点的相对位移进行描述,因此需要 6 个自由度(每个点 3 个自由度)。Kirchhoff 和 FSDT 理论均将所需自由度简化为每个节点 3 个平动自由度和 2 个面内转动自由度。这两个转动自由度 ϕ_x 和 ϕ_y 均是垂直于参考面法向轴的转动。因为它们的转动向量均位于壳的面内,因此被称为面内转动。对于平滑的曲面壳和平面壳,不需要考虑绕曲面法向轴自身的转动自由度 ϕ_z,这个自由度被称作面外转动。然而,如果壳有弯折,那么壳的一侧有面内转动,就会对应另一侧有面外转动(图 3.6)。此时,为了位移的协调,必须对面外转动进行约束。因此,对于无面外转动自由度的 5 自由度单元,仅适用于分析平滑的曲面壳,而不适用于弯折壳。

Abaqus/Standard 在下述情况下会自动将任何节点处的 5 自由度单元转换为 6 自由度单元。

(1)节点的转动自由度上施加运动学(位移、速率)边界条件。

(2)节点被用于包括转动自由度的多点约束中。

(3)节点被一个在所有节点处均应用了 3 个全局转动分量的梁单元或壳单元共用。

(4)节点在壳的弯折线上。

(5)节点上施加了弯矩。

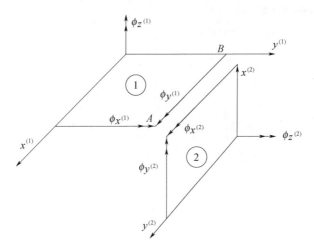

图 3.6　平板沿 AB 线弯折的分解图(单元 1 和 2 的面外转动分别是 $\phi_z^{(1)}$ 和 $\phi_z^{(2)}$)

3.2.9　层合板 FEA 中 A、B、D、H 参数的输入

如前所述,如果仅分析结构变形、模态和屈曲而不进行详细的应力分析,结

构在宏观尺度(层合板层次)上的分析精度已经足够。因此,不需要输入层合板的铺层信息(LSS)、厚度和层合板每一单层的弹性性能,仅需要输入式(3.9)中定义的层合板弹性参数(A、B、D、H 矩阵)。由于仅需要一些整体参数,因此计算会方便很多。层合板输入参数的简化,使得仅需要 4 个矩阵参数就可以模拟铺层数可以无限多的层合板结构。

当 A、B、D、H 矩阵用于层合板有限元计算时,计算模型拥有结构刚度性能,但没有层合板铺层信息(LSS)。因此,软件可以计算结构变形响应(包括屈曲和振动),甚至可以计算层合板厚度方向的应变分布,但无法计算层合板应力分布,因为软件中无各铺层的材料性能。

A、B、D、H 参数的计算可参考式(3.9)或文献[12]。A、B、D、H 数据的输入过程参见例 3.1。

Abaqus 中 A、B、D 矩阵通过 Edit General Stiffness Section 对话框(图 3.7)输入。调用步骤如下所示:

  Module: Property
    Menu: Section, Edit,[Section - name]

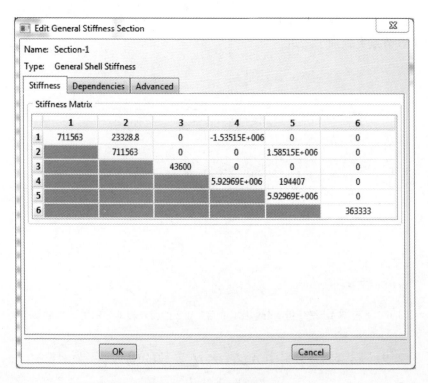

图 3.7 A、B、D 矩阵的输入

A、B、D 矩阵的数据通过图 3.7 的 Stiffness 选项卡输入。H 矩阵数据需转换为横向剪切刚度值,并通过图 3.8 的 Advanced 选项卡输入。横向剪切刚度系数 H_{44}、H_{45} 和 H_{55} 的定义参见式(3.9)。考虑到 Abaqus 符号缩写规则,见表 1.1,与本书中使用的标准(Voigt)规则对比,可得到图 3.8 中横向剪切系数表示为

$$\begin{cases} K_{11} = H_{55} \\ K_{22} = H_{44} \\ K_{12} = H_{45} \end{cases} \quad (3.16)$$

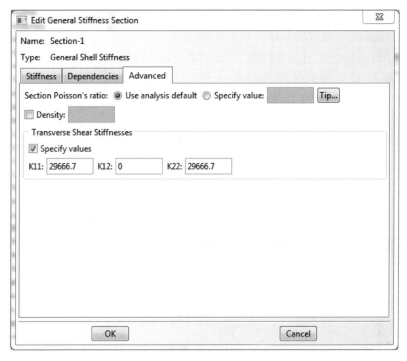

图 3.8 H 矩阵数据的输入

例 3.1 边长 $a_x = a_y = 2000\text{mm}$ 的正方形简支板,采用 $[0/90]_n$ 的铺层顺序,材料为 AS4D/9310(表 3.1)。板的总厚度为 $t = 10\text{mm}$。板单边受 $N_x = 1\text{N/mm}$ ($N_y = N_{xy} = M_x = M_y = M_{xy} = 0$)的压缩载荷。当铺层 $n = 1,5,10,15$ 和 20 时,试分别计算板中心的挠度。计算 A、B、D、H 矩阵,并输入 Abaqus,使用对称性对 1/4 的板进行建模。

解 求解 $n = 1$ 时 $[0/90]_n$ 铺层的层合板过程如下:

ⅰ. 确保文件保存在所选文件夹中。
Menu: File, Save as,[C:\SIMULIA\User\Ex_3.1\Ex_3.1.cae]
Menu: File, Set Work Directory,[C:\SIMULIA\User\Ex_3.1]

ii. 创建部件。

由于板的对称性,仅需对 1/4 板进行建模。

Module: Part

Menu: Part, Create,

　　[Part-1], 3D, Deformable, Shell, Planar, Approx. size[4000], Cont

Menu: Add, Line, Rectangle,

　　# Enter the points[0,0],[1000,1000], X # to end the command, Done

iii. 创建材料和截面,并且将截面赋予部件。

在 Abaqus 中,A、B、D、H 矩阵以截面属性的形式输入。由于 A、B、D、H 矩阵包含了层合板的所有弹性性能,因此不需要定义材料。

A、B、D、H 矩阵的计算参见式(3.9),详细的说明见文献[12]。对于 $[0/90]_1$ 层合板,A、B、D、H 矩阵计算如下:

$$[A] = \begin{bmatrix} 711541 & 23317 & 0 \\ & 711541 & 0 \\ & & 43060 \end{bmatrix} (\mathrm{MPa \cdot mm})$$

$$[B] = \begin{bmatrix} -1585193 & 0 & 0 \\ & 1585193 & 0 \\ & & 0 \end{bmatrix} (\mathrm{MPa \cdot mm^2})$$

$$[D] = \begin{bmatrix} 5929510 & 194306 & 0 \\ & 5929510 & 0 \\ & & 358833 \end{bmatrix} (\mathrm{MPa \cdot mm^3})$$

$$[H] = \begin{bmatrix} 29442 & 0 \\ 0 & 29442 \end{bmatrix} (\mathrm{MPa \cdot mm})$$

将它们输入 Abaqus/CAE 中的步骤代码如下:

Module: Property

Menu: Section, Create,[Section - 1],

　　Shell, General Shell Stiffness, Cont

　　# In the pop - up window fill the stiffness matrix as follows

　　Tab: Stiffness,

　　# Upper part of matrix A in rows 1 to 3, columns 1 to 3

　　# matrix B in rows 1 to 3, columns 4 to 6

　　# Upper part of matrix D in rows 4 to 6, columns 4 to 6

　　# The Stiffness matrix should be

　　# 711541　　23317　　0　　-1585193　　0　　0

　　#　　　　　711541　　0　　　0　　1585193　　0

```
#           43060        0       0      0
#                     5929510  194306   0
#                             5929510   0
#                                     358833
```
Tab: Advanced,
The values corresponding to the H matrix are entered here
Transverse Shear Stiffness, # checkmark: Specify values,
K11:[29442], K12:[0], K22:[29442], OK
Menu: Assign, Section, # pick the part, Done, OK

ⅳ．创建装配体。

Module: Assembly
Menu: Instance, Create, Independent, OK

ⅴ．创建分析步。

Module: Step
Menu: Step, Create, Name[Step-1], Static/General, Cont, OK

ⅵ．施加边界条件和载荷。

Module: Load
Menu: BC, Manager,
    Create, Name[X-symm], Step: Initial,
    Mechanical, Symm/Anti/Enca, Cont
    # pick the left vertical line, Done, XSYMM, OK
    Create, Name[Y-symm], Step: Initial,
    Mechanical, Symm/Anti/Enca, Cont
    # pick the upper horizontal line, Done, YSYMM, OK
    Create, Name[Simple-Support],
    Step: Initial, Mechanical, Disp/Rota, Cont
    # pick right vertical and lower horizontal lines, Done,
    # checkmark: U3, OK, # close BC Manager
Menu: Load, Create, Step: Step-1, Mechanical, Shell edge load, Cont
    # pick the right vertical line, Done, Magnitude[1], OK

ⅶ．创建网格。

Module: Mesh
Menu: Seed, Instance, Approximate global size[50], Apply, OK

Menu: Mesh, Controls, Element Shape[Quad], Technique[Structured], OK
Menu: Mesh, Instance, # At the bottom of the WS select[Yes]

ⅷ. 求解并可视化结果。
Module: Job
Menu: Job, Manager
  Create, Name[0-90_1], Cont, OK
  # Job name should be changed for each laminate so the user can
  # access results of different models without re-running
  Data Check # To check the model for errors,
  Submit # To run the model,
  Results # To visualize the solution
Toolbar: Views, Apply Front View
Menu: Plot, Contours, On Deformed Shape,
  Results, Field Output,
  Output Variable[U], Component[U3], Apply, OK
  # Note that maximum deflection should be -2.196e-01
  # at the upper left corner of the plate

ⅸ. 保存模型数据。
Menu: File, Save,
Menu: File, Save As[Ex_3.1(1).cae], OK
Menu: File, New Model Database, With Standard/Explicit Model

ⅹ. 通过"Save As",用户可保存一个$[0/90]_1$铺层的层合板模型,同时通过对模型修改完成不同铺层方式的分析。

模型结果的读取步骤如下:
Menu: File, Open, # Browse to C:\SIMULIA\User\Ex_3.1
  File Filter, Output Database(*.odb*), # Select[0-90_1.odb], OK

对于 $n>1$ 的层合板,求解过程与上述步骤一致。改变铺层的数量,保持层合板厚度不变,此时只有拉-弯耦合刚度阵 B 改变。对于其他铺层的层合板,B 矩阵计算参考文献[12],得

对于 $[0/90]_5$:

$$[B] = \begin{bmatrix} -317039 & 0 & 0 \\ & 317039 & 0 \\ & & 0 \end{bmatrix} (\text{MPa} \cdot \text{mm}^2)$$

对于$[0/90]_{10}$:
$$[B] = \begin{bmatrix} -158519 & 0 & 0 \\ & 158519 & 0 \\ & & 0 \end{bmatrix} (\mathrm{MPa \cdot mm^2})$$

对于$[0/90]_{15}$:
$$[B] = \begin{bmatrix} -105468 & 0 & 0 \\ & 105468 & 0 \\ & & 0 \end{bmatrix} (\mathrm{MPa \cdot mm^2})$$

对于$[0/90]_{20}$:
$$[B] = \begin{bmatrix} -79260 & 0 & 0 \\ & 79260 & 0 \\ & & 0 \end{bmatrix} (\mathrm{MPa \cdot mm^2})$$

然后,对于已保存的模型 Ex_3.1.cae,改变 $n>1$ 时不同的拉－弯耦合刚度阵。

ⅰ. 修改截面属性,提交计算并可视化。

Menu: File, Open, # Browse to... \Ex_3.1,[Ex_3.1.cae], OK

Module: Property
Menu: Section, Edit, Section‐1
　　# In the pop‐up window modify the cells of the new matrix B
　　# rows 1 to 3, columns 4 to 6. The new B sub‐matrix is
　　# ‐317039　　　 0　　　 0
　　#　　 0　　 317039　　 0
　　#　　 0　　　 0　　　 0
　　OK
Module: Job
Menu: Job, Manager, Create, Name[0‐90_5], Cont, OK
　　Data Check # To check the model for errors,
　　Submit # To run the model,
　　Results # To visualize the solution

Toolbar: Views, Apply Front View
Menu: Plot, Contours, On Deformed Shape,
　　Results, Field Output,
　　Output Variable[U], Component[U3], Apply, OK

最大变形出现在平板的左上角,为 -2.109×10^{-2} mm。

ⅱ. 保存模型。

Menu: File, Save,

Menu: File, Save As[Ex_3.1(5).cae], OK

Menu: File, New Model Database, With Standard/Explicit Model

表 3.2 中列出了所有计算结果,拉 – 弯耦合刚度导致了横向变形,且随着铺层数的增加,横向变形减少。注意拉 – 弯耦合系数 $-B_{11}$ 随着铺层的增加的比例关系。

表 3.2 例 3.1 中横向变形与铺层数量的关系

n	δ/mm	$-B_{11}$	%
1	0.2196	1585193	100
5	0.0211	317039	1/5
10	0.0104	158518	1/10
15	0.0069	105468	1/15
20	0.0052	79260	1/20

3.2.10 层合板 FEA 中等效正交各向异性参数的输入

部分有限元软件没有层合板单元,因此无法像 3.2.9 节中一样输入 A、B、D、H 矩阵参数。然而,如果软件中有正交各向异性单元,仍可以进行层合板的变形、振动、屈曲分析,本章将进行讲解。

1. 单向层合板 FEA

可采用标准的壳单元构建单向板模型,即使不是层叠壳单元属性,但使用它们依然可以计算获得位移、应变和应力结果。壳的几何选取在壳的中面,中面位于壳厚度方向上的中间位置。壳厚度坐标在垂直于壳中面的垂线上。这也是壳单元法线的定义。例 3.2 中有详细的描述。

例 3.2 在 Abaqus 中建立简支约束的矩阵板。$a_x = 4000\mathrm{mm}$,$a_y = 2000\mathrm{mm}$,厚度 $t = 10\mathrm{mm}$。施加均布横向载荷 $q_0 = 0.12\times10^{-3}$ MPa。材料为单向的 AS4D/9310(表 3.1)。纤维方向为 x – 向。计算板中心的变形,利用式(1.102)~式(1.104)计算正交各向异性刚度阵,并通过 Type:Orthotropic 输入材料属性。本例在例 3.8 中将继续。

解 控制方程中消除单向板的厚度坐标,因此 3D 问题简化为 2D 问题。在这个过程中,厚度参数作为截面参数通过软件输入,模型的建立和求解如下:

Menu: File, Save as[C:\SIMULIA\User\Ex_3.2\Ex_3.2.cae]
Menu: File, Set Work Directory[C:\SIMULIA\User\Ex_3.2]

ⅰ. 创建部件。

考虑到板的对称性,只建立 1/4 模型。

Module: Part
Menu: Part, Create,
    [Part-1], 3D, Deformable, Shell, Planar, Approx. size[4000], Cont
Menu: Add, Line, Rectangle,
    # Enter the points[0,0],[2000,1000], X # ends the command, Done

ⅱ. 创建材料和截面,将截面赋予部件。

Module: Property
Menu: Material, Create,
    Mechanical, Elasticity, Elastic, Type: Orthotropic,
    # Enter the values of the components of the stiffness matrix, OK

注意 Abaqus 中刚度矩阵的建立与式(1.68)中定义的顺序不一致。在 Abaqus 中为

$$\begin{Bmatrix} \sigma_1 \\ \sigma_2 \\ \sigma_3 \\ \sigma_6 \\ \sigma_5 \\ \sigma_4 \end{Bmatrix} = \begin{bmatrix} C_{11} & C_{12} & C_{13} & 0 & 0 & 0 \\ \cdot & C_{22} & C_{23} & 0 & 0 & 0 \\ \cdot & \cdot & C_{33} & 0 & 0 & 0 \\ \cdot & \cdot & \cdot & C_{66} & 0 & 0 \\ \cdot & \cdot & \cdot & \cdot & C_{55} & 0 \\ \cdot & \cdot & \cdot & \cdot & \cdot & C_{44} \end{bmatrix} \begin{Bmatrix} \epsilon_1 \\ \epsilon_2 \\ \epsilon_3 \\ \gamma_6 \\ \gamma_5 \\ \gamma_4 \end{Bmatrix}$$

$D_{1111}=C_{11}, D_{1122}=C_{12}, D_{2222}=C_{22}, D_{1133}=C_{13}, D_{2233}=C_{23}, D_{3333}=C_{33}, D_{1212}=C_{66}$, $D_{1313}=C_{55}, D_{2323}=C_{44}$。使用 AS4D/9310 材料性能计算柔度矩阵式(1.104),并求逆得到刚度矩阵,如下:

$$\begin{bmatrix} 136222.40 & 3908.70 & 3908.70 & 0 & 0 & 0 \\ \cdot & 9251.35 & 3731.31 & 0 & 0 & 0 \\ \cdot & \cdot & 9251.35 & 0 & 0 & 0 \\ \cdot & \cdot & \cdot & 4306.00 & 0 & 0 \\ \cdot & \cdot & \cdot & \cdot & 4306.00 & 0 \\ \cdot & \cdot & \cdot & \cdot & \cdot & 2760.00 \end{bmatrix} (\text{MPa})$$

Menu: Section, Create,

Shell, Homogeneous, Cont
　　　　Shell thickness[10], Material[Material-1], OK
Menu: Assign, Section, # pick the part, Done, OK

另一个材料性能输入的方法是通过 Abaqus/CAE 中 Edit Materials 窗口的 Engineering Constants(9 个常数值)选项卡或 Lamina(5 个常数值[①])选项卡。

ⅲ. 创建装配体。
Module: Assembly
Menu: Instance, Create, Independent, OK

ⅳ. 创建分析步。
Module: Step
Menu: Step, Create, Name[Step-1], Static/General, Cont, OK

ⅴ. 创建约束和载荷。
Module: Load
Menu: BC, Manager,
　　Create, Name[X-symm], Step: Initial,
　　Mechanical, Symm/Anti/Enca, Cont
　　# pick the left vertical line, Done, XSYMM, OK
　　Create, Name[Y-symm], Step: Initial,
　　Mechanical, Symm/Anti/Enca, Cont
　　# pick the upper horizontal line, Done, YSYMM, OK
　　Create, Name[Simple-supp], Step: Initial,
　　Mechanical, Disp/Rota, Cont
　　# pick the right vertical and the lower horizontal lines, Done,
　　# checkmark: U3, OK, # close BC Manager

Menu: Load, Create
　　Step: Step-1, Mechanical, Pressure, Cont
　　# pick the part, Done,
Select side for the shell, Brown, Magnitude[0.12e-3], OK

ⅵ. 创建网格。
Module: Mesh

---

① 对于横观各向同性材料，$G_{12} = G_{13}$。

Menu: Seed, Instance, Approximate global size[50], Apply, OK
Menu: Mesh, Controls, Element Shape[Quad], Technique[Structured], OK
Menu: Mesh, Instance, # At the bottom of the WS select[Yes]

ⅶ. 求解并可视化结果。
Module: Job
Menu: Job, Manager
    Create, Cont, OK
    Data Check # To check the model for errors,
    Submit # To run the model
    Results # To visualize the solution
    Toolbar: Views, Apply Front View
Menu: Plot, Contours, On Deformed Shape,
    Results, Field Output,
    Output Variable[U], Component[U3], Apply, OK

平板中心最大位移为 17.43mm。

2. 对称层合板 FEA

多向铺层的层合板如果是对称平衡的,层合板的表观正交各向异性性能可以通过 1.15 节介绍的方法获得。表观的层合板性能表现为等效的(假想的)正交各向异性刚度,受面内载荷作用时,其响应与实际的层合板一致,但表观性能不能用于弯曲响应分析。当结构仅为弯曲响应时,例如一个厚的悬臂板,需要采用文献[1,式(6.36)]中的公式计算获得层合板的表观性能。对于大部分复合材料壳体结构,层合板仅受面内载荷作用,因此可采用 1.15 节中的计算公式。

如果层合板对称但并不平衡,正交异性轴相对于层合板坐标系旋转一个角度,但按照 1.15 节内容,该层合板依然可以认为是正交各向异性材料。例如,一个单向板的铺层方向与全局坐标夹角为 θ,其单向板应在含纤维方向的坐标系中进行建模(见 3.2.14 节)。

例 3.3 使用 Abaqus 建立简支约束的矩形板模型,其 $a_x = 2000$mm, $a_y = 2000$mm,铺层顺序为 $[\pm 45/0]_S$。施加边拉力 $N_x = 200$N/mm,计算最大水平位移,单层厚度为 1.0mm,性能如下:

$$E_1 = 37.88\text{GPa} \quad G_{12} = 3.405\text{GPa} \quad \nu_{12} = 0.299$$
$$E_2 = 9.407\text{GPa} \quad G_{23} = 3.308\text{GPa} \quad \nu_{23} = 0.422$$

解 因为层合板是平衡对称的,使用 1.15 节中方法计算层合板平均性能

E_x、E_y、E_z,如下:

$$E_x = 20.104 \text{GPa} \quad G_{xy} = 8.237 \text{GPa} \quad \nu_{xy} = 0.532$$
$$E_y = 12.042 \text{GPa} \quad G_{yz} = 3.373 \text{GPa} \quad \nu_{yz} = 0.203$$
$$E_z = 10.165 \text{GPa} \quad G_{xz} = 3.340 \text{GPa} \quad \nu_{xz} = 0.307$$

模型的建立步骤如下:

Menu: File, Save as,[C:\SIMULIA\User\Ex_3.3\Ex_3.3.cae]

Menu: File, Set Work Directory,[C:\SIMULIA\User\Ex_3.3]

i. 创建部件。

由于平板是对称的,因此只需建立1/4模型。

Module: Part

Menu: Part, Create,

　　[Part-1], 3D, Deformable, Shell, Planar, Approx. size[4000], Cont

Menu: Add, Line, Rectangle,

　　# Enter the points[0,0],[1000,1000], X # ends the command, Done

ii. 创建材料和截面,将截面赋予部件。

Module: Property

Menu: Material, Create,

　　Mechanical, Elasticity, Elastic, Type[Orthotropic],

　　# Enter the values of the components of the stiffness matrix, OK

注意 Abaqus 中刚度矩阵的建立与式(1.68)中定义的顺序不一致。在 Abaqus 中为

$$\begin{Bmatrix} \sigma_1 \\ \sigma_2 \\ \sigma_3 \\ \sigma_6 \\ \sigma_5 \\ \sigma_4 \end{Bmatrix} = \begin{bmatrix} C_{11} & C_{12} & C_{13} & 0 & 0 & 0 \\ \cdot & C_{22} & C_{23} & 0 & 0 & 0 \\ \cdot & \cdot & C_{33} & 0 & 0 & 0 \\ \cdot & \cdot & \cdot & C_{66} & 0 & 0 \\ \cdot & \cdot & \cdot & \cdot & C_{55} & 0 \\ \cdot & \cdot & \cdot & \cdot & \cdot & C_{44} \end{bmatrix} \begin{Bmatrix} \epsilon_1 \\ \epsilon_2 \\ \epsilon_3 \\ \gamma_6 \\ \gamma_5 \\ \gamma_4 \end{Bmatrix}$$

$D_{1111} = C_{11}, D_{1122} = C_{12}, D_{2222} = C_{22}, D_{1133} = C_{13}, D_{2233} = C_{23}, D_{3333} = C_{33}, D_{1212} = C_{66}, D_{1313} = C_{55}, D_{2323} = C_{44}$。使用前面给出的材料性能计算柔度矩阵式(1.104),并求逆得到刚度矩阵。

$$\begin{bmatrix} 26677.59 & 9531.84 & 5736.79 & 0 & 0 & 0 \\ \cdot & 15897.74 & 4216.73 & 0 & 0 & 0 \\ \cdot & \cdot & 11813.63 & 0 & 0 & 0 \\ \cdot & \cdot & \cdot & 8264.46 & 0 & 0 \\ \cdot & \cdot & \cdot & \cdot & 3378.38 & 0 \\ \cdot & \cdot & \cdot & \cdot & \cdot & 3344.48 \end{bmatrix} (\text{MPa})$$

其通过下述步骤输入 Abaqus/CAE 中：

Menu: Section, Create,
  Shell, Homogeneous, Cont
  Shell thickness[10], Material[Material-1], OK

Menu: Assign, Section, # pick the part, Done, OK

另一个方法是通过 Abaqus/CAE 中的 Edit Materials 对话框中 Engineering Constant(9 个常量)选项卡输入材料属性。

ⅲ. 创建装配体。

Module: Assembly
Menu: Instance, Create, Independent, OK

ⅳ. 创建分析步。

Module: Step
Menu: Step, Create, Name[Step-1], Static/General, Cont, OK

ⅴ. 创建约束和载荷。

Module: Load
Menu: BC, Manager,
  Create, Name[X-symm], Step: Initial, Mechanical,
  Symm/Anti/Enca, Cont
  # pick the left vertical line, Done, XSYMM, OK
  Create, Name[Y-symm], Step: Initial, Mechanical,
  Symm/Anti/Enca, Cont
  # pick the upper horizontal line, Done, YSYMM, OK
  Create, Name[Simple-supp], Step: Initial, Mechanical,
  Disp/Rota, Cont
  # pick the right vertical and the lower horizontal lines, Done,
  # checkmark: U3, OK, # close BC Manager

```
Menu: Load, Create
    Step: Step-1, Mechanical, Shell edge load, Cont
    # pick the right vertical line, Done, Magnitude[-200], OK
```

ⅵ. 创建网格。
```
Module: Mesh
Menu: Seed, Instance, Approximate global size[50], Apply, OK
Menu: Mesh, Controls, Element Shape[Quad], Technique[Structured], OK
Menu: Mesh, Instance, # At the bottom of the WS select[Yes]
```

ⅶ. 求解并可视化结果。
```
Module: Job
Menu: Job, Manager
    Create, Cont, OK
    Data Check # To check the model for errors,
    Submit # To run the model
    Results # To visualize the solution
    # 'Apply Front View' from the 'Views toolbar',
Menu: Plot, Contours, On Deformed Shape,
    Results, Field Output,
    Output Variable[U], Component[U1], Apply, OK
```

最大位移出现在平板的右侧边上,为 1.667mm。

1/4 模型计算得到的最大水平位移为 1.667mm。铺层为 [±45/0]$_S$ 的层合板,其 $x=0$ 和 $y=0$ 平面为非对称平面。然而一旦采用等效正交各向异性材料描述该层合板,像本例中的操作一样,虽然单层板层面不对称,但对于层合板中面的位移求解没有影响。因此,如果不进行应力的求解,1/4 板模型可以很好地模拟整板。此外,采用正交各向异性材料参数还可以进行层合板位移和中面应变的分析。但即使采用整板建模,该方法计算的应力也不是层合板的真实应力。本例中材料不是均质材料,而是多向铺层材料,而例 3.2 中材料为单向均质材料,因此,本例中应力值不正确,但例 3.2 中应力值是正确的。

3. 非对称层合板 FEA

如果层合板是非对称的,拉 - 弯耦合效应就需要考虑。严格来讲,非对称层合板并不是正交各向异性,因此不能按等效的层合板材料性能来分析。尽管如此,如果只能采用正交各向异性壳单元进行分析,且拉 - 弯耦合效应较小时,就可采用正交各向异性材料性能进行近似分析,同时忽略拉 - 弯耦合刚度阵 ***B*** 和

弯曲刚度阵 D 的影响。文献[1,式(6.37)~式(6.39)]中定义的比率可以用来评估使用层合板表观弹性参数获得近似值的准确性。必须对非平衡层合板多加注意,其 A、B 矩阵的计算坐标系应与层合板正交轴一致。

3.2.11 多向层合板 FEA 中的 LSS

对于中尺度下(单层板层次)的应变和应力的计算,必须要知道层合板每单层的属性和铺层的规则。多向层合板的描述信息中包括 LSS 信息,即每单层相对层合板 x-轴的角度、厚度和每单层材料的弹性性能。然后,软件可以根据铺层的属性计算出层合板的 A、B、D、H 矩阵。进一步,软件可以计算出每层铺层的应力。该方法在例 3.4 中进行了应用。在 Abaqus/CAE 中,LSS 信息可以通过 Composite Section 或者 Composite Layup 工具定义。

1. Composite Section

Composite Section 是用户定义复合材料 LSS 信息的基本工具。它的使用方法与典型截面(实体、壳、梁、桁架等)的定义和赋予不同。首先,用户需要定义一个或多个材料,这时需要指定各向同性、工程常数、单层板、正交各向异性或者各向异性类型。在 CAE 模块中,按照如下方式进行:

Module: Property
Menu: Material, Create, Mechanical, Elasticity, Elastic, Type: Lamina

然后,定义层合板截面,如下:
Module: Property
Menu: Section, Create, Category: Solid, Composite, Cont
    Material:[select], Thickness:[value], Orientation Angle:[value],
    Integration Points:[value], Ply name:[value]

注意:类别(category)选项中可以选 shell 也可以选 solid,其取决于使用的单元类型。连续体壳单元(如 SC8R)、薄壳单元(如 S8R5)和通用壳单元(如 S8R)需要 shell、composite 截面。实体单元,例如 C3D20,需要 solid、composite 截面。

实体单元也可以进行层叠,但是位移变化的最大阶次是二次的。因此,应变变化最多是线性的,这导致模拟层合板厚度方向上应变变化的精度不足。为缓解这个问题,实体单元可以进行堆叠,甚至可以每单层铺层使用一层单元。然而,由于实体纵横比的限制,建立的单层铺层实体单元其他两个尺寸不能大于厚度的 10 倍。因此单元需要划分得很小,而这又导致计算量的增加。

对于每个截面的定义,可通过 Eidt Section 对话框,用户可以定义层合板的 LSS,包括层合板中每个铺层的 Material(材料)、Thickness(厚度)、Orientation Angle(方向角),Integration Points(积分点)和 Ply Name(名称)。铺层名称是可选项,但推荐填写以方便理解分析结果。材料是从预先定义的材料库中进行选择。厚度和铺层方向需要输入。积分点最好使用默认的 3 个 Simpson 积分点,其分别位于为单层铺层的两个表面和一个中面上。用户可以通过编辑 Field Output Requests 对话框对输出结果进行优化编辑,具体如下:

```
Module: Step
Menu: Output, Field Output Requests, Edit, F-Output-1,
      Output at shell, Specify: 1,2,3
```

其中,1,2,3,…,3N,是壳厚度方向的所有 Simpson 积分点,N 表示层合板中铺层数。积分点结果的可视化是可选的:

```
Module:Visualization
Menu:  Results, Section Points, Selection method: Plies,
       Plies:[select],
       Ply result location:[Bottommost, Middle/Single, Topmost,
         Top & Bottom]
```

如果选择 Top & Bottom,那么层合板表面的计算值将显示出来。如果表面不在视图中,只需通过旋转模型即可。

最后,用户需要采用常规的方法将截面信息赋予相应的区域(Region)。如果使用的是常规壳单元和通用壳单元,该区域是平面或者曲面区域,即 2D 单元。如果使用的是连续体壳单元,该区域是 3D 单元。

如果使用常规壳单元和通用壳单元,层合板的铺层信息(LSS)通过 composite section 进行定义,并赋予相应单元区域。连续体壳单元可以在横跨壳的整个厚度上使用一个单元,这时截面的赋予方式类似,即对每个平/曲面赋予一个复合材料截面信息。但连续体壳单元的可以采用堆叠的形式,也就是说在 3D 壳的厚度方向上可以分割为多层连续体壳单元,这时每层连续体壳单元可以被赋予不同的复合材料截面信息。每个连续体壳单元将形成层合形式,其与复合材料截面定义的一致。

2. Composite Layup

Composite Layup 是另一种定义层合板铺层信息的方法。这里面,区域的概念与典型 CAE 中的通常所说的区域不同。一个 Composite Layup 必须与某一区域厚度上的一个单独的单元相关联。如果在厚度方向上进行细化多个网格,再通过 Composite Layup 方式对其多个网格进行定义,会导致在厚度上每个单元均被赋予完整层合板的铺层信息,而这将导致错误的计算结果。

第 3 章 层合板的刚度和强度

采用这个方法,用户不需要创建截面再将其赋予给某个区域。使用 Composite Layup 工具可以同时完成上述两件工作。如果误操作,对同一区域同时采用了 Composite Section 和 Composite Layup 两种定义方法,Composite Layup 的定义会覆盖前者。

Composite Layup 对话框可通过下述操作开启:

Module: Property

Menu: Composite, Create

这时用户必须确定需要创建的单元类型:常规壳、连续体壳或实体。连续体壳单元(如 SC8R)、薄壳单元(如 S8R5)和通用壳单元(如 S8R)需要的是 shell、composite 截面。实体单元,例如 C3D20 需要的是 solid、composite 截面。

Edit Composite Layup 对话框(图 3.9)有 4 个选项卡:Plies、Offset、Shell Parameters 和 Display。LSS 信息在 Plies 选项卡中定义。参考面的位置在 Offset 选项卡中定义。厚度模量(thickness modulus)参数在 Shell Parameters 选项卡中定义。

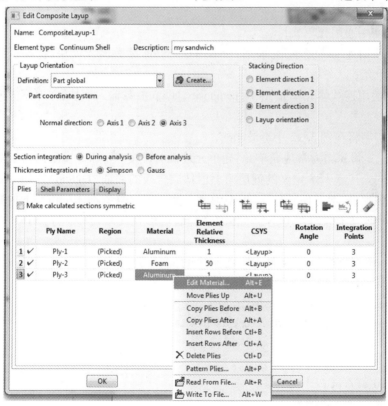

图 3.9　使用 Composite Layup 方式输入层合板铺层属性

Piles 选项卡中有 7 列待输入内容：Ply Name、Region、Material、Element Relative Thickness、CSYS、Rotation Angle 和 Integration Points。Ply Name 可选填。Region 为赋予铺层信息的区域。Material 中选择预先已定义好的材料。Element Relative Thickness 为用来确定铺层厚度的各铺层之间的比例系数，其值不必是实际的铺层厚度参数，并且系数之和也不必为 1。CSYS 为额外的可用于定义铺层的方向与模型全局坐标系关系的功能键。用户可以通过 CSYS 整体旋转铺层坐标系。Rotation Angle 是每个单层与层合板坐标系之间的夹角 θ_k。对于 Integration points，最好使用默认值，即 3 个 Simpson 积分点，分别为单层的两个表面积分点和一个中面积分点。结果可视化可以像 3.2.11 节中阐述的一样通过编辑 field output requests 进行优化定义。

例 3.4 一个简支约束的正方形平板 $a_x = a_y = 2000\text{mm}$，厚度 $t = 10\text{mm}$，采用 AS4D/9310 材料（表 3.1），铺层顺序为 $[0/90/\pm 45]_S$。板承受的拉力载荷为 $N_x = 100\text{N/mm}(N_y = N_{xy} = M_x = M_y = M_{xy} = 0)$。采用 4 种不同的单元模型计算单层板坐标系下的面内剪切应力 σ_6：

a. 使用常规壳单元；
b. 使用通用壳单元；
c. 使用连续体壳单元，采用 composite section 参数输入模块，在厚度方向只有一个单元；
d. 使用连续体壳单元，采用 composite layup 参数输入模块，在厚度方向有多个单元。

本实例将在例 3.12 中继续求解。

解 a 和 b 均为常规壳单元，壳单元均没有厚度。a 和 b 模型的区别是单元类型不同。注意 LSS 定义的铺层第 1 层从层合板的底面（bottom）开始。建立平板的整体模型，不考虑对称性，模型中节点被约束以避免刚体位移。下述操作步骤代码说明了整个建模过程。如果有疑问，可以在文献[5]中打开 Ex_3.4.cae 文件。

ⅰ. 确保这个 .cae 文件保存在正确的位置，并且将工作目录设置在文件保存的文件夹中。

Menu: File, Save as[c:\simulia\user\Ex_3.4\Ex_3.4.a.cae]
Menu: File, Set Work Directory[c:\simulia\user\Ex_3.4]

ⅱ. 创建部件。
Module: Part,
Menu: Part, Create,
Name[Part-1], 3D, Deformable, Shell, Planar, Approx size[2000], Cont

Menu: Add, Line, Rectangle,[-1000,-1000],[1000,1000], X, Done

ⅲ. 创建材料和截面,并且将截面赋予部件。

Module: Property

Menu: Material, Create, Name[Material-1], Mech. , Elast. , Elastic,
  Type[Engineering Constants],
  # Note: you can cut and paste an entire row of values from Excel
  [1.3386E+05 7.7060E+03 7.7060E+03 3.0100E-01 3.0100E-01
  3.9600E-01 4.3060E+03 4.3060E+03 2.7600E+03], OK

Menu: Section, Create, Shell, Composite, Cont
  # checkmark: Symmetric layers
  Integration[Simpson]
  # insert rows with right click until you have 4 rows
  Material[Material-1] # for all
  Thickness[1.25] #for all
  Orientation[0,90,45,-45]
  Integration points[3,3,3,3] # results at bot/mid/top of each ply
  Ply Names[k1,k2,k3,k4], OK, # k1:bottom, k4:middle
  # due to symmetry, CAE adds[-45,45,90,0] to the model,

Menu: Assign, Section, # pick part, Done, OK

ⅳ. 创建装配体。

Module: Assembly

Menu: Instance, Create, Independent, OK

ⅴ. 创建分析步。

Module: Step

Menu: Step, Create, Name[Step-1], Static/General, Cont, OK

Menu: Output, Field Output Requests, Edit, F-Output-1
  Step: Step-1, Output variables:
  # unselect everything, # select only what you need: S, E, U, RF
  # request at 8x3 = 24 points (8 layers, 3 Simpson points per layer)
  # Output at shell and layered section points:
  Specify:[1,...,24], OK # list all #'s from 1 to 24

ⅵ. 施加载荷。

Module: Load

```
# since the part is a shell, an edge load is: Shell edge load
Menu: Load, Create, Name[Load-1], Step: Step-1, Mechanical,
Shell edge load, Cont
    # pick the edges x = -1000 and x = 1000, Done, Magnitude[-100], OK
```

为了在创建边界条件时方便选择节点,需要预先进行网格划分。

ⅶ. 创建网格。

```
Module: Mesh
Menu: Seed, Instance, Approx global size[100], Apply, OK
Menu: Mesh, Controls, Element Shape[Quad], Technique[Structured],
    OK
Menu: Mesh, Element Type, Std, Quad, DOF[5], OK # element S8R5
Menu: Mesh, Instance, Yes
Menu: Tools, Set, Manager
    Create,[center], Node, Cont, # center node, Done
    Create,[y-hold], Node, Cont, # nodes at (-1000,0) and (1000,0),
    Done
    Create,[edges], Node, Cont, # nodes at the edges of the plate,
    Done
    # close Set Manager
```

ⅷ. 施加边界条件。

```
Module: Load
Menu: BC, Manager
    Create, Name[edges], Step: Initial, Mechanical, Disp/Rota, Cont
    Sets, # in the Region Selection pop-up window,[edges], Cont
    # checkmark: U3, OK
    Create, Name[center], Step: Initial, Mechanical, Disp/Rota, Cont
    # in the Region Selection pop-up window,[center], Cont
    # checkmark: U1, U2, and U3, OK
    Create, Name[y-hold], Step: Initial, Mechanical, Disp/Rota, Cont
    # in the Region Selection pop-up window,[y-hold], Cont
    # checkmark: U2, and U3, OK
    # close BC Manager
```

ⅸ. 求解并结果可视化。

```
Module: Job
```

```
Menu: Job, Manager
    Create, Cont, OK,
    Submit # to run the model
    Results # to visualize the solution
Menu: Viewport, Viewport Annotation Options, Tab: State Block,
    Set font, Size[12], Apply to[# checkmark all], OK, OK
Menu: Plot, Undeformed Shape,
Menu: Plot, Allow Multiple Plot States,
Menu: Plot, Deformed Shape,
Menu: Plot, Material Orientations, On Deformed Shape,
    # by default shows the 0-deg layer on bottom
Menu: Result, Section Points, Plies, # select K1, Apply
    # repeat for all the plies to visualize their orientation, OK
Menu: Plot, Contour, On Deformed Shape,
    Field Output Toolbox, Primary, S, S12 # laminate stress_6
Menu: Result, Section Points, Plies, # select K1, Apply
    # repeat for all the plies to visualize S12, OK
    # select ply K1, Apply # sigma_6(K1) = 0.0
    # select ply K2, Apply # sigma_6(K2) = 0.0
    # select ply K3, Apply # sigma_6(K3) = -1.112
    # select ply K4, Apply # sigma_6(K4) = 1.122, OK
```

a. 采用薄壳单元求解。

在上述步骤中,通过下述步骤选择常规壳单元 S8R5：

```
Module: Mesh
Menu: Mesh, Element Type, Standard, Quadratic, DOF[5], OK
```

本例可以方便地对模型进行保存并且修改模型以求解(b)。

```
Menu: File, Save as[c:\simulia\user\Ex_3.4\Ex_3.4.a.cae], OK
```

在单层板坐标系下可视化单层板的应力,

$$\sigma_6 = 0, 0, 1.112, 1.112 \text{MPa}$$

其值分别为 $0, 90, 45°, -45°$ 铺层的剪切应力。这些结果可以与文献[12]中通过经典层合理论的计算值进行检验和比较。

b. 采用通用壳单元求解。

仅需选择 DOF[6] 改变单元类型,即可将薄壳单元改变为通用壳单元,如下：

Menu: File, Save as[c:\simulia\user\Ex_3.4\Ex_3.4.b.cae], OK
Module: Mesh
Menu: Mesh, Element Type, Standard, Quadratic, DOF[6], OK # S8R

求解使用的单元为 S8R,该单元为二次的减缩积分通用(厚/薄)壳单元,模型修改后需重新提交计算。因为本例中没有弯曲,因此结果与(a)中求解的结果一样。

c. 采用连续体壳单元求解,铺层通过 composite section 模块定义。

要使用连续体壳单元,几何模型必须建成实体,尺寸为 2000mm × 2000mm × 10mm。

ⅰ. 保存模型,设置工作目录。
Menu: File, Save as[c:\simulia\user\Ex_3.4\Ex_3.4.c.cae]
Menu: File, Set Work Directory[c:\simulia\user\Ex_3.4]

ⅱ. 创建部件。
Module: Part
Menu: Part, Create,
　　Name[Part-1], 3D, Deformable, Solid, Extrusion, size[2000], Cont
Menu: Add, Line, Rectangle,[-1000,-1000],[1000,1000], X, Done,
　　Depth[10], OK

ⅲ. 创建材料和截面,并且将截面赋予部件。
Module: Property
Menu: Material, Create, Name[Material-1], Mech. , Elast. , Elastic,
　　Type[Engineering Constants],
　　# Note: you can cut and paste an entire row of values from Excel
　　[1.3386E+05 7.7060E+03 7.7060E+03 3.0100E-01 3.0100E-01
　　3.9600E-01 4.3060E+03 4.3060E+03 2.7600E+03], OK
Menu: Section, Create,[Section-1], Shell, Composite, Cont
　　# insert rows with right click until you have 8 rows
　　Material[Material-1] # for all, Thickness[1.25] # for all
　　Orientation[0,90,45,-45,-45,45,90,0],
　　Integration points[3] # for all
　　Ply Names[PLY-1,PLY-2,PLY-3,PLY-4,PLY-5,PLY-6,PLY-7,PLY-8], OK
Menu: Assign, Section, # pick the part, Done, OK

ⅳ. 创建装配体。
Module: Assembly

Menu: Instance, Create, Independent, OK

ⅴ. 创建分析步。

Module: Step

Menu: Step, Create, Name[Step-1], Static/General, Cont, OK

Menu: Output, Field Output Requests, Edit, F-Output-1

　　Step: Step-1, Output variables:

　　# unselect everything, # select only what you need: S, E, U, RF

　　# request at 8x3 = 24 points (8 layers, 3 Simpson points per layer)

　　# Output at shell and layered section points:

　　Specify:[1,...24], OK # 1,2,3,4,...

ⅵ. 施加载荷。

Module: Load

Menu: Load, Create

　　Name[Load-1], Step: Step-1, Mechanical, Surface traction, Cont

　　# an edge load of 100 N/mm on a surface 10 mm-thick equals 10 N/mm2

　　# Zoom into the upper-right corner of the plate

　　# pick the surface x = 1000, Done,

　　Distribution[Uniform], Traction[General],

　　Vector, Edit,[0,0,0],[1,0,0], Magnitude[10], OK

　　# repeat the procedure to apply a surface traction of 10 N/mm2

　　# on the x = -1000 surface, # Name[Load-2] and direction[-1,0,0]

为了在创建边界条件时方便选择节点,需要预先进行网格划分。

ⅶ. 创建网格。

Module: Mesh

Menu: Seed, Instance, Approx global size[100], OK

Menu: Mesh, Controls, # if required select the whole plate, Done,

　　Element Shape[Hex], Technique[Sweep], OK

Menu: Mesh, Element Type, # if required select whole plate, Done,

　　Family[Continuum Shell], # element SC8R has been selected, OK

Menu: Mesh, Instance, Yes

　　# create sets of nodes to facilitate applying the BC to the model

　　Views Toolbar, Front View # to visualize better

Menu: Tools, Set, Manager

　　Create, Name[edges], Type[Node], Cont

```
        # select all the nodes along the edges of the plate, Done
        Create, Name[center], Type[Node], Cont
        # select the nodes at the center of the plate (x = 0,y = 0), Done
        Create, Name[y-hold], Type[Node], Cont
        # select the nodes at (x = -1000,y = 0) and (x = 1000,y = 0), Done
        # close the Set Manager pop-up window
```

ⅷ. 施加边界条件。
```
Module: Load
Menu: BC, Manager
    Create, Name[edges], Step: Initial, Mechanical, Disp/Rota, Cont
    Sets, # in the Region Selection pop-up window,[edges], Cont
    # checkmark: U3, OK
    Create, Name[center], Step: Initial, Mechanical, Disp/Rota, Cont
    # in the Region Selection pop-up window,[center], Cont
    # checkmark: U1, U2, and U3, OK
    Create, Name[y-hold], Step: Initial, Mechanical, Disp/Rota, Cont
    # in the Region Selection pop-up window,[y-hold], Cont
    # checkmark: U2, and U3, OK
    # close BC Manager
```

ⅸ. 求解并结果可视化。
```
Module: Job
Menu: Job, Manager
    Create, Cont, OK,
    Submit # to run the model
    Results # to visualize the solution
Menu: Viewport, Viewport Annotation Options, Tab: State Block,
    Set font, Size[12], Apply to[# checkmark all], OK, OK
Menu: Plot, Undeformed Shape,
Menu: Plot, Allow Multiple Plot States,
Menu: Plot, Deformed Shape,
Menu: Plot, Material Orientations, On Deformed Shape
Menu: Result, Section Points, Plies, # select PLY-1, Apply
    # repeat for all the plies to visualize their orientation, OK
Menu: Plot, Contour, On Deformed Shape,
    Field Output Toolbox, Primary, S, S12 # laminate stress_6
Menu: Result, Section Points, Plies, # select PLY-1, Apply
```

repeat for all the plies to visualize S12 stress, OK

d. 采用连续体壳单元求解，复合材料铺层通过 Composite Layup 模块定义。

为了使用 Composite Layup 模块模拟层合板，其实体模型必须在厚度方向上划分为多个区域，每个区域对应层合板一层。部件的创建过程需要做如下修改：

ⅰ. 创建部件。

Module: Part
Menu: Part, Create,
    Name[Part-1], 3D, Deformable, Solid, Extrusion, size[2000], Cont
Menu: Add, Line, Rectangle, [-1000,-1000], [1000,1000], X, Done,
    Depth[10], OK
    # zoom into the upper-right corner of the plate
    # divide the thickness of the plate in 8 segments
Menu: Tools, Partition, Type[Edge], Method[Enter parameter]
    # pick edge, Done, Normalized edge param[0.5], Create Partition
    # repeat procedure until the edge is divided in 8 segments, Done
    # in the Create Partition pop-up window select
    Type[Cell], Method[Define cutting plane], Point & Normal,
    # select one point defined previously (start at top of the plate),
    # select edge used as normal direction (along positive z-axis),
    Create Partition # create first region representing the top ply
    # pick region to partition, Done,
    Point & Normal, # repeat procedure,
    # once 8 regions have been created, Done, # close pop-up window

材料属性定义需做如下修改：

ⅱ. 创建材料和截面，并且将截面赋予部件。

Module: Property
Menu: Material, Create, Name[Material-1], Mech., Elast., Elastic,
    Type[Engineering Constants],
    # Note: you can cut and paste an entire row of values from Excel
    [1.3386E+05 7.7060E+03 7.7060E+03 3.0100E-01 3.0100E-01
    3.9600E-01 4.3060E+03 4.3060E+03 2.7600E+03], OK
Menu: Composite, Create
    Initial ply count[8], Element Type[Continuum Shell], Cont
    Thickness integration rule[Simpson]

```
# in column named Region double-click cell corresponding Ply-1,
# pick top region of laminate, Done, # repeat for other plies
# in column named Material double-click cell corresponding Ply-1,
Material[Material-1], OK
# repeat to assign Material-1 to the other plies
Element Relative Thickness[1] # for all plies
Rotation Angle[0, 90, 45, -45, -45, 45, 90, 0] # one per ply
Integration points[3] # for all plies, OK
```

后续的步骤与上述(c)中的步骤一样。

3.2.12 含丢层层合板的 FEA

有时为了方便,特别是当层合板含有丢层时(图3.10),最好将壳的参考面定义在底面(或顶面)。当设计需要将层合板局部的厚度降低时,可以将层合板的铺层数量逐渐减少,从而使壳的厚度由厚变薄。通常丢层应该控制在 1∶16～1∶20比率(Th∶L,见图3.10)之间,除非有细致的分析和(或)试验表明可以使用更大的比率。对于丢层层合板,最好指定一个几何光滑的表面,或模具面(tool surface)作为参考面。

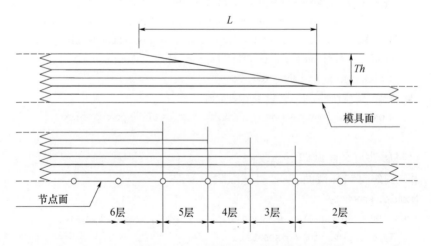

图 3.10 丢层区域的长度(L)和厚度(Th)以及有限元模型的简化

由于层合板铺层的减少,与其相关的单元材料和厚度也要改变,这在下面例子中将进行说明。并不是所有分析软件都能进行复合材料层合板的丢层分析,进行分析时需假设层合板中面是不变的,而实际上对于丢层结构只有基准面是不变的。只要层合板厚度与长宽相比很小,该假设并不会对整体分析结果造成

第3章 层合板的刚度和强度

较大影响,如变形、屈曲和膜应力等。当需要采用3D应力分析方法对丢层区域进行分析时,那么对该区域厚度方向的几何准确描述非常重要,此时采用局部3D模型进行分析更合适。

例3.5 长120mm、宽100mm的复合材料层合板,在A区域:$[90/0]_s$和B区域:$[90/0]$之间有一个丢层区域,单层厚度为0.75mm。丢层的比率为1:20,也就是说丢层区域的横向长度是减少厚度的20倍。层合板底边受$N=10$N/mm的拉力,材料为AS4D/9310(表3.1)。采用1/2对称模型进行分析。求解并可视化最大变形(z-向,垂直于板面),试列出每个铺层顶面和底面处的最大应力σ_{11}(不是最小值,也不是绝对值最大值)。

解 对于区域A、B和丢层区域(称AB),分别定义3个不同的截面。丢层区域仅最顶层的90°铺层去除。丢层区域的厚度为两个单层的厚度,$0.75\times2=1.5$mm。由于丢层的比率为1:20,丢层区域的长度为30mm。如果区域A跨度为$0<x<60$mm,过渡区跨度为$60<x<90$mm区域,那么区域B跨度为$90<x<120$mm。

每个区域层合板的中面与底面的偏移是不同的。因此,最好将载荷施加于共用的底边上。由于拉力面与中面的偏心,会带来附加的弯矩。这个附加弯矩的影响可以通过施加一个相反的弯矩抵消,本例中为了着重分析变形结果,并未施加抵消弯矩。

底部铺层被定义为#1铺层,其余的铺层沿单元坐标系的z-向逐层进行堆叠。层合板平面位于x-y平面内,x-z平面为对称面(YSYMM)。在模型左侧的较厚的一侧(A区域)根部施加Encastre-condition约束,这可以保证模型关于y-z平面对称。

材料属性通过Edit Material对话框进行编辑定义(图3.11),但这种方法比较麻烦。如果将材料属性列在Excel中或者其他制表文件中,那么就可以通过复制粘贴直接将材料属性输入到Edit Material对话框。此外,也可以通过网络服务直接使用CADEC数据库[18]。

层合板铺层顺序信息(LSS)在Edit Section对话框中完成定义,具体操作如下:

```
Module: Property
Menu: Section, Create, Shell, Composite, Cont
```

在Edit Section对话框中(图3.12),第一行为#1铺层,位于层合板的底部。其余行用于定义层合板中的第n铺层,最后一行为层合板的顶面铺层。对铺层分别命名为k1、k2、k3、k4。

图 3.11　Edit Material 对话框

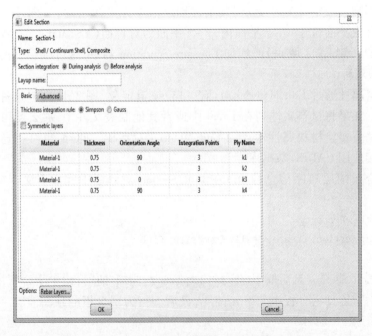

图 3.12　在 Edit Section 对话框中定义铺层顺序

计算完成后,用户需要选择计算结果积分点的位置才能显示,即选择需要显示结果的铺层以及该铺层厚度上积分点的位置。这些操作都是通过 Section Points 对话框完成,可通过下述步骤打开对话框:

Module: Visualization

Menu: Results, Section Points, Selection method:[Plies]

在 Section Points 对话框(图 3.13)中,选择了 k4,Topmost,表示选择显示 k4 铺层的顶面的计算结果。类似地,k1,Bottommost 表示选择显示 k1 铺层的底面结果。

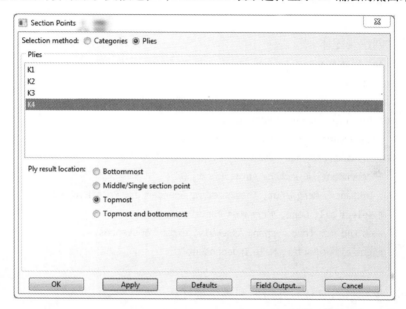

图 3.13 在 Section Points 对话框中选择需要显示的积分点位置

下面的操作代码可执行整个过程。如果有疑问,可以参考文献[5]中的 EX_3.5.cae。为了简化代码,可以通过 .py 文件创建材料和截面,利用之前例子中的方法可以方便地完成定义。

Units: N, mm, MPa

Menu: File, Save as[c:\simulia\user\Ex_3.5\Ex_3.5.cae]

Menu: File, Set Work Directory[c:\simulia\user\Ex_3.5]

i. 创建部件。

Module: Part

Menu: Part, Manager

    Create,[Part-A], 3D, Deformable, Shell, Planar,

```
        Approx size[500], Cont
Menu: Add, Line, Rectangle,[0,0],[60,50], X, Done
    Create,[Part-AB], 3D, Deformable, Shell, Planar,
    Approx size[500], Cont
Menu: Add, Line, Rectangle,[60,0],[90,50], X, Done
    Create,[Part-B], 3D, Deformable, Shell, Planar,
    Approx size[500], Cont
Menu: Add, Line, Rectangle,[90,0],[120,50], X, Done
    # close the Part Manager
```

ⅱ. 创建装配体。

```
Module: Assembly
Menu: Instance, Create,[Part-A],
    Independent, # uncheck Auto-offset, OK
Menu: Instance, Create,[Part-AB],
    Independent, # uncheck Auto-offset, OK
Menu: Instance, Create,[Part-B],
    Independent, # uncheck Auto-offset, OK
Menu: Instance, Merge/Cut, Intersecting Boundaries: Retain, Cont
    # select all, Done, # creates Part-1
    # on the.mdb tree, expand Assembly, expand Instances,
    right-click Part-1, Make Independent
```

ⅲ. 创建材料、截面,并将截面赋予部件。

```
Module: Property
    # to display the whole Part-1, set the tabbed menus as follows:
Module:[Property], Model:[Model-1], Part:[Part-1]
    # create materials by running.py scripts
Menu: File, Run Script,[AS4D--9310.py], OK
    # create sections by running.py scripts
Menu: File, Run Script,[laminate-A.py], OK
Menu: File, Run Script,[laminate-AB.py], OK
Menu: File, Run Script,[laminate-B.py], OK
    # recall that from bottom to top, section A is[90/0/0/90],
    # section B is[90/0], and section AB is[90/0/0]
Menu: Section, Assignment Manager
    Create, # pick the region/part Section-A, Done
Section:[Section-A], Shell Offset Definition:[Bottom Surface], OK
```

　　　　Create, # pick the region/part Section-AB, Done
Section:[Section-AB], Shell Offset Definition:[Bottom Surface], OK
　　　　Create, # pick the region/part Section-B, Done
Section:[Section-B], Shell Offset Definition:[Bottom Surface],
　　　OK # close the Section Assignment Manager,
　　　Done # close current command

ⅳ. 定义分析步。
Module: Step
Menu: Step, Create, Name[Step-1], Static/General, Cont, OK
Menu: Output, Field Output Request, Edit, F-Output-1
　　　Output at shell, Specify[1,2,3,4,5,6,7,8,9,10,11,12], OK

ⅴ. 施加边界条件和载荷。
Module: Load
Menu: BC, Manager
　　　Create,[BC-1], Step: Initial, Mechanical, Symm/Anti/Enca, Cont
　　　# pick edge @ x = 0, Done, Encastre, OK
　　　Create,[BC-2], Step: Initial, Mechanical, Symm/Anti/Enca, Cont
　　　# pick the 3 segments @ y = 0, Done, YSYMM, OK, # close BC Manager
　　　# since the part is a Shell, an edge load is: Shell edge load
　　　# load applies on the offset defined during Assign
Menu: Load, Create
　　　[Load-1], Step: Step-1, Mechanical, Shell edge load, Cont
　　　# pick the edge at x = 120, Done, Magnitude[-10], OK

ⅵ. 创建网格。
Module: Mesh
Menu: Seed, Instance, Approx global size:[5], Apply, OK
Menu: Mesh, Controls, # select all the regions, Done
　　　Element Shape[Quad], Technique[Structured]
Menu: Mesh, Element Type, # select all the regions, Done
　　　Standard, Quadratic, DOF[6], OK # element selected S8R
　　　Mesh, Instance, Yes

ⅶ. 求解并结果可视化。
Module: Job

Menu: Job, Manager

    Create, Cont, OK, Submit # when completed, Results

    Module: Visualization

Menu: Viewport, Viewport Annotation Options,

    Tab: State Block, Set Font, Size[10], Apply to[checkmark all],

    OK, OK

Menu: File, Save Options, OK # to save options for future sessions

Menu: Plot, Undeformed Shape

Menu: Plot, Allow Multiple Plot States

Menu: Plot, Deformed Shape # you will see both shapes

Menu: Plot, Contours, On Deformed Shape

    Field Output Toolbar,[Primary],[U],[U3]

    # read the maximum value from the legend on top left, U3 = 2.578 mm

    Field Output Toolbar,[Primary],[S],[S11]

    # next to select the top or bottom surface of each lamina

Menu: Result, Section Points

    Selection method:[Plies], K1, Bottommost, Apply

    # repeat for the other plies

    Selection method:[Plies], K1, topmost, Apply

    # repeat for the other plies

    # a gray region means the ply is not modeled in that region

最大的变形 $w=2.578\mathrm{mm}$,最大的应力在表 3.3 中列出。

表 3.3 例 3.5 的应力计算

层号#	z/mm	$\theta(°)$	σ_{11}/MPa
1(bot)	0.00	90	1.233
1(top)	0.75	90	0.6206
2(bot)	0.75	0	3.17
2(top)	1.50	0	6.596
3(bot)	1.50	0	6.596
3(top)	2.25	0	-19.15
4(bot)	2.25	90	0.1907
4(top)	3.00	90	0.4094

例 3.6 建立图 3.14 中的层合板有限元模型。其中区域 A 的铺层顺序为 [+45/-45/0/90/0]。单层铺层的厚度为 1.2mm。丢层区域的丢层比率为 1:10。

层合板长 120mm,宽 100mm。在层合板的底边施加 $N_x = 100\text{N/mm}$ 的拉力。使用 1/2 对称模型。材料为 AS4D/9310(表 3.1)。计算并显示最大横向位移 U_3。同时,可视化和列表显示各个铺层(k1、k2、k3、k4、k5)顶面的最大应力 σ_{11}。对于丢层区域没有的铺层,对应的结果将缺失。

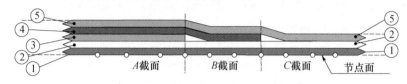

图 3.14 含有丢层区域的层合板

解 本例中,我们使用分区方法建立 3 个不同的区域。丢层区域丢层比率为 1∶10,因此总的丢层区域长度为 $2 \times 1.2 \times 10 = 24\text{mm}$。丢层区域在层合板中心线两侧各 12mm,而中心线据左侧边界 60mm 处,所以丢层过渡区分别位于 48~72mm 范围内。底部铺层被定义为#1。其余铺层按单元坐标系的正法向逐层堆叠在底层之上。丢层的铺层号为#3 和#4。下述操作代码可以执行整个过程。如果有疑问,可参考文献[5]中的 Ex_3.6.cae。

```
# Units: N, mm, MPa
Menu: File, Set Work Directory[c:\simulia\user\Ex_3.6]
Menu: File, Save As[Ex_3.6.cae]
```

ⅰ. 创建部件。

```
Module: Part
Menu: Part, Create
    [Part-1], 3D, Deformable, Shell, Planar, Approx size[300], Cont
Menu: Add, Line, Rectangle,[0,0],[120,50], X, Done
    # creating 3 partitions for the different regions
Menu: Tool, Datum, Plane, Offset from principal plane
    [YZ Plane],[48], Enter,[YZ Plane],[72], Enter, X
    # close the Create Datum pop-up window
Menu: Tools, Partition, Face, Use datum plane,
    # pick a datum plane, Create Partition
    # pick face to partition (area not partitioned yet), Done
    # pick datum plane, Create Partition, Done
    # close the Create Partition pop-up window
```

ⅱ. 创建装配体。

Module: Property

Menu: Material, Create
    Name[AS4D--9310], Mechanical, Elasticity, Elastic, Type[Lamina]
    # you can cut and paste an entire row of values from Excel
    [1.3386E5, 7.706E3, 0.301, 4.306E3, 4.306E3, 2.76E3], OK
    # or use a.py file to create materials as in previous example
Menu: Section, Manager
    Create, Name[Section-A], Shell, Composite, Cont
    # insert rows with right click until you have 5 rows
    Material[AS4D--9310] # for all, Thickness[1.2] # for all
    Orientation[45, -45,0,90,0], Integration points[3,3,3,3,3]
    Ply names[k1,k2,k3,k4,k5], OK
    Copy,[Section-AB], OK, Edit, # delete the center ply (k3), OK
    Copy,[Section-B], OK, Edit, # delete the center ply (k4), OK
    # close Section Manager
Menu: Section, Assignment Manager
    Create, # pick the region on the left, Done
    Section[Section-A], Shell Offset[Bottom surface], OK
    Create, # pick the region in the middle, Done
    Section[Section-AB], Shell Offset[Bottom surface], OK
    Create, # pick the region on the right, Done
    Section[Section-B], Shell Offset[Bottom surface], OK, Done
    # close Section Assignment Manager

iii. 创建材料、截面,并将截面赋予部件。
Module: Assembly
Menu: Instance, Create, Independent, OK

iv. 定义分析步。
Module: Step
Menu: Step, Create
    Name[Step-1], Procedure type: General, Static/General, Cont, OK
Menu: Output, Field Output Request, Edit, F-Output-1
    Specify[1,2,3,4,5,6,7,8,9,10,11,12,13,14,15], OK

v. 施加边界条件和载荷。
Module: Load
Menu: BC, Create

Name[BC-1], Step: Initial, Mechanical, Symm/Anti/Enca, Cont
　　# pick edge @ x = 0, Done, Encastre, OK
Menu: BC, Create
　　Name[BC-2], Step: Initial, Mechanical, Symm/Anti/Enca, Cont
　　# pick 3 segments @ y = 0, Done, YSYMM, OK
　　# since the part is a Shell, an edge load is: Shell edge load
　　# load applied with an offset defined during Assign
Menu: Load, Create
　　Name[Load-1], Step: Step-1, Mechanical, Shell edge load, Cont
　　# pick edge @ x = 120, Done, Magnitude[-10], OK

ⅵ. 创建网格。
Module: Mesh
Menu: Seed, Instance, Approx global size:[5], Apply, OK
Menu: Mesh, Controls, # select all regions, Done,
　　Element Shape[Quad], Technique[Structured], OK
Menu: Mesh, Element Type, # select all regions, Done
　　[Standard],[Quadratic], DOF[6], OK, # element used S8R
Menu: Mesh, Instance, Yes

ⅶ. 求解并结果可视化。
Module: Job
Menu: Job, Manager
　　Create, Cont, OK, Submit, # when completed, Results
Module: Visualization
Menu: Viewport, Viewport Annotation Options
　　Tab: State Block, Set Font, Size[10],
　　Apply to[checkmark all], OK, OK
Menu: File, Save Options, OK # to save options for future sessions
Menu: Plot, Undeformed shape
Menu: Plot, Allow Multiple Plot States
Menu: Plot, Deformed Shape # you will see both shapes
Menu: Plot, Material Orient. , On Def. Shape # identify the 1-axis
Menu: Result, Section Points,[Plies], K1, Bottommost
　　# pick one ply at a time and comment on the results, OK
　　Field Output Toolbar,[Primary],[S],[S11]
Menu: Result, Section Points,[Plies], K1, Topmost
　　# pick one ply at a time and tabulate the results, OK

最大的横向变形为0.6075mm,最大的应力在表3.4中列出。

表3.4 例3.6中各铺层顶面的最大应力

层号#	z/mm	θ/(°)	σ_{11}/MPa
1	1.2	45	10.1
2	2.4	-45	17.88
3	3.6	0	17.63
4	4.8	90	4.303
5	6.0	0	9.632

3.2.13 夹层壳的FEA

当结构由两个薄面板和一个较厚但刚度较弱的夹芯(蜂窝、泡沫、轻质木材等)组成时,其层合结构应模拟为夹层壳。薄面板承担几乎所有的弯矩和面内载荷,并假设夹芯承受几乎所有的横向剪力。例3.7说明了如何定义和计算一个夹层悬臂梁。

对于夹层壳结构通常有如下的假设:

(1)式(3.9)中的H_{ij}项仅取决于中间层(夹芯层),按下式计算:

$$H_{ij} = (\overline{Q}_{ij}^*)_{core} t_{core} \quad (i,j=4,5) \tag{3.17}$$

(2)对于上、下面板而言,横向剪切模量G_{23}和G_{13}为0,因此面板对夹层壳横向刚度无贡献。

(3)面板的横向剪切应变和应力为0,可以忽略。

(4)夹芯的横向剪切应变和应力在厚度方向上保持不变。

例3.7 四周固支约束的夹层板,在上表面施加均匀压力[①]。夹层板由上、下面板和夹芯组成,具体厚度和性能以及压力载荷见表3.5。使用对称边界条件,建立1/4模型。采用6×6个壳单元建模。采用如下4种单元计算夹层板中心变形:

(a)常规壳单元S4R。

(b)连续体壳单元SC8R,厚度方向只有一层单元,定义了夹层板的上、下面板和夹芯。

(c)连续体壳单元SC8R,在厚度方向堆叠单元,上、下面板分别由一层单元定义,夹芯由一层或多层单元定义。

---

① 本例由文献[16, Test R0031/3]修改而来,其简支边界更改为固支边界。做出这样的改变是因为在连续体壳边缘施加简支边界条件较为困难。若确实需要施加简支边界的话,可以使用约束方程。

(d) 夹芯层采用连续体壳单元 SC8R，上、下面板采用常规壳单元 S4R。

表 3.5　例 3.7 中的计算参数

参数	单位	面板	夹芯
E_1	MPa	68947.57	0.068948
E_2	MPa	27579.03	0.068948
v_{12}		0.3000	0.0000
G_{12}	MPa	12927.67	0.068948
G_{13}	MPa	12927.67	209.6006
G_{23}	MPa	12927.67	82.73708
厚度	mm	0.7112	19.0500
1/2 板宽	mm	127.00	
压力	MPa	0.689476	
厚度模量	MPa	689475.7	689475.7

解　我们采用 4 种方法进行求解，结果列在表 3.6 中。堆叠的连续体壳模型(c)和堆叠的连续体壳+面板模型(d)比(a)、(b)模型更柔软，(c)、(d)模型中在夹芯和两个面板上分别单独施加了 FSDT 约束(直法线)。因此，每个单层(夹芯和每个面板)都能独立进行转动，这意味着每个单层均会达到自身应有的横向剪切变形。因此，夹芯的剪切模量(G_{13}，G_{23})较小，那么夹芯的变形会更大。这额外的弹性会影响夹层壳的整体横向变形，尤其是横向剪切应变和应力。对于夹层壳，因为夹芯本身是匀质的(不是层合结构)，因此增加夹芯厚度方向单元的数量不会改变计算结果。因此，它的法线变形是一条直线，与厚度方向单元数量无关。使用常规壳单元模型和使用厚度上只有一个单元的连续体壳单元模型会得到一样的计算结果，因为厚度上只有一个单元的连续体壳单元 SC8R 和常规壳单元 S4R 具有一样的动力学行为(假设 SC8R 单元通过选取很大的厚度模量实现横向不可压缩)。

在 Abaqus 中，连续体壳单元需要输入厚度模量参数 E_3，该参数不能通过之前给定的材料性能中读取，即使通过 Engineering Properties 或者 Isotropic($E_3 = E$) 和 Lamina($E_3 = E_2$) 定义也无法读取 E_3。因此，用户必须通过 Edit Composite Layup 对话框(图 3.15)中的 Shell Parameters 选项卡定义厚度模量。因为壳在厚度方向假设是不可压缩的，我们可以选取面板最大弹性模量的 10 倍作为厚度模量值[①]，即 $E_3 = 10 \times E_1 = 689475.7$ MPa。常规壳单元本身具有厚度方向不可压缩

---

① 这与文献([19]4.9.3 节 R0031(3))中的 r313_std_sc8r_composite.inp 数据一致。

属性,因此不需要定义厚度模量。

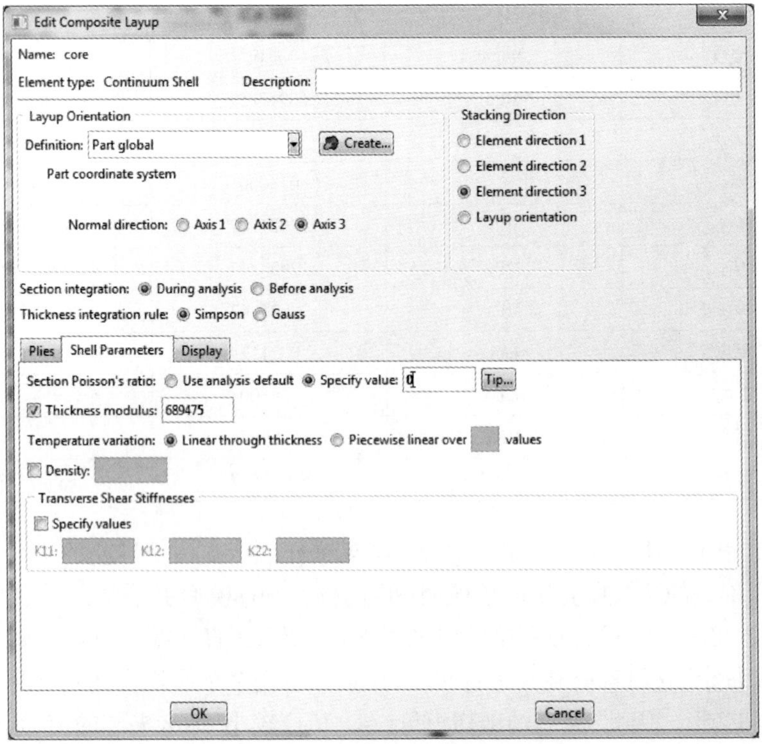

图 3.15 对壳截面进行厚度模量赋值

a. 常规壳单元方法求解。

下面将夹层板几何建立为 2D 平面模型,然后通过在常规壳单元下定义层合板属性。整个操作过程代码如下。如果有疑问,可以文献[5]中的 EX_3.7.a.cae。

```
# NAFEMS sandwich clamped metric conventional shell elements
Menu: File, Set Work Directory[c:\simulia\user\Ex_3.7]
Menu: File, Save As[Ex_3.7.a.cae]
```

ⅰ. 创建部件。

Module: Part

Menu: Part, Create, 3D, Deformable, Shell, Planar, Cont

Menu: Add, Line, Rectangle,[0,0],[127,127], X, Done

  # name the part to make it easier to use Composite Modeler later

Menu: Tools, Set, Create, Name[plate], Cont, # select all, Done

ⅱ. 创建材料、截面,并将截面赋予部件。

Module: Property

Menu: Material, Create

Name[face], Mechanical, Elasticity, Elastic, Type[Lamina]

　　[68947.57, 27579.028, 0.3000, 12927.66938, 12927.66938,
　　12927.66938], OK

Menu: Material, Create

Name[core], Mechanical, Elasticity, Elastic, Type[Lamina]

　　[0.06894757, 0.06894757, 0.0000, 0.06894757, 209.6006128,
　　82.737084], OK

use Composite Modeler instead of Create + Assign Sections

Menu: Composite, Create, Name[NAFEMS-sandwich],

　　Conventional Shell, Cont

　　# Name, region, Material, Thickness, CSYS, Angle, Integ Points

　　Ply-1, plate, face, 0.7112, < Layup >, 0, 3

　　Ply-2, plate, core, 19.050, < Layup >, 0, 3

　　Ply-3, plate, face, 0.7112, < Layup >, 0, 3

　　OK

　　# to assign Region double click the cell corresp. to desired ply

　　Sets # to open Region Selection pop-up window, plate, Cont

　　# to assign Material double click the cell corresp. to desired ply

　　# in Select Material pop-up window, Material[face or core], OK

ⅲ. 创建装配体。

Module: Assembly

Menu: Instance, Create, Independent, OK

ⅳ. 定义分析步。

Module: Step

Menu: Step, Create

　　Procedure type: Linear perturbation, Static/General, Cont, OK

ⅴ. 施加边界条件和载荷。

Module: Load

Menu: Load, Create

　　Name[Load-1], Step: Step-1, Mechanical, Pressure, Cont,

　　# pick the part, Done, Choose a side for the shell[Brown]

Magnitude[0.689476], OK
Menu: BC, Manager
   Create, Step: Initial, Mechanical, Symm/Anti/Enca, Cont
   # pick edges @ x = 0 and y = 0, Done, Encastre, OK
   Create, Step: Initial, Mechanical, Symm/Anti/Enca, Cont
   # pick edge @ x = 127, Done, XSYMM, OK
   Create, Step: Initial, Mechanical, Symm/Anti/Enca, Cont
   # pick edge @ y = 127, Done, YSYMM, OK
   # close BC Manager

ⅵ. 创建网格。
Module: Mesh
Menu: Seed, Instance, Approx global size[21], OK
Menu: Mesh, Controls, Element Shape[Quad], Tech[Structured], OK
Menu: Mesh, Element Type, # select all surfaces, Done
   Element Library[Standard], Geometric Order[Linear], OK # S4R
Menu: Mesh, Instance, Yes

ⅶ. 求解并结果可视化。
Module: Job
Menu: Job, Manager
   Create, Cont, OK, Submit # when completed, Results
Module: Visualization
   Field Output Toolbar,[Primary],[U],[U3]

模型右上角的位移为 −1.667mm。
b. 单层连续体壳单元方法求解。
下面将夹层板建立为 3D 体模型，然后在连续体壳单元下定义层合板铺层属性。注意，相对厚度项里面的数据为铺层之间的厚度比例，其每个铺层的实际厚度由单元厚度计算得到。为了说明，尽管模型的尺寸为国际单位制，本例中我们将铺层的厚度按英寸单位定义为(0.028″,0.75″,0.028″)。下面为整个过程的操作代码。如果有疑问，可以参考文献[5]中的 Ex_3.7.b.cae。
   # NAFEMS sandwich clamped metric laminated continuum shell elements
Menu: File, Set Work Directory[c:\simulia\user\Ex_3.7]
Menu: File, Save As[Ex_3.7.b.cae]

ⅰ. 创建部件。
Module: Part

Menu: Part, Create, 3D, Deformable, Solid, Extrusion, Cont
Menu: Add, Line, Rectangle,[0,0],[127,127], X, Done
    Depth[20.4724], OK # core + two faces
    # name the part to make it easier to use Composite Modeler later
    Menu: Tools, Set, Create, Name[plate], Cont, # select all, Done

ⅱ. 创建材料、截面,并将截面赋予部件。
Module: Property
Menu: Material, Create
    Name[face], Mechanical, Elasticity, Elastic, Type[Lamina]
    [68947.57, 27579.028, 0.3, 12927.66938, 12927.66938,
    12927.66938], OK
Menu: Material, Create
    Name[core], Mechanical, Elasticity, Elastic, Type[Lamina]
    [0.06894757, 0.06894757, 0.0, 0.06894757, 209.6006128,
    82.737084], OK
use Composite Modeler instead of Create + Assign Section
Menu: Composite, Create, Name[sandwich], Type: Continuum Shell, Cont
    # Name, region, Material, Thickness, CSYS, Angle, Integ Points
    Ply-1, plate, face, 0.028, <Layup>, 0, 3
    Ply-2, plate, core, 0.75, <Layup>, 0, 3
    Ply-3, plate, face, 0.028, <Layup>, 0, 3
    # to assign Region double click the cell corresp. to desired ply
    Sets # to open Region Selection pop-up window, plate, Cont
    # to assign Material double click the cell corresp. to desired ply
    # in Select Material pop-up window, Material[face or core], OK
    Tab: Shell Parameters
    Section Poisson's ratio: Specify value[0.0]
    # checkmark: Thickness modulus[689475.7], OK
    # the last two instructions are very important!

ⅲ. 创建装配体。
Module: Assembly
Menu: Instance, Create, Independent, OK

ⅳ. 定义分析步。
Module: Step

Menu: Step, Create
    Procedure type: Linear perturbation, Static/General, Cont, OK
Menu: Output, Field Output Requests, Edit, F-Output-1
    Output variables: Edit variables[CF,CTSHR,LE,RF,S,SE,TSHR,U]
    Output at shell: Specify[1,2,3,4,5,6,7,8,9], OK

Ⅴ. 施加边界条件和载荷。

Module: Load
Menu: Load, Create
    Step: Step-1, Mechanical, Pressure, Cont
    # pick the top surface (z = 20.4724), Done, Magnitude[0.689476], OK
Menu: BC, Manager
    Create, Step: Initial, Mechanical, Symm/Anti/Enca, Cont
    # pick surfaces @ x = 0 and y = 0, Cont, Encastre, OK
    Create, Step: Initial, Mechanical, Symm/Anti/Enca, Cont
    # pick surface @ x = 127, Cont, XSYMM, OK
    Create, Step: Initial, Mechanical, Symm/Anti/Enca, Cont
    # pick surface @ y = 127, Cont, YSYMM, OK
    # close BC Manager

Ⅵ. 创建网格。

Module: Mesh
Menu: Seed, Instance, Approx global size[21], OK
Menu: Mesh, Controls, Element Shape[Hex], Technique[Sweep], OK
Menu: Mesh, Element Type, # select all, Done
    Family[Continuum Shell], OK # notice the element chosen is SC8R
Menu: Mesh, Instance, Yes

Ⅶ. 求解并结果可视化

Module: Job
Menu: Job, Manager
    Create, Cont, OK, Submit # when completed, Results
Module: Visualization
    Field Output Toolbar,[Primary],[U],[U3]

模型右上角的位移为 -1.667 mm。

c. 多层连续体壳单元方法求解。

下面将夹层板建立为 3D 体模型,并且厚度上分为两层面板和夹芯。面板和夹芯区域分别用连续体壳单元建模。为了避免计算时自锁,采用 3D 实体单元离散时,网格必须划分很细。但实际上,面板厚度很小,其网格单元尺寸可能大于厚度的 10 倍。本例中我们将学习如何利用 Local Seeds 对话框(图 3.16)布种子,如何对一条边重新布种子。同时需要通过 Edit Composite Layup 对话框中的 Shell Parameters 选项卡定义厚度模量(图 3.15)。下面为整个过程的操作代码,如果有疑问可以参考文献[5]中的 Ex_3.7.c.cae。如果将夹芯在厚度方向上重新划分为 4 个单元,计算结果保持不变。

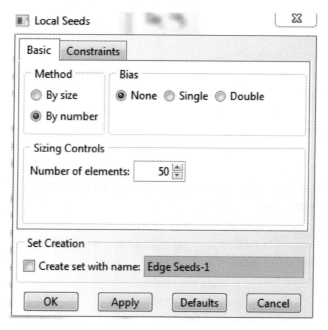

图 3.16 使用 By number 方法在边上布种子,边上有 50 个单元

```
# NAFEMS sandwich clamped metric stack 3 continuum shell elements
Menu: File, Set Work Directory[c:\simulia\user\Ex_3.7]
Menu: File, Save As[Ex_3.7.c.cae]
```

i. 创建部件。

```
Module: Part
Menu: Part, Create, 3D, Deformable, Solid, Extrusion, Cont
Menu: Add, Line, Rectangle,[0,0],[127,127], X, Done
    Depth[20.4724], OK # core + two faces
```

```
# partition the faces from the core
Menu: Tool, Datum, Plane, Offset from principal plane
    [XY],[0.7112], Enter,[XY],[19.7612], Enter,
    X, # close Create Datum
Menu: Tools, Partition, Cell, Use datum plane
    # the entire part is automatically selected
    Select datum plane # back datum plane, Create Partition
    # pick front cell to create the new partition, Done
    Select datum plane, Create Partition, Done
```

ii. 创建材料、截面,并将截面赋予部件。

```
Module: Property
Menu: Material, Manager
    Create, Name[face], Mechanical, Elasticity, Elastic,
    Type[Lamina]
    [68947.57, 27579.028, 0.3, 12927.66938, 12927.66938,
    12927.66938], OK
    Create, Name[core], Mechanical, Elasticity, Elastic,
    Type[Lamina]
    [0.06894757, 0.06894757, 0.0, 0.06894757, 209.6006128,
    82.737084], OK
    # close Material Manager, # zoom in to pick the regions easily
    # use Composite Modeler instead of Create + Assign Section
Menu: Composite, Create
    Name[front face], Initial ply count[1],
    Element Type[Continuum Shell], Cont
    # Name, region, Material, Rel. Thick., CSYS, Angle, Integ. Points
    Ply-1, <picked>, face, 1, <Layup>, 0, 3
    Tab: Shell Param., Section Poisson's ratio: Specify value[0.0]
    # checkmark: Thickness modulus:[689475.7], OK
    # the last two instructions are very important!
Menu: Composite, Create
    Name[back face], Initial ply count[1],
    Element Type[Continuum Shell], Cont
    # Name, region, Material, Rel. Thick., CSYS, Angle, Integ. Points
    Ply-1, <picked>, face, 1, <Layup>, 0, 3
    Tab: Shell Param., Section Poisson's ratio: Specify value[0.0]
    # checkmark: Thickness modulus:[689475.7], OK
```

```
# the last two instructions are very important!
Menu: Composite, Create
    Name[core], Initial ply count[1], Type[Continuum Shell], Cont
    # Name, region, Material, Rel. Thick. , CSYS, Angle, Integ. Points
    Ply-1, <picked>, core, 1, <Layup>, 0, 3
    Tab: Shell Param. , Section Poisson's ratio: Specify value[0.0]
    # checkmark: Thickness modulus:[689475.7], OK
```

iii. 创建装配体。

```
Module: Assembly
Menu: Instance, Create, Independent, OK
```

iv. 定义分析步。

```
Module: Step
Menu: Step, Create
    Procedure type: Linear perturbation, Static/Linear pert. , Cont, OK
Menu: Output, Field Output Requests, Edit, F-Opuput-1
    Output variables: Edit variables[CF,CTSHR,LE,RF,S,SE,TSHR,U]
    Output at shell: Specify[1,2,3], OK
```

v. 施加边界条件和载荷。

```
Module: Load
Menu: Load, Create
    Step: Step-1, Mechanical, Pressure, Cont
    # pick the top surface of the part (z = 20.4724), Done
    Magnitude[0.689476], OK
Menu: BC, Manager
    Create, Step: Initial, Mechanical, Symm/Anti/Enca, Cont
    # pick 3 surfaces @ x = 0 and the 3 @ y = 0, Done, Encastre, OK
    Create, Step: Initial, Mechanical, Symm/Anti/Enca, Cont
    # pick the three surfaces @ x = 127, Done, XSYMM, OK
    Create, Step: Initial, Mechanical, Symm/Anti/Enca, Cont
    # pick the three surfaces @ y = 127, Done, YSYMM, OK
    # close the BC Manager pop-up window
```

vi. 创建网格。

```
Module: Mesh
```

Menu: Seed, Instance, Approx global size[21], OK
Menu: Seed, Edges
    # select the four vertical edges of the core part, Done
    Method[By number], Number of elements[1], OK, X
Menu: Mesh, Controls, # select all regions, Done, Sweep, OK
Menu: Mesh, Element Type, # select all regions, Done
    Family[Continuum Shell], OK # SC8R
Menu: Mesh, Instance, Yes

ⅶ. 求解并结果可视化。
Module: Job
Menu: Job, Manager
    Create, Cont, OK, Submit # when completed, Results
Module: Visualization
    Field Output Toolbar,[Primary],[U],[U3]

模型右上角的位移为 $-1.892\mathrm{mm}$。将夹芯网格细化,程序进行如下修改：
Module: Mesh
Menu: Seed, Edges
    # select the four vertical edges of the core part, Done
    Method[By number], Number of elements[4], OK, X
Menu: Mesh, Instance, Yes

d. 连续体壳单元和常规壳单元混合方法求解。

下面将夹层板夹芯建立为3D体模型,将夹层板面板建立为表层(skin)2D模型。夹芯采用连续体壳单元离散,面板采用常规壳单元离散。需要注意面板的参考面位置必须进行准确偏移,以使面板表面与夹芯表面重合,否则计算结果将出现与面板厚度相关的偏差。为了对面板正确进行偏移,需将参考面定义为面板的顶面或底面,这取决于壳法向的方向,需要首先进行法向方向的显示。如果法向朝外,那么需将面板底面定义为参考面,此时面板底面与夹芯表面可以重合。下面为整个过程的操作代码,如果有疑问,可以参考文献[5]中的 Ex_3.7.d.cae。

NAFEMS sandwich clamped metric using skins
Menu: File, Set Work Directory[c:\simulia\user\Ex_3.7]
Menu: File, Save As[Ex_3.7.d.cae]

第3章 层合板的刚度和强度

ⅰ. 创建部件。
Module: Part
Menu: Part, Create, 3D, Deformable, Solid, Extrusion, Cont
Menu: Add, Line, Rectangle,[0,0],[127,127], X, Done
    Depth[19.05], OK # notice you created only the core part
    # name the part to make it easier to use Composite Modeler later
Menu: Tools, Set, Create, Name[plate], Cont, # select all, Done

ⅱ. 创建材料、截面,并将截面赋予部件。
Module: Property
Menu: Special, Skin, Create, # pick top face, Done
Menu: Special, Skin, Create, # pick back face, Done, X
Menu: Material, Create
    Name[face], Mechanical, Elasticity, Elastic, Type[Lamina]
    [68947.57, 27579.028, 0.3, 12927.66938, 12927.66938,
    12927.66938], OK
Menu: Material, Create
    Name[core], Mechanical, Elasticity, Elastic, Type[Lamina]
    [0.06894757, 0.06894757, 0.0, 0.06894757, 209.6006128,
    82.737084], OK
    # Notice the type of element used for the core (Continuum Shell)
Menu: Composite, Create
    Name[core], Initial ply count[1], Type[Continuum Shell], Cont
    # Name, Region, Mat., Relat. Thick., CSYS, Angle, Integ. Points
    Ply-1, plate, core, 1, < Layup >, 0, 3
    Tab: Shell Parameters,
    Section Poisson's ratio: Specify value[0.0]
    # checkmark: Thickness modulus:[689475.7], OK
    # Notice type of element used for skins/faces (Conventional Shell)
Menu: Composite, Create
    Name[back face], Initial ply count[1], Type[Conv. Shell], Cont
    # Name, Region, Material, Thick., CSYS, Angle, Integ. Points
    Ply-1, < picked >, face, 0.7112, < Layup >, 0, 3
    # use actual thickness of the laminate
    Tab: Offset,
    Shell Reference Surface and Offsets[Bottom surface], OK
Menu: Composite, Create, Name[front face],
    Initial ply count[1], Type[Conventional Shell], Cont

    # Name, Region, Material, Thick. , CSYS, Angle, Integ. Points
    Ply-1, <picked>, face, 0.7112, <Layup>, 0, 3
    # use the actual thickness of the laminate
    Tab: Offset,
    Shell Reference Surface and Offsets[Bottom surface], OK

ⅲ．创建装配体。

Module: Assembly

Menu: Instance, Create, Independent, OK

ⅳ．定义分析步。

Module: Step

Menu: Step, Create

    Procedure type: Linear perturbation, Static/Linear pert. , Cont, OK

Menu: Output, Field Output Requests, Edit, F-Opuput-1

    Output variables: Edit variables[CF,CTSHR,LE,RF,S,SE,TSHR,U]

    Output at: Specify[1,2,3,4,5,6,7,8,9], OK # important step!

ⅴ．施加边界条件和载荷。

Module: Load

Menu: Load, Create

    Step: Step-1, Mechanical, Pressure, Cont

    # pick the top surface of the part (z = 19.05), Done

    Magnitude[0.689476], OK

Menu: BC, Manager

    Create, Step: Initial, Mechanical, Symm/Anti/Enca, Cont

    # pick the surfaces @ x = 0 and @ y = 0, Done, Encastre, OK

    Create, Step: Initial, Mechanical, Symm/Anti/Enca, Cont

    # pick the surface @ x = 127, Done, XSYMM, OK

    Create, Step: Initial, Mechanical, Symm/Anti/Enca, Cont

    # pick the surface @ y = 127, Done, YSYMM, OK

    # close BC Manager pop-up window

ⅵ．创建网格。

Module: Mesh

Menu: Seed, Instance, Approx global size[21], OK

Menu: Mesh, Controls, Sweep, OK

```
Menu: Mesh, Element Type, # pick the core part, Done
    Family[Continuum Shell], OK # element SC8R was applied
    # pick the front and back faces (the skins), Done
    Family[Shell], OK # element S4R was applied
Menu: Mesh, Instance, Yes
```

ⅶ. 求解和结果可视化。
```
Module: Job
Menu: Job, Manager
    Create, Cont, OK, Submit # when completed, Results
Module: Visualization
    Field Output Toolbar,[Primary],[U],[U3]
```

模型右上角的位移为 −1.892mm。

例 3.7 中的中心位移计算结果如表 3.6 所列。

表 3.6 例 3.7 中的中心位移计算结果

计算方法	中心位移/mm
常规壳	1.667
单一的连续体壳	1.667
多个层叠的连续体壳	1.892
连续体壳 + 表层(skin)	1.892

3.2.14 单元坐标系

在前处理器中,定义层合板过程中,首先要知道层合板的坐标系方向。材料参数、每个铺层相对于层合板轴线方向以及其他参数和属性的定义都需要参考层合板坐标系,除非另有其他规定。同时,也可以获得在层合板坐标系下的计算派生结果(应变和应力)。在有限元分析中,层合板坐标系与单元坐标系相关,通过唯一的右手正交法则与每个单元关联。

单元坐标系方向与单元类型有关。对于杆和梁单元,x-轴方向一般为单元两个端点连线的方向。对于实体单元,x-轴方向与全局坐标系方向一致。对于壳单元,x-轴和 y-轴在单元平面内,z-轴为单元的法向。x-轴和 y-轴的默认方向取决于软件类型和单元类型。

对于壳单元,定义单元坐标系方向的方法有多种。在 ANSYS 中(图 3.17(a)),x-轴与单元中由第 1、第 2 个节点定义的边的方向一致,z-轴为壳平面法向(右手法则定义的向外方向),y-轴正交于 x-轴和 z-轴。在 MSC-MARC

中,x-轴与单元对边中点的连线方向一致(图 3.17(b))。在 Abaqus 中(图 3.17(c)),单元 x-轴方向由全局坐标系 X-轴方向在单元平面上的投影计算得到,即

$$\begin{cases} \hat{i} = (\hat{I} - \hat{i}^*) / \parallel \hat{I} - \hat{i}^* \parallel \\ \hat{k} = \hat{n} \\ \hat{j} = \hat{k} \times \hat{i} \end{cases} \quad (3.18)$$

式中:$\hat{i}^* = (\hat{I} \cdot \hat{n})\hat{n}$ 为全局坐标系 X-轴方向的 $\hat{I}$ 在单元法向方向 $\hat{n}$ 上的投影。如果全局坐标系 X-轴方向,$\hat{I}$,与单元法向 $\hat{n}$ 的夹角小于 0.1°,即它们几乎平行,那么使用全局坐标系的 Z-轴方向 $\hat{K}$ 代替 $\hat{I}$ 进行投影计算得到单元局部坐标 x-轴的方向。单元局部坐标系也可以通过 Abaqus 帮助文档中描述的相关过程进行定义(文献[7],2.2.5 节)。

如果没有定义附加旋转,层合板坐标系与单元坐标系是重合的。附加旋转可在 ANSYS 中的 ESYS 模块或者 Abaqus 中的 CSYS 模块中定义,参见例 3.8 和例 3.9。

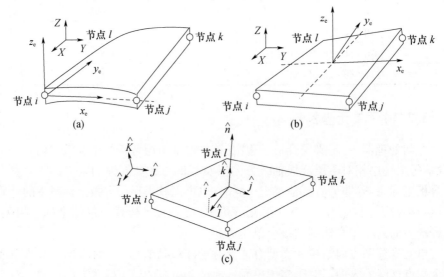

图 3.17 壳单元的单元坐标系默认方向
(a) ANSYS;(b) MSC-MARC;(c) Abaqus。

在例 3.2~例 3.7 中,矩形板仅进行了矩形单元的网格划分并进行了分析。所有单元的第 1 和第 2 个节点连线均与全局坐标系的 x-轴对齐。因此,材料坐标系的轴也与全局坐标系的轴平行。但是这样定义并不是必须的。大部分商业

第3章 层合板的刚度和强度

软件都可以改变单元坐标系。例 3.8 中说明了如何定义单元局部坐标系,例 3.9 说明了在曲面壳上如何定义单元坐标系,例 3.10 说明了如何在结构不同区域定义不同的坐标方向。

例 3.8 使用局部坐标系建立例 3.2 中的平板有限元模型,其中正交各向异性材料绕 x – 轴方向转动 $+30°$。本例将在例 3.13 中继续分析。

解 本例是例 3.2 的延续。然而,材料不再对称,因此不能再使用 Ex_3.2.cae 程序,需要重新建立一个新的完整的模型。在下面的操作过程中,材料的力学属性使用 Lamina 定义,而不是采用 [ABDH] 矩阵定义。

Menu: File, Set Work Directory, [C:\SIMULIA\User\Ex_3.8], OK
Menu: File, Save as, [Ex_3.8.cae], OK

ⅰ. 创建部件。
Module: Part
Menu: Part, Create
    3D, Deformable, Shell, Planar, Approximate size[4000], Cont
Menu: Add, Line, Rectangle
    # Enter the points[-2000,-1000],[2000,1000], X, Done

ⅱ. 创建材料。
Module: Property
Menu: Material, Create, Mechanical, Elast., Elastic, Type: Lamina
    [133860, 7706, 0.301, 4306, 4306, 2760], OK

ⅲ. 创建局部坐标系,绕全局坐标系转动 $30°$,将材料方向与局部坐标系匹配。
Menu: Tools, Datum
    # enter: origin, point along new x-axis, point on xy-plane
    # cos(30) = 0.866, sin(30) = 0.5
    CSYS, 3 Points, Name[csys-1], Rectangular, Cont
    [0,0,0],[0.866,0.5,0],[1,1,0], # close pop-up windows
Menu: Assign, Material Orientation, # pick the region, Done
    Datum CSYS List, Names: csys-1, OK, # look at the figure, OK

ⅳ. 创建截面,并将截面赋予部件。
Menu: Section, Create
    Shell, Homogeneous, Cont

Section integration: Before analysis, Thickness: Value[10], OK
Menu: Assign, Section, # pick part, Done, OK

ⅴ. 创建装配体。
Module: Assembly
Menu: Instance, Create, Independent, OK
 # the global and user-defined coordinate systems are displayed

ⅵ. 定义分析步。
Module: Step
Menu: Step, Create, Name[Step-1], Static/General, Cont, OK

ⅶ. 施加边界条件和载荷。
Module: Load
Menu: BC, Manager
 Create, Name[SS1-xx], Step: Initial, Mech., Disp/Rota, Cont
 # pick the lines @ x = -2000 and x = 2000, Done, U2, U3, RU1, OK
 Create, Name[SS1-yy], Step: Initial, Mech., Disp/Rota, Cont
 # pick the lines @ y = -1000 and y = 1000, Done, U1, U3, RU2, OK
 # close BC Manager
Menu: Load, Create
 Step[Step-1], Mech., Pressure, Cont, # pick the part, Done
 # Choose a side for the shell: Brown, Magnitude[0.12e-3], OK

ⅷ. 创建网格。
Module: Mesh
Menu: Seed, Instance, Approximate global size[500], Apply, OK
Menu: Mesh, Controls, Elem. Shape: Quad, Technique: Structured, OK
Menu: Mesh, Element Type, Geometric Order: Quadratic, DOF: 5, OK
Menu: Mesh, Instance, Yes

ⅸ. 求解和结果可视化。
Module: Job
Menu: Job, Manager
 Create, Cont, OK, Submit # when completed, Results
Module: Visualization
Menu: Plot, Contours, On Deformed Shape

Field Output Toolbox, Primary, U, U3

模型的最大变形为 8.704mm。

例 3.9 将 3D 曲面壳上的单元坐标系 x-轴(层合板坐标系)与全局坐标系 Y-轴在曲面上的投影保持一致。

解 材料为 AS4D/9310,铺层顺序为 [90/45/-45]。该例分两步进行求解。首先,层合板坐标系需要与单元坐标系对齐,在 Composite Layup 窗口中的 CSYS:<Layup> 内填写。单元坐标系方向的解释参见 3.2.14 节。然后,定义用户定义坐标系,该坐标系绕单元坐标系 z-轴转 90°,因此将层合板坐标系的 x-轴与全局坐标系 Y-轴在曲面上的投影进行了对齐。

Menu: File, Set Work Directory, [C:\SIMULIA\User\Ex_3.9], OK
Menu: File, Save as, [Ex_3.9.cae], OK

i. 创建部件。
Module: Part
Menu: Part, Create, 3D, Deformable, Point, Approx. size[500], Cont
　　[0,0,0] # creates a reference point denoted by RF
Menu: Tools, Datum, Point, Enter coordinates
　　[300,0,135], [0,0,235], [-300,0,135], [200,200,0],
　　[0,200,135], [-200,200,0], X, # close pop-up, Views: Iso View
Menu: Shape, Wire, Spline, # pick the two rightmost points, Done
Menu: Shape, Wire, Spline, # leftmost and topmost points, Done
Menu: Shape, Wire, Spline, # pick the lower three points, Done
Menu: Shape, Wire, Spline, # pick the upper three points, Done
Menu: Shape, Shell, Loft
　　Insert Before, # pick the leftmost straight line, Done
　　Insert After, # pick the rightmost straight line, Done
　　Tab: Transition, Method: Select path
　　Add, # pick lower curve, Done, Add, # pick upper curve, Done, OK

例 3.9 中的部件几何如图 3.18 所示。

ii. 创建材料,截面,并将截面赋予部件。
Module: Property
Menu: Material, Create, Mechanical, Elast., Elastic, Type: Lamina
　　[133860, 7706, 0.301, 4306, 4306, 2760], OK
Menu: Tools, Datum, # create a CSYS to be used later
　　CSYS, 3 Points, Name[user-csys], Rectangular, Cont

```
    [0,0,0],[0,1,0],[-1,1,0], # close pop-up windows
Menu: Composite, Create, Element Type: Conventional Shell, Cont
    Region # double click the cell corresponding to each ply
    # pick the part, Done # the cell must read (Picked)
    Material # double click the cell corresponding to each ply, OK
    Thickness[1.05] # all plies, CSYS # <Layup> all plies
    Rotation Angle[90,45,-45], Integ. Points[3] # all plies, OK
```

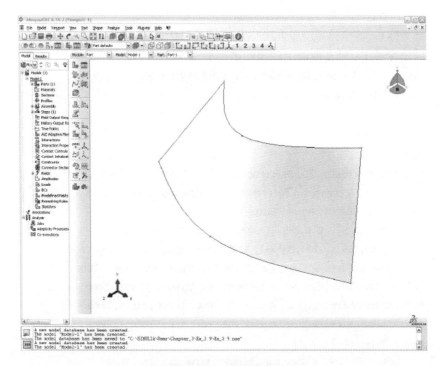

图 3.18　例 3.9 中的部件几何

ⅲ. 创建装配体。

Module: Assembly

```
Menu: Instance, Create, Independent, OK
    # the global and user-defined coordinate systems are displayed
```

ⅳ. 定义分析步。

Module: Step

```
Menu: Step, Create, Name[Step-1], Static/General, Cont, OK
Menu: Output, Field Output Requests, Edit, F-Output-1
    Output at shell: Specify[1,2,3,4,5,6,7,8,9], OK
```

ⅴ. 创建网格。
Module: Mesh
Menu: Seed, Instance, Approximate global size[100], Apply, OK
Menu: Mesh, Controls, Element Shape: Quad, Technique: Structure, OK
Menu: Mesh, Instance, Yes

由于本例主要是分析单元和层合板坐标系方向是如何定义的,所以无需施加载荷和边界,也不需求解。只需对模型进行 Data Check,显示出单元坐标系的方向即可。

ⅵ. 可视化层合板坐标系。
Module: Job
Menu: Job, Manager
  Create, Cont, OK, Data Check # when completed, Results
  # to visualize the coordinate system of the elements
Module: Visualization
Toolbar: Render Style, Render Model: Wireframe
Menu: Plot, Material Orientations, On Undeformed Shape
  # select different plies and visualize their orientations
Menu: Results, Section Points, Selection method: Plies

在 Composite Layup 对话框中,通过选择 CSYS 数据列中的 <Layup>,用户可以将层合板坐标系与单元坐标系对齐。若使用用户定义坐标系,那么更改 Composite Layup 对话框中 CSYS。例 3.9 中的求解过程中,定义了一个名为 user-csys 的用户定义坐标系。下面的操作代码说明了如何进行修改并使用用户定义坐标系。

```
# Defining materials, sections, and assigning sections to parts
Module: Property
Menu: Composite, Edit, CompositeLayup-1
    Region # must read (Picked) for all the plies
    Material # must read Material-1 for all the plies
    Thickness[1.05] # for all plies
    CSYS # double the cell corresponding to each ply
    Base Orientation: CSYS, Select, Datum CSYS List, user-csys, OK, OK
    # must read user-csys.3 for all the plies
    Rotation Angle[90,45,-45], Integ. Points[3] # all plies, OK
```

90°方向铺层的单层板坐标系在旋转前后分别显示在图 3.19 和图 3.20 中。通过对不同铺层的单层板坐标系进行可视化,用户可以了解用户定义坐标系对铺层坐标系的影响。

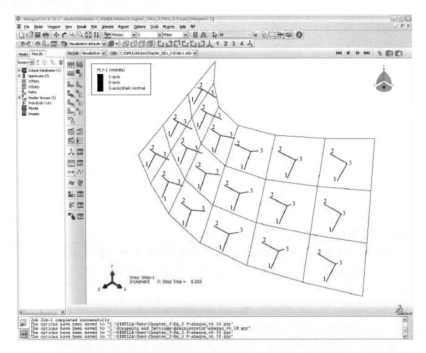

图 3.19　使用默认坐标系对[90/45/-45]层合板中的 90°层进行单元坐标系定义

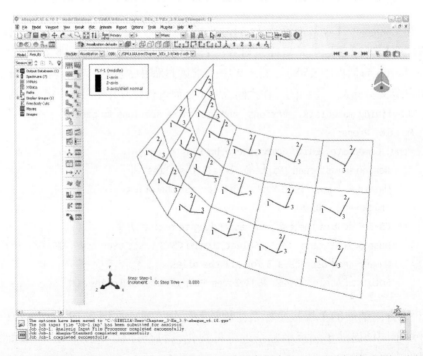

图 3.20　使用用户定义的坐标系对[90/45/-45]层合板中的 90°层进行单元坐标系定义

例 3.10 在 Abaqus 中建立一个具有轴向和径向层合板方向的法兰管模型。在圆柱管部分,参考轴为纵向方向;在法兰部分,参考轴为径向方向(图 3.21)。

解 材料选用 AS4D/9310,其铺层顺序为[0/45/-45]。为了方便定义方向,需要建立一个柱坐标系。下面的操作过程代码描述了如何将圆柱管上的单元坐标系主方向与柱坐标系轴向(柱坐标系的 z-轴)对齐,将法兰上单元的坐标系主方向与柱坐标系径向对齐。

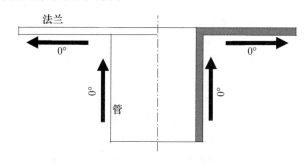

图 3.21 法兰管参考轴

```
Menu: File, Set Work Directory,[C:\SIMULIA\User\Ex_3.10], OK
Menu: File, Save as,[Ex_3.10.cae], OK
```

ⅰ. 创建部件。
```
Module: Part
Menu: Part, Create, 3D, Solid, Extrusion, Approx size[600], Cont
Menu: Add, Circle,[0,0],[0,175],[0,0],[0,275], X, Done
    Depth[350], OK
Menu: Shape, Shell, From Solid, # pick the part, Done, X
Menu: Tools, Geometry Edit, Category: Face, Method: Remove
    # pick the external cylindrical surface, Done
    # pick the flat surface at the back of the model, Done
    X, # close the Geometry Edit pop-up window
Menu: Tools, Set, Manager
    Create, Name[cylinder], Cont, # pick cylindrical surface, Done
    Create, Name[flange], Cont, # pick the flat surface, Done
    # close the Set Manager pop-up window, # sets will be used later
```

ⅱ. 创建材料、截面,并将截面赋予部件。
```
Module: Property
Menu: Material, Create, Mechanical, Elasticity, Elastic, Type: Lamina
```

[133860, 7706, 0.301, 4306, 4306, 2760], OK
Menu: Tools, Datum, CSYS, 3 Points, Name[cyl-csys], Cylindrical, Cont
[0,0,0],[1,0,0],[1,1,0], # close pop-up windows
Menu: Composite, Manager
Create, Name[cyl-zone], Element Type: Conventional Shell, Cont
Region # double click the cell corresponding to each ply
Sets, cylinder, Cont, # the cell must read cylinder for all plies
Material # double click the cell corresponding to each ply, OK
must read Material-1 for all plies, Thickness[1.05] # all plies
CSYS # double click the cell corresponding to each ply
Base Orientation: CSYS, Select, Datum CSYS List, cyl-csys, OK
Normal Direction: Axis 2, OK, # must read cyl-csys.2 for all plies
Rotation Angle[0,45,-45], Integration Points[3] # all plies, OK
Create, Name[flange-zone], Element Type: Conventional Shell, Cont
Region # double click the cell corresponding to each ply
Sets, flange, Cont, # the cell must read flange for all plies
Material # double click the cell corresponding to each ply, OK
must read Material-1 for all cells, Thickness[1.05] # all plies
CSYS # double click the cell corresponding to each ply
Base Orientation: CSYS, Select, Datum CSYS List, cyl-csys, OK
Normal Direction: Axis 3, OK, # must read cyl-csys.3 for all plies
Rotation Angle[0,45,-45], Integration Points[3] # all plies, OK
close Composite Layup Manager

ⅲ. 创建装配体。
Module: Assembly
Menu: Instance, Create, Independent, OK
the global and user-defined coordinate systems are displayed

ⅳ. 定义分析步。
Module: Step
Menu: Step, Create, Name[Step-1], Static/General, Cont, OK
Menu: Output, Field Output Requests, Edit, F-Output-1
Output at shell: Specify[1,2,3,4,5,6,7,8,9], OK

ⅴ. 创建网格。
Module: Mesh

```
Menu: Seed, Instance, Approximate global size[75], Apply, OK
Menu: Mesh, Controls, # pick all the surfaces, Done
     Element Shape: Quad, Technique: Free, OK
Menu: Mesh, Instance, Yes
```

本例主要是展示如何利用柱坐标系确定单元坐标,所以无需施加载荷和边界,也不需求解。只需对模型进行 Data Check,再对单元坐标进行可视化即可。

ⅵ. 单元坐标可视化。

```
Module: Job
Menu: Job, Manager
     Create, Cont, OK, Data Check # when completed, Results
Module: Visualization
     Toolbar: Render Style, Render Model: Shaded
     # to easily visualize the coordinate system of all the elements
Menu: Plot, Material Orientations, On Undeformed Shape
Menu: Results, Section Points, Selection method: Plies
     # select the different plies and visualize their orientations
```

3.2.15 约束

约束以多种方式用来限制模型局部运动,总结如下:

(1) Tie Constraint(绑定约束):绑定约束可以将两个区域固连在一起,不论两个区域网格是否相同。见例 6.3。

(2) Rigid Body Constraint(刚体约束):刚体约束是将一个组件中多个区域的运动耦合到一参考点的运动上。在分析过程中,不同区域之间的相对位置被设置为刚体,保持不变。在刚体约束定义中,pin(nodes)方式使连接区域的所有的平动位移与参考点保持一致,tie(nodes)方式使连接区域的所有平动和转动与参考点保持一致。刚体约束能够方便地施加铰接约束,见例 3.11。

(3) Coupling Constraint(耦合约束):耦合约束将单个面的运动与一个点的运动耦合在一起。虽然对于 3D 实体单元,每个节点有 3 个自由度,但是参考点会有 6 个自由度。因此,如果想使得被耦合面作为刚性面只有平动而没有旋转,参考点必须约束转动自由度。见例 7.5。

(4) Multi-point Constraint(多点约束)(MPC):多点约束将选取的从节点的运动与参考点的运动耦合在一起。

(5) Shell-to-solid Coupling Constraint(壳 - 实体耦合约束):壳 - 实体耦合约

束将壳边运动与相连的实体面的运动进行耦合,该耦合方式可用于局部进行精细实体单元划分而其他整体采用壳单元简化之间的约束。

(6) Embedded Region Constraint(区域嵌入约束):区域嵌入约束将一个模型的区域(小)嵌入到一个(大)模型的主区域中。该功能在局部-整体分析、多尺度分析和多物理场分析中用处较多。

例 3.11 建立一个拉挤成型的复合材料梁的有限元模型,其受轴向压缩载荷 $P = 11452\text{N}$[20],计算柱体端部的轴向位移 $u(L/2)$,$x=0$ 为梁的中点。梁两端 $(x=(-L/2,L/2))$ 采用简支约束,因为梁在长度方向上对称,可以只建立 1/2 有限元模型。梁的长度 $L = 1.816\text{m}$。梁截面为工字型,并且 $H = W = 304.8\text{mm}$。工字型梁的缘板和腹板厚度 $t_f = t_w = 12.7\text{mm}$。材料属性以 A、B、D 和 H 矩阵的形式赋值,单位分别为 mm·MPa、mm^2·MPa、mm^3·MPa 和 mm·MPa。

对于缘板:

$$[A] = \begin{bmatrix} 335035 & 47658 & 0 \\ 47658 & 146155 & 0 \\ 0 & 0 & 49984 \end{bmatrix}; \quad [B] = \begin{bmatrix} -29251 & -1154 & 0 \\ -1154 & -5262 & 0 \\ 0 & 0 & -2274 \end{bmatrix}$$

$$[D] = \begin{bmatrix} 4261183 & 686071 & 0 \\ 686071 & 2023742 & 0 \\ 0 & 0 & 67544 \end{bmatrix}; \quad [H] = \begin{bmatrix} 34216 & \\ & 31190 \end{bmatrix}$$

对于腹板:

$$[A] = \begin{bmatrix} 338016 & 44127 & 0 \\ 44127 & 143646 & 0 \\ 0 & 0 & 49997 \end{bmatrix}; \quad [B] = \begin{bmatrix} -6088 & -14698 & 0 \\ -14698 & -6088 & 0 \\ 0 & 0 & 0 \end{bmatrix}$$

$$[D] = \begin{bmatrix} 4769538 & 650127 & 0 \\ 650127 & 2155470 & 0 \\ 0 & 0 & 739467 \end{bmatrix}; \quad [H] = \begin{bmatrix} 34654 & \\ & 31623 \end{bmatrix}$$

解 工字梁的几何由缘板的底面和腹板的中面定义,因此下面程序中工字梁高度为 292.1mm,宽度为 304.8mm。主体结构在长度方向(Z-轴)上对称,因此只需建立 1/2 模型。载荷以集中力形式进行施加。载荷施加点与梁的载荷端节点之间需建立 tie(nodes)方式的刚体约束,以使梁受载边上的所有节点的平动和转动与载荷施加点保持一致。

```
Menu: File, Set Work Directory,[C:\SIMULIA\User\Ex_3.11], OK
Menu: File, Save as,[Ex_3.11.cae], OK
```

i. 创建部件。

Module: Part

第 3 章　层合板的刚度和强度

Menu: Part, Create
　　　3D, Deformable, Shell, Extrusion, Approx size[500], Cont
Menu: Add, Line, Connected Lines,[-152.4,-146.05],
　　　[152.4,-146.05], X
Menu: Add, Line, Connected Lines,[-152.4,146.05],
　　　[152.4,146.05], X
Menu: Add, Line, Connected Lines,[0,-146.05],
　　　[0,146.05], X, Done
　　　Depth[908], OK

ⅱ. 创建材料、截面,并将截面赋予部件。
Module: Property
Menu: Section, Create, Name[flange],
　　　　Shell, General Shell Stiffness, Cont
　　# In the pop-up window fill the stiffness matrix as follows
　　Tab: Stiffness
　　# Upper part of matrix A in rows 1 to 3, columns 1 to 3
　　# matrix B in rows 1 to 3, columns 4 to 6
　　# Upper part of matrix D in rows 4 to 6, columns 4 to 6
　　# The Stiffness matrix should be
　　# 335053　　47658　　　0　　　-29251　　-1154　　　0
　　#　　　　　146155　　　0　　　-1154　　-5262　　　0
　　#　　　　　　　　　49984　　　　0　　　　0　　　-2274
　　#　　　　　　　　　　　　　4261183　　686071　　　0
　　#　　　　　　　　　　　　　　　　　2023742　　　0
　　#　　　　　　　　　　　　　　　　　　　　　677544
　　Tab: Advanced
　　# The values corresponding to the H matrix are entered here
　　Transverse Shear Stiffness: # checkmark: Specify values
　　　K11[34216], K12[0], K22[31190], OK
Menu: Section, Create, Name[web],
　　　　Shell, General Shell Stiffness, Cont
　　Tab: Stiffness
　　# The Stiffness matrix should be
　　# 338016　　44127　　　0　　-6088　　-14698　　　0
　　#　　　　　143646　　　0　　-14698　　-6088　　　0
　　#　　　　　　　　　49997　　　0　　　　0　　　　0
　　#　　　　　　　　　　　　　4769538　　650127　　　0

```
    #                              2155470    0
    #                                         739467
    Tab: Advanced
    Transverse Shear Stiffness: # checkmark: Specify values
      K11[34654], K12[0], K22[31623], OK
Menu: Assign, Section
    # pick the four regions forming the flanges of the beam, Done
    Section: flange, OK
    # pick the region forming the web of the beam, Done
    Section: web, OK, X # to end the command
    # Create orientations to align the material accordingly
Menu: Tools, Datum, Type: CSYS, Method: 2 lines
    Name[web-csys], Rectangular, Cont
    # pick a line along Z-global, pick a line along Y-global
    Name[flange-csys], Rectangular, Cont
    # pick a line along Z-global, pick a line along X-global
    # close the pop-up windows
Menu: Assign, Material Orientation
    # pick the web, Done, Datum CSYS List, web-csys, OK,
    Normal: Axis 3, OK
    # pick the four regions forming the flange, Done
    Datum CSYS List, # pick flange-csys, OK, Normal: Axis 3, OK
```

ⅲ. 创建装配体。

```
Module: Assembly
Menu: Instance, Create, Independent, OK
```

ⅳ. 定义分析步。

```
Module: Step
Menu: Step, Create, Cont, OK
```

ⅴ. 创建约束。

```
Module: Interaction
    # create a reference point (RP) at the loaded end of the I-beam
Menu: Tools, Reference Point,[0,0,0]
    # define rigid body constraint of the loaded end
    # Tie nodes makes all disp and rota equal to those of the RP
```

Menu: Constraint, Create, Rigid Body, Cont
  Tie (nodes), Edit, # pick the 5 lines at the loaded end, Done
  Reference Point: Edit, # pick the RP created, OK

vi. 施加载荷和边界条件。
Module: Load
Menu: Load, Create
  Step: Step-1, Mechanical, Concentrated force, Cont
  # select the RP, Done, CF3[11452], OK
Menu: BC, Create
  Step: Initial, Mechanical, Symm/Anti/Enca, Cont
  # pick lines opposite to loaded end, Done, ZSYMM, OK

vii. 创建网格。
Module: Mesh
Menu: Seed, Edges
  # pick the six longitudinal lines, Done
  Method: By Number, Bias: None, Number of elements[10], Apply, OK
  # pick edges of the flanges at both ends of the beam, Done
  Method: By Number, Bias: None, Number of elements[3], Apply, OK
  # pick the edges of the web at both ends of the beam, Done
Method: By Number, Bias: None, Number of elements[4], Apply, OK
Menu: Mesh, Controls, # select all, Done, Technique: Structured, OK
Menu: Mesh, Element Type, # select all, Done
  Geometric Order: Quadratic, DOF: 6, Family: Shell, OK # S8R in use
Menu: Mesh, Instance, Yes

创建一个集合，其包含未受载端截面的中心点，该中心点上施加一个额外边界，阻止对称平面上的位移。
Module: Mesh
Menu: Tools, Set, Create, Name[Cpoint], Type: Node, Cont
  # pick the point at the center of unloaded end, Done
Module: Load
Menu: BC, Create
  Step: Initial, Mechanical, Disp/Rota, Cont
  Sets, Cpoint, Cont, # checkmark: U1,U2,UR3, OK

ⅷ. 求解并可视化结果。
Module: Job
Menu: Job, Manager
    Create, Cont, OK, Submit, # when completed, Results
Module: Visualization
Menu: Plot, Allow Multiple Plot States
Menu: Plot, Deformed Shape
    Toolbar: Field Output, Primary, U, U3

载荷施加端的位移为 0.03584mm。

3.3 失效准则

失效准则是基于单轴应力试验数据来预测结构在多轴应力条件下破坏的拟合曲线。本章中所述的破坏准则均为首层失效准则,无法获得层合板的失效扩展和最终破坏。第 8 章和第 9 章中使用的损伤力学方法可以获得层合板的损伤演变直至层合板的最终破坏过程。

本节中,失效准则通过失效指数(failure index)的形式进行表示,失效指数在很多有限元软件中均使用,其定义为

$$I_F = \frac{应力}{强度} \tag{3.19}$$

当 $I_F \geq 1$ 时,出现失效。强度比(strength ratio)(文献[1],7.1.1 节)定义为失效指数的倒数

$$R = \frac{1}{I_F} = \frac{强度}{应力} \tag{3.20}$$

当 $R \leq 1$ 时,出现失效。

3.3.1 2D 失效准则

在有限元分析中,通常使用基于强度的失效准则预测复合材料结构的失效。对于平面应力($\sigma_3 = 0$)状态的单向板目前有多种失效准则。常用的失效准则可参见文献[1],如下:

—最大应力准则
—最大应变准则
—修正的最大应变准则
—相互影响失效准则

本节中还将介绍一些其他的失效准则。

1. Hashin 失效准则

Hashin 失效准则(HFC)提出了 4 种独立的失效模式：

(1) 纤维拉伸失效

(2) 纤维压缩失效

(3) 基体拉伸失效

(4) 基体压缩失效

其由 4 个独立的方程式描述,即①

$$I_{Fft}^2 = \left(\frac{\sigma_1}{F_{1t}}\right)^2 + \alpha\left(\frac{\sigma_6}{F_6}\right)^2 \quad (\sigma_1 \geqslant 0) \tag{3.21}$$

$$I_{Ftc}^2 = \left(\frac{\sigma_1}{F_{1c}}\right)^2 \quad (\sigma_1 < 0) \tag{3.22}$$

$$I_{Fmt}^2 = \left(\frac{\sigma_2}{F_{2t}}\right)^2 + \left(\frac{\sigma_6}{F_6}\right)^2 \quad (\sigma_2 \geqslant 0) \tag{3.23}$$

$$I_{Fmc}^2 = \left(\frac{\sigma_2}{2F_4}\right)^2 + \left[\left(\frac{F_{2c}}{2F_4}\right)^2 - 1\right]\frac{\sigma_2}{F_{2c}} + \left(\frac{\sigma_6}{F_6}\right)^2 \quad (\sigma_2 < 0) \tag{3.24}$$

式中:α 为权重系数,主要用于表示纤维失效中剪切的影响大小。当 $\alpha = 0$ 时,Hashin 失效准则和最大应力失效准则评估单向带纵向拉伸破坏的方程式一致。注意,式(3.22)预测纵向压缩失效没有考虑剪切应力的影响,尽管剪切应力状态对纵向压缩失效有较大影响[21]。

根据 Hashin 失效准则,式(3.21)~式(3.24)定义的失效指数为平方形式。Abaqus中 Hashin 准则、Tsai-Hill 准则、Azzi-Tsai-Hill 准则和 Tsai-wu 准则②均采用失效指数的平方形式,而最大应力准则(MSTRS)和最大应变准则(MSTRN)的失效指数不同(不是平方形式)。

本书和大部分教科书中的压缩强度 F_{1c} 和 F_{2c} 均考虑成正值,而 Abaqus 中不同的输值位置需要不同的符号。例如,在 Edit Material, Materail Behavior: Elastic, Suboption: Fail Stress 中输入 F_{1c} 和 F_{2c} 均为负值；而在 Edit Material, Material Behavior: Hashin Damage 中输入 F_{1c} 和 F_{2c} 作为损伤初始阈值时,必须是正值。

在 Abaqus 中,虽然损伤初始阈值通过 Hashin 准则计算得到,但随后的损伤累积分析是通过独立于初始损伤准则的逐步软化的方法进行分析。尽管如

---

① 式(3.21)~式(3.24)右侧式子计算得到失效指数的 I_F 的平方值,与最大应力准则($I_{Fft} = \sigma_1/F_{1T}$)等不同。

② Tsai-Hill 准则、Azzi-Tsai-Hill 准则和 Tsai-wu 准则并不推荐,因为其过度强调纵向纤维(σ_1)和横向基体(σ_2)之间的相互影响关系。

此,Abaqus 文档和 Abaqus/CAE 中均认为整个破坏的分析方法为 Hashin 损伤方法。

2. Puck 失效准则

Puck 失效准则[22]分为纤维失效(FF)和基体失效(MF)2 种。对于平面应力状态,MF 又分为 3 种模式。模式 A:横向拉伸应力下(有/无面内剪切应力),产生横向开裂。模式 B:仍为发生横向开裂,但在面内剪切应力和较小的横向压缩应力下。模式 C:受较大的横向压缩应力作用,发生斜向开裂(典型的是碳纤维/环氧层合板 53°角度方向)。

FF 和 3 种 MF 失效模式得到了不同的失效指数。Puck 准则假设 FF 主要取决于纵向拉伸。因此,FF 模式的失效指数定义为

$$I_{FF} = \begin{cases} \sigma_1/F_{1t} & (\sigma_1 > 0) \\ -\sigma_1/F_{1c} & (\sigma_1 < 0) \end{cases} \quad (3.25)$$

MF 具体的失效指数与激活的 MF 失效模式有关。当承受正的横向应力时,模式 A 激活。模式 A 的失效指数计算如下

$$I_{MF,A} = \sqrt{\left(\frac{\sigma_6}{F_6}\right)^2 + \left(1 - p_{6t}\frac{F_{2t}}{F_6}\right)^2 \left(\frac{\sigma_2}{F_{2t}}\right)^2} + p_{6t}\frac{\sigma_2}{F_6} \quad (\sigma_2 \geq 0) \quad (3.26)$$

式中:p_{6t}为拟合参数,若缺乏相关试验数据,一般建议 $p_{6t}=0.3$[22]。

当承受负的横向应力时,模式 B 或 C 激活,主要取决于面内剪切应力和横向应力的关系。模式 B 和 C 之间的选择取决于 F_{2A}/F_{6A},其中

$$F_{2A} = \frac{F_6}{2p_{6c}}\left[\sqrt{1 + 2p_{6c}\frac{F_{2c}}{F_6}} - 1\right] \quad (3.27)$$

$$F_{6A} = F_6\sqrt{1 + 2p_{2c}} \quad (3.28)$$

其中 p_{2c} 定义为

$$p_{2c} = p_{6c}\frac{F_{2A}}{F_6} \quad (3.29)$$

其中,p_{6c}是另一个拟合参数,若缺乏试验数据,一般建议 $p_{6c}=0.2$[22]。

最后,模式 B 失效指数为

$$I_{MF,B} = \frac{1}{F_6}\left[\sqrt{\sigma_6^2 + (p_{6c}\sigma_2)^2} + p_{6c}\sigma_2\right] \left(\begin{cases} \sigma_2 < 0 \\ \left|\frac{\sigma_2}{\sigma_6}\right| \leq \frac{F_{2A}}{F_{6A}} \end{cases}\right) \quad (3.30)$$

模式 C 失效指数为

$$I_{MF,C} = -\frac{F_{2c}}{\sigma_2}\left[\left(\frac{\sigma_6}{2(1+p_{2c})F_6}\right)^2 + \left(\frac{\sigma_2}{F_{2c}}\right)^2\right] \left(\begin{cases} \sigma_2 < 0 \\ \left|\frac{\sigma_2}{\sigma_6}\right| \leq \frac{F_{2A}}{F_{6A}} \end{cases}\right) \quad (3.31)$$

3.3.2 3D失效准则

本节主要介绍的3D广义失效准则与文献[1,7.1节]内容类似。3.3.1节中所述的2D失效准则,可以认为是3D失效准则中当$\sigma_4=\sigma_5=0$时的情况。使用有限元软件分析时需注意:部分软件仅采用面内应力分量计算失效指数(如Abaqus),尽管所有6个应力分量在分析中均可获得。在这种情况下,层间应力和厚度方向应力分量需单独进行失效评估。

本节中,数字下标表示方向:1—纤维,2—面内的纤维横向,3—单层板厚度方向。字母下标表示:t—拉,c—压。剪切项使用符号缩写形式,见1.5节。

1. 最大应变准则

失效指数定义为

$$I_F = \max \begin{cases} \epsilon_1/\epsilon_{1t} & (\epsilon_1>0) \quad 或 \quad -\epsilon_1/\epsilon_{1c} \quad (\epsilon_1<0) \\ \epsilon_2/\epsilon_{2t} & (\epsilon_2>0) \quad 或 \quad -\epsilon_2/\epsilon_{2c} \quad (\epsilon_2<0) \\ \epsilon_3/\epsilon_{3t} & (\epsilon_3>0) \quad 或 \quad -\epsilon_3/\epsilon_{3c} \quad (\epsilon_3<0) \\ \text{abs}(\gamma_4)/\gamma_{4u} \\ \text{abs}(\gamma_5)/\gamma_{5u} \\ \text{abs}(\gamma_6)/\gamma_{6u} \end{cases} \quad (3.32)$$

式中的分母均为单向板的极限应变,式(3.32)中的极限压缩应变均为正值。

2. 最大应力准则

失效指数定义为

$$I_F = \max \begin{cases} \sigma_1/F_{1t} & (\sigma_1>0) \quad 或 \quad -\sigma_1/F_{1c} \quad (\sigma_1<0) \\ \sigma_2/F_{2t} & (\sigma_2>0) \quad 或 \quad -\sigma_2/F_{2c} \quad (\sigma_2<0) \\ \sigma_3/F_{3t} & (\sigma_3>0) \quad 或 \quad -\sigma_3/F_{3c} \quad (\sigma_3<0) \\ \text{abs}(\sigma_4)/F_4 \\ \text{abs}(\sigma_5)/F_5 \\ \text{abs}(\sigma_6)/F_6 \end{cases} \quad (3.33)$$

字母F表示单向板的强度值[23],式(3.33)中压缩强度为正值。

3. Tsai-wu 失效准则

Tsai-wu准则的失效指数为

$$I_F = \frac{1}{R} = \left[-\frac{B}{2A} + \sqrt{\left(\frac{B}{2A}\right)^2 + \frac{1}{A}} \right]^{-1} \quad (3.34)$$

式中

$$A = \frac{\sigma_1^2}{F_{1t}F_{1c}} + \frac{\sigma_2^2}{F_{2t}F_{2c}} + \frac{\sigma_3^2}{F_{3t}F_{3c}} + \frac{\sigma_4^2}{F_4^2} + \frac{\sigma_5^2}{F_5^2} + \frac{\sigma_6^2}{F_6^2} \quad (3.35)$$

$$+ c_4 \frac{\sigma_2\sigma_3}{\sqrt{F_{2t}F_{2c}F_{3t}F_{3c}}} + c_5 \frac{\sigma_1\sigma_3}{\sqrt{F_{1t}F_{1c}F_{3t}F_{3c}}} + c_6 \frac{\sigma_1\sigma_2}{\sqrt{F_{1t}F_{1c}F_{2t}F_{2c}}}$$

$$B = (F_{1t}^{-1} - F_{1c}^{-1})\sigma_1 + (F_{2t}^{-1} - F_{2c}^{-1})\sigma_2 + (F_{3t}^{-1} - F_{3c}^{-1})\sigma_3 \quad (3.36)$$

式中：c_i，$i=4,\cdots,6$，为 Tsai-Wu 耦合系数，一般取 -1。注意，式(3.35)和式(3.36)中的压缩强度为正值。

在公开文献中，厚度方向强度值 F_{3t} 和 F_{3c} 很少可获得数据，因此通常采用相应的面内横向强度值代替。另外，层间强度 F_5 一般假设与面内剪切强度相等。层间剪切强度 F_4 可由横向压缩强度 F_{2c} 和断裂面夹角 α_0 计算（文献[1]，式(4.109)），该夹角一般为 $53°$。

例 3.12 分别采用最大应力失效准则和 Tsai-Wu 失效准则计算例 3.4 中每层铺层的失效指数 I_F 和强度比 R。单层板的弹性模量和强度见表 3.1。试分别使用这两种准则计算层合板的强度比。

解 在 Abaqus 中，Hashin 准则、最大应变准则和最大应力准则是较为适用的失效准则，而 Tsai-Hill 和 Tsai-Wu 失效准则由于过高估计了 $\sigma_1(S_{11})$ 和 $\sigma_2(S_{22})$ 之间的关联，并不推荐。本例中采用的 Tsai-Wu 失效准则与文献[8]中的例 3.12 一致。与本书中式(3.34)中的定义不一样，Abaqus 中 Tsai-Wu 失效指数定义为

$$\text{TSAIW} = f_1 \cdot S_{11} + f_2 \cdot S_{22} + f_{11} \cdot S_{11}^2 + f_{22} \cdot S_{22}^2 + f_{66} \cdot S_{12}^2 + 2f_{12} \cdot S_{11} \cdot S_{22}$$

上式为失效指数的二次式，因此 TSAIW 的值与 MSTRN（式(3.32)）和 MSTRS（式(3.33)）失效指数不一样，除非对 TSAIW 值开二次方。Abaqus 中 Tsai-Hill 失效指数 TSAIH 和 4 个 Hashin 失效指数（见式(3.21)~式(3.24)）一样也是二次形式。

强度值 F_{1T}，F_{1C}，F_{2T}，F_{2C}，F_6，f_{12} 和 Slim 可通过 Abaqus/CAE 中 Edit Material，Material Behavior：Elastic，Suboption：Fail Stress 输入。注意，对于计算 SAIW，F_{1C} 和 F_{2C} 应赋予负值，但对于 Hashin 损伤计算，F_{1C} 和 F_{2C} 应赋予正值（见 3.3.1 节）。如果不赋值，f_{12} 和双轴应力 Slim 默认为 0。因为 f_{12} 过高估计 $\sigma_1(S_{11})$ 和 $\sigma_2(S_{22})$ 之间相互影响，因此 f_{12} 和 Slim 默认为 0 可以减少这种影响。本例中，我们取 $f_{12} = -1$ 以重现文献[8，例 3.12]中的计算结果。

为求解本例，打开例 3.4 中 Ex_3.4.a.cae 并且另存为 Ex_3.12.cae。然后，按如下步骤操作：

ⅰ. 定义工作目录。

Menu: File, Set Work Directory,[C:\SIMULIA\User\Ex_3.12], OK

ii. 定义材料破坏参数。
Module: Property
Menu: Material, Manager
    Edit, Suboptions, Fail Stress, # you can copy/paste from excel
    [1830 -1096 57 -228 71 -1 0], OK, OK
    # close Material Manager pop - up window

iii. 定义分析步,创建一个失效指数的场输出。
Module: Step
Step: Step-1
Menu: Output, Field Output Requests, Edit, F-Output-1
    # expand: Failure/Fracture, # checkmark: CFAILURE
    # uncheck: everything else we don't need, OK

iv. 求解并可视化结果。
Module: Jobs
Menu: Job, Manager
    Submit, # when completed, Results
Module: Visualization
Menu: Plot, Contours, On Deformed Shape
Menu: Result, Field Output, MSTRS, OK
Menu: Results, Section Points, Method: Plies
    # pick a ply, Apply, # read the value, OK
Menu: Result, Field Output, TSAIW, OK
Menu: Results, Section Points, Method: Plies
    # pick a ply, Apply, # read the value, OK

结果见表 3.7,可将结果与文献[12]中使用经典层合理论(CLT)的求解值进行对比。

表 3.7 例 3.12 中采用最大应力准则和 Tsai – Wu 准则计算的失效指数

层#	MSTRS		TSAIW	
	I_F	R	I_F	R
1	0.01442	69.34813	0.01447	69.10850
2	0.02430	41.15226	0.03107	32.18539
3	0.01566	63.85696	0.01873	53.39028
4	0.01566	63.85696	0.01873	53.39028

例 3.13 采用 UMAT 子程序(文献[5],umatS8R5.for)计算例 3.8 中每层铺层的 Tsai-Wu 失效指数。弹性模量和强度值见表 3.1。

解 执行含有用户子程序的计算时,需先正确配置环境变量(附录 C)。求解本例,先打开例 3.8 中的 Ex_3.8.cae,另存为 Ex_3.13.cae。将工作目录设置为新建的文件夹内,以便所有文件便于读取。

```
Menu: File, Set Work Directory,[C:\SIMULIA\User\Ex_3.13], OK
Menu: File, Open,[C:\SIMULIA\User\Ex_3.8\Ex_3.8.cae], OK
Menu: File, Save As,[C:\SIMULIA\User\Ex_3.13\Ex_3.13.cae], OK
```

仅需对 .mdb 文件的小部分内容进行修改。记住,Abaqus 中壳单元的应力/应变分量的顺序为:11,22,12,13,23。那么,截面剪切刚度可通过文献[12]中的程序计算。另外,对于厚度为 t 的单层壳,横向剪切刚度系数计算如下

$$H_{44} = (5/6)tG_{23}$$
$$H_{45} = 0$$
$$H_{55} = (5/6)tG_{13}$$

或者对于一个层合板,横向剪切刚度也利用文献[1,式(6.16)]计算,或者利用 Abaqus 计算这几个值,然后通过下述步骤从 .dat 文件中读取。通过 Abaqus 计算,首先打开例 3.8 的 .cae 文件,并且进行如下修改:

ⅰ. 修改材料定义,增加失效参数。

```
Module: Property
# unidirectional material from Ex. 3.8 used to calculate H44, H45, H55
Menu: Material, Create
    Name[unidirectional], Mechanical,
    Elasticity, Elastic, Type: Lamina
    [133860 7706 0.301 4306 4306 2760]
    # or right-click on Data field, Read from File
    Select,[uni_lam_prop.txt], OK, OK
    # add Fail Stress values
    Suboptions, Fail Stress
    [1830 -1096 57 -228 71 -1], OK
    # or right-click on Data field, Read from File
    Select,[uni_fail_stress.txt], OK, OK, OK,
    OK # to close Edit Material
Menu: Section, Edit, Section-1
    Tab: Basic, Material: unidirectional, OK
```

ⅱ. 提交计算并可视化结果。
Module: Job
Menu: Job, Manager
 Edit, Tab: General, # checkmark: Print model definition data, OK
 Submit, # when completed, Results
 Field Output Toolbar: Primary, U, U3, # U3max = 8.704 mm

使用文本编辑器打开 job-1.dat 文件。该文件必须保存在工作目录中(C:\SIMULIA\User\Ex_3.13)。读取截面的横向剪切刚度数据,如下:

$$K_{ts11} = K_{11} = H_{55} = 35883.0$$
$$K_{ts22} = K_{22} = H_{44} = 23000.0$$
$$K_{ts12} = K_{12} = H_{45} = 0.0$$

为运行用户材料子程序,按下述步骤进行:
ⅰ. 通过 UMAT 子程序创建一个用户自定义材料,并且定义截面。
Module: Property
Menu: Material, Create
 Name[ud-user], General, User Material,
 User material type: Mechanical
 Mechanical constants,[133860 7706 0.301 0.396
 4306 1830 1096 57 228 71 -1]
 # or right-click on Data field, Read from File
 Select,[user_mat_props.txt], OK, OK
 # see umats8r5.for for interpretation of Mechanical constants
 General, Depvar,
 Number of solution-dependent state variables[2], OK
Menu: Section, Edit, Section-1
 Tab: Basic
 Section integration: During analysis
 Thickness: Shell thickness: Value[10]
 Material: ud-user
 Thickness integration rule: Simpson
 Thickness integration points[3]
 Tab: Advanced
 Transverse Shear Stiffness: # checkmark: Specify values
 K11:[35883], K12:[0], K22:[23000], OK

ⅱ. 在分析步中定义输出变量。
Module: Step
Step: Step-1
Menu:　Output, Field Output Requests, Edit, F-Output-1
　　Output variables: Edit variables:[S,E,U,RF,SDV], OK

ⅲ. 提交计算并可视化结果。
　　在提交计算前,先确认系统是否支持运行 UMAT 子程序。如果系统不支持,那么提交计算时将报错。
Module: Job
Menu: Job, Manager
　　Edit, Tab: General,
　　User subroutine file: Select,[umatS8R5.for], OK, OK
　　Submit, # when completed, Results
Module: Visualization
Menu: Options, Contour, Contour Type: Quilt, OK
Menu: Plot, Contours on Deformed Shape
　　Field Output Toolbar: Primary, SDV1
Menu: Result, Section Points, Selection method: Plies
　　Ply result location: Bottommost, Apply
　　Ply result location: Topmost, Apply
　　OK # close the Section Points pop-up window

　　如果报错,先确认配置文件 C:\SIMULIA\Abaqus\Commands\abq6102.bat 是否正确。同时保证.inp 文件中包含下面的代码:
　　** Section: Section-1
　　　* Shell Section, elset=_PickedSet2, material=uni-user, orientation=Ori-1
　　　10., 3
　　　* Transverse Shear
　　　35883., 23000., 0.

　　　和
　　* Material, name=uni-user
　　　* Depvar
　　　2,
　　　* User Material, constants=11
　　　133860., 7706., 0.301, 0.396, 4306., 1830., 1096., 57.
　　　228., 71., -1.

从 SDV1 可视化云图可知,底面失效指数为 $I_F = 1.767 \times 10^{-2}$,顶面失效指数为 $I_F = 9.444 \times 10^{-3}$。

3.4 预定义场

下面实例给出了考虑热膨胀的分析方法,并且说明了如何引入预定义场(Predefined Field)。

例3.14 计算并显示铺层顺序为$[0/90]_T$层合板的变形情况以说明热膨胀耦合效应(在文献[1],6.1.4节,图6.7中讨论)。每层铺层厚为1.2mm,材料为AS4D/9310(表3.1)。板的几何形状为正方形,$a_x = a_y = 1000$mm。热载荷为$\Delta T = -125$℃。单元类型为S8R5。

解 铺层顺序为$[0/90]_T$的层合板承受均匀热载荷ΔT,几何形状为在$x-y$平面内的正方形($a_x = a_y$),其关于$x-z$和$y-z$平面对称,因此只需建立1/4对称模型。注意,本例中LLS并不对称,但是不影响模型离散的对称性。模型离散的对称性主要考虑3个方面因素:(a)几何对称;(b)载荷对称;(c)材料对称。下面的操作过程代码详细说明了求解过程。平板的变形图见图3.22。CAE文件可从网站[5,Ex_3_15.cae]获取。

 i. 几何。
```
Module: Part
Menu: Part, Create,[Part-1], 3D, Deformable,
    Shell, Planar, Approximate size[4000], Cont
Menu: Add, Line, Rectangle,
    # Enter the points[0,0],[1000,1000], X, Done
```

 ii. 材料和截面属性。
```
Module: Property
Menu: Material, Create,
    Name[Material-1], Mechanical, Elasticity, Elastic, Lamina,
    [13386000 7706 0.301 4306 4306 2760]
    Mechanical, Expansion, Type, Orthotropic,
    [13.23672E-007 2.58908E-005 2.58908E-0051], OK
Menu: Section, Create, Name[Section-1], Shell, Composite, Cont
    Material: Material-1 # for all
    Thickness[1.2] # for all
    Orientation[0,90]
Menu: Assign, Section
```

171

pick part, Done, OK

ⅲ. 装配和载荷步。
Module: Assembly
Menu: Instance, Independent, OK
Module: Step
Menu: Step, Create, Name[Step-1], Static/General, Cont, OK

ⅳ. 载荷和边界条件。
Module: Load
Menu: BC, Create, Name[BC-1], Step: Initial,
　　　Mechanical, Symmetry, Cont
　　　# pick left vertical line, Done, XSIMM, OK
Menu: BC, Create, Name[BC-2], Step: Initial,
　　　Mechanical, Symmetry, Cont
　　　# pick lower horizontal line, Done, YSIMM, OK
Menu: BC, Create, Displacement/Mechanical, Cont,
　　　# pick bottom left point, Done
　　　# checkmark all (U1 to UR3), OK
Menu: Predefined Field, Create, Other, Temperature, Cont
　　　# Select rectangle, Done, Magnitude[0], OK
Menu: Predefined Field, Manager
　　　# Double click cell Predefined Field-2, Step-1 (Propagated)
　　　Status, Modified, Magnitude[-150], OK

ⅴ. 划分网格。
Module: Mesh
Menu: Seed, Instance, Apply, OK
Menu: Mesh, Element type, Std, Quad, DOF[5], OK # S8R5
Menu: Mesh, Instance, Yes

ⅵ. 计算并可视化。
Module: Jobs
Menu: Job manager
　　　Create, Cont, OK
　　　Submit,
　　　Results

Module: Visualization
Menu: Plot,
　　Allow multiple plot states, Undeformed shape, Deformed shape
Menu: Options, Common,
　　Tab: Basic, Render style: Shaded, Visible Edges: Free edges,
　　Tab: Other, Scale coordinates, Z:[3.0], OK
Menu: Options, Superimpose plot options,
　　Visible Edges: Feature edges, OK
Menu: View, Odb display options,
　　Tab: Mirror/Pattern, Mirror planes: XZ and YZ

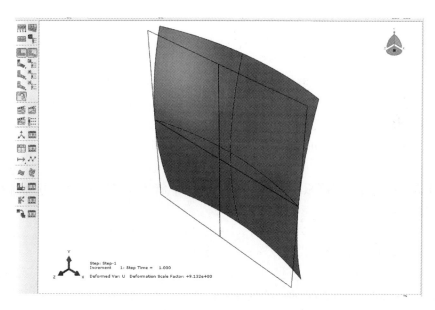

图 3.22　例 3.14 中热载荷作用下的变形

习题

习题 3.1　计算外径 r_0 和内径 r_i 的圆管梁单位截面积能够承受的最大弯矩 m_u，载荷为纯弯矩，无剪切。材料是均质材料，当最大应力达到材料的强度极限 σ_u 时发生破坏。圆管中填充泡沫以防止屈曲。在相同外径情况下，使用圆管最大弯矩 m_u 与实芯圆柱最大弯矩 m_u 之比推导截面效率表达式。当使用强度大，但价格较贵的材料时，是选圆管还是实芯圆柱比较合理？

习题 3.2　对于一个长度为 L，在端部由施加载荷 P，其他信息与习题 3.1

中的一致的悬臂梁,计算悬臂梁的最大外径。材料剪切强度为 $\tau_u = \sigma_u/2$,仅考虑剪切。薄壁梁的屈曲效应对实际壁厚有一定限制。

习题 3.3 计算外径 r_0 和内径 r_i 的圆管梁每单位体积 δV 可以承受的最大挠度。悬臂梁长度为 L,在端部施加载荷 P。由均质材料制成,材料弹性模量为 E 和 $G = E/2.5$,圆管中填充泡沫以防止屈曲。在相同外径情况下,使用圆管与实芯圆柱 δV 之比推导截面效率表达式。当使用刚度不大,但价格较贵的材料时,推荐小的还是大的半径比较合理?

习题 3.4 编写相关程序计算式(3.9)。输入数据为层合板铺层顺序(LSS)、每层铺层的厚度以及材料的弹性模量,结果输出在一个文件中。

习题 3.5 使用习题 3.4 中的程序计算下面层合板的 A, B, D 和 H 矩阵。材料为 AS4D/9310,每层铺层均为 0.85mm。分析下面几种情况本构方程的耦合情况:(a) 单层 $[0]$;(b) 单层 $[30]$;(c) $[0/90]_2$;(d) $[0/90]_S$;(e) $[0/90]_8$;(f) $[\pm 45]_2 = [+45/-45/+45/-45]$;(g) $[\pm 45]_S = [+45/-45/-45/+45]$;(h) $[+45/0/90/\pm 30]$。

习题 3.6 计算例 3.2 中绝对值最大的横向剪切应变 γ_{23} 值及其位置。在该位置绘制 γ_{23} 随平板厚度的变化曲线。应变的分布是否合理?

习题 3.7 在双重正弦载荷作用下,重新计算例 3.2
$$q(x,y) = q_0 \sin(\pi x/2a)\sin(\pi x/2b)$$
其中 $2a$ 和 $2b$ 分别是平板在 x 和 y 方向的尺寸。将计算结果与平板中心的精确解对比,即
$$\omega_0 = 16q_0 b^4/[\pi^4(D_{11}s^4 + 2(D_{12} + 2D_{66})s^6) + D_{22}]$$
其中 $s = b/a$(文献 [14],式(5.2.8)~式(5.2.10))。

习题 3.8 计算层合板振动解析解为 $\overline{\omega}_{mn}^2 = \pi^4[D_{11}m^4s^4 + 2(D_{12} + 2D_{66})m^2n^2s^2 + D_{22}n^4]/(16\rho h b^4)$ 的一阶频率 $\overline{\omega}_{11}$,其中 ρ 和 h 分别为层合板的密度和厚度(文献 [14],式(5.7.8))。

习题 3.9 考虑一个矩形平板 $a_x = 1000mm$,$b_x = 1000mm$。每层铺层厚度为 1.2mm,材料为 AS4D/9310(表 3.1)。

建立 3 种不同的铺层顺序,每一种产生下述 3 种耦合效应(显示在文献 [1],图 6.7]):(a) 弯曲-拉伸耦合;(b) 扭转-拉伸耦合;(c) 剪切-拉伸耦合。

考虑对称边界简化有限元模型:(1) 采用 1/2 模型 500mm×100mm,或 (2) 采用 1/4 模型 500mm×50mm。分别使用合适的最小有限元模型(1/2、1/4 或整体)计算 (a)~(c) 3 种铺层层合板。施加面内拉力 $N_x = 1000N/mm$(其他为 0)。单元类型为 S8R5。

习题 3.10 当 $I_F = 1$ 时,失效包络是3D面。失效包络面与坐标平面的交线

为 2D 曲线。使用 MATLAB 类似的程序,在 $\sigma_1 - \sigma_2$ 平面内绘制最大应力失效准则、Tsai – Wu 失效准则和 Puck 失效准则的失效包络曲线。类似地,在 $\sigma_1 - \sigma_2$ 平面内绘制失效包络曲线。

习题 3.11 分别使用最大应力失效准则和 Hashin 失效准则计算例 3.12 中每层的失效指数 I_F,单向板的强度值见表 3.1。

习题 3.12 采用 UMAT 子程序,使用 Puck 失效准则计算例 3.12 中每层的失效指数 I_F。单向板的模量和强度值见表 3.1。Puck 失效准则的参数值见 3.3.1 节的推荐值。

习题 3.13 将习题 3.11 中每个铺层上、下表面的节点应力写入一个文本文件。然后,使用 MATLAB 类似的外部程序,依据平板中心的节点应力值计算失效指数,并与习题 3.11 的结果进行对比。

第4章 屈　　曲

大部分复合材料结构都是薄壁结构,主要有如下原因:

(1)复合材料比常规材料强度高,可以采用较小截面积的结构承受较大的载荷,因此,大部分复合材料构件都很薄。

(2)复合材料比常规材料贵,因此,有很强的减少材料用量的需求,复合材料结构会被设计的尽可能薄。

(3)聚合物基复合材料的成本随刚度的增加而增加。纤维方向刚度值可通过纤维刚度和纤维含量估算,$E_1 = E_f V_f$。例如,玻璃纤维聚合物基复合材料的刚度比铝的刚度低。而芳纶纤维聚合物基复合材料的刚度与铝相当。碳纤维聚合物基复合材料刚度比钢低。因此,有很强的需求在不增加截面积的情况下,尽量增加梁和加强筋的惯性矩。最好的选择就是通过增大截面尺寸和减少壁厚的方法增加惯性矩。

上述原因导致复合材料结构设计时,一般采用大尺寸的薄壁截面,因此结构的失效模式主要是屈曲。

4.1　特征值屈曲分析

屈曲是由于几何效应而不是材料失效导致结构丧失承载的现象。如果屈曲后结构的变形无法抑制,那么将进一步导致发生材料的失效和破坏。大部分结构在线弹性范围内工作,也就是说,如果去除载荷,结构可以恢复初始未变形状态。当超出弹性范围时,将导致结构产生永久变形,如复合材料中出现基体裂纹。

考虑一个受简支约束的圆柱梁,截面积为 A,长度为 L,惯性矩为 I,由弹性模量为 E,强度为 F 的均质材料构成。圆柱梁受到位于截面形心的压力 P 的作用[3]。如果圆柱梁的几何、载荷和材料均无缺陷,那么圆柱梁的轴向变形为

$$u = PL/EA \tag{4.1}$$

其侧向变形 $w=0$。结构屈曲前的变形 (u,v,w) 称为主路径(图4.1)。细微的缺陷将导致结构发生屈曲,屈曲载荷为

$$P_{cr} = \pi^2(EI)/L^2 \tag{4.2}$$

细长梁的承载能力由屈曲限制,与受材料的压缩强度控制不同。梁达到屈曲临界载荷后的行为主要取决于支撑条件。对于简支梁,当载荷刚刚超过临界载荷P_{cr}时,其侧向变形①为

$$w = A\sin(\pi x/L) \tag{4.3}$$

将趋向无穷大($A \to \infty$)。如此大的侧向变形将导致材料失效和梁的最终破坏。结构在屈曲之后的行为称为后屈曲。

图4.1所示的简支梁在屈曲发生前,不会产生屈曲模态形式的变形(式4.3)。这就是说,对于无缺陷结构,在理想载荷作用下,结构具有简单的变形主路径。对于这样的结构,屈曲一般发生在分叉点上,即变形主路径和次路径的交叉点。次路径即为后屈曲路径[24]。

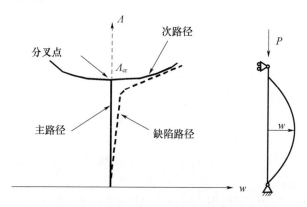

图4.1 无缺陷圆柱梁的平衡路径

对于不同屈曲模态的分叉载荷,通过商业软件可以比较容易计算得到。结构几何没有初始变形和缺陷,承受名义载荷作用,材料为弹性材料。此类问题计算对于分析者来说比较简单。商业软件一般通过特征值屈曲分析(eigenvalue buckling analysis)计算屈曲载荷,因为屈曲临界载荷值为系统离散方程组的特征向量λ_i

$$([K] - \lambda[K_s])\{v\} = 0 \tag{4.4}$$

式中:K,K_s分别为刚度和应力刚度矩阵;v为特征值向量(屈曲模态)[24]。

例4.1 考虑一个简支约束平板,$a_x = 1000\text{mm}$,$a_y = 500\text{mm}$,在平板边缘施加压缩载荷 $N_x = N_y = 1\text{N/mm}$。平板铺层为$[(0/90)_3]_S$,材料为 AS4/9310(表4.1),纤维体积含量0.6,总厚度$t_T = 10.2\text{mm}$。采用特征值方法计算平板前4阶屈曲载荷,并可视化前4阶屈曲模态。

---

① 使用 x 表示梁的长度方向位置。

表 4.1　AS4/9310 单层板性能,$V_f = 0.6$

弹性模量	剪切模量	泊松比
$E_1 = 145880 \text{MPa}$	$G_{12} = G_{13} = 4386 \text{MPa}$	$\nu_{12} = \nu_{13} = 0.263$
$E_2 = E_3 = 13312 \text{MPa}$	$G_{23} = 4529 \text{MPa}$	$\nu_{23} = 0.470$

解 建立平板整体模型计算屈曲模态,而不是建立对称模型。选用 S8R5 单元模拟薄板。层合板为对称铺层,计算屈曲时不需计算每层铺层的应力,所以可采用 3 种不同的方法计算屈曲载荷。对于 S8R5 单元,如果壳单元截面为 Homogeneous 或 Composite,那么不需要指定 Transverse Shear Stiffness(见(a)和(c)情况),如果壳单元截面为 General Stiffness((b)情况),那么需要指定 Transverse Shear Stiffness。

a. 使用层合板等效模量求解。

在第一种方法中,先计算得到层合板等效弹性模量并与正交各向异性壳单元一起使用。此时,层合板弹性模量代表了正交各向异性板的刚度,其受面内载荷作用时行为与真实层合板等效(见 3.2.10 节)。层合板等效弹性模量可参考 1.15 节。将单层板材料性能(表 4.1)输入式(1.91),每层按照式(1.53)进行角度转换,再根据式(1.102)将所有铺层逐层叠加,按照式(1.105)计算出层合板模量,见表 4.2。对于本例题,使用层合板弯曲模量(文献[1],式(6.36))可能会带来更好的结果。

下面的操作步骤代码利用 Abaqus/CAE 中的 GUI 进行建模、求解和结果可视化。Ex_4.1.a.cae 可通过文献[5]查找。求解结果见表 4.3。

```
# Ex.4.1.a using equivalent laminate moduli
Menu: File, Set Work Directory[C:\SIMULIA\user\Ex_4.1], OK
Menu: File, Save As[C:\SIMULIA\user\Ex_4.1\Ex_4.1.a.cae], OK
```

表 4.2　$[(0/90)_3]_S$ 层合板的等效模量

弹性模量	剪切模量	泊松比
$E_x = 79985 \text{MPa}$	$G_{xy} = 4386 \text{MPa}$	$\nu_{xy} = 0.044$
$E_y = 79985 \text{MPa}$	$G_{yz} = 4458 \text{MPa}$	$\nu_{yz} = 0.415$
$E_z = 16128 \text{MPa}$	$G_{xz} = 4458 \text{MPa}$	$\nu_{xz} = 0.415$

ⅰ. 创建部件。

```
Module: Part
Menu: Part, Create
    3D, Deformable, Shell, Planar, Approx size[2000], Cont
```

Menu: Add, Line, Rectangle,[-500,-250],[500,250]
　　X # to close command, Done

ⅱ. 定义材料、截面,并将截面赋予部件。

Module: Property
Menu: Material, Create
　　Name[Material-1], Mechanical, Elasticity, Elastic
　　Type: Engineering Constants
　　[79985, 79985, 16128, 0.044, 0.415, 0.415, 4386, 4458, 4458], OK
Menu: Section, Create
　　Name[Section-1], Shell, Homogeneous, Cont
　　Shell thickness: Value[10.2], Material: Material-1, OK
　　Menu: Assign, Section
　　# pick the part, Done, OK

ⅲ. 创建装配体。

Module: Assembly
Menu: Instance, Create, Independent, OK

ⅳ. 定义分析步。

Module: Step
Menu: Step, Create
　　Procedure type: Liner Perturbation, Buckle, Cont
　　Number of eigenvalues requested[10], Vectors/iteration[10], OK

ⅴ. 施加边界条件和载荷。

Module: Load
Menu: Load, Create
　　Step: Step-1, Mechanical, Shell Edge Load, Cont
　　# pick: edges @ x = 500 & y = 250, Done, Magnitude[1.0], OK
Menu: BC, Manager
　　Create, Step: Initial, Mechanical, Disp/Rota, Cont
　　# pick: the 4 edges, Done, # checkmark: U3, OK
　　Create, Step: Initial, Mechanical, Disp/Rota, Cont
　　# pick: edge @ x = -500, Done, # checkmark: U1, OK
　　Create, Step: Initial, Mechanical, Disp/Rota, Cont
　　# pick: edge @ y = -250, Done, # checkmark: U2, OK, # close Manager

ⅵ. 网格划分。

Module: Mesh

Menu: Seed, Instance, Size:[100], OK

Menu: Mesh, Controls, Technique: Structured, OK

Menu: Mesh, Element Type, Geometric Order: Quadratic, DOF: 5, OK

Menu: Mesh, Instance, Yes

ⅶ. 求解并可视化结果。

Module: Job

Menu: Job, Manager

  Create, Name[Ex-4-1-a], Cont, OK, Submit, # when completed, Results

Module: Visualization

Menu: Plot, Deformed Shape # buckling mode 1 will be displayed

Menu: Result, Step/Frame, # pick a mode, Apply # to visualize

  # eigenvalues are shown in the WS and recorded in Ex-4-1-a. dat

b. 使用 $A-B-D-H$ 矩阵法求解。

在第二种方法中,将使用 $A-B-D-H$ 矩阵。将单层板材料性能(表4.1)代入式(3.9),即可获得层合板属性(A、B、D 和 H 矩阵)。计算结果如下:

$$\begin{bmatrix} A & B \\ B & D \end{bmatrix} = \begin{bmatrix} 817036 & 35937.6 & 0 & 0 & 0 & 0 \\ 35937.6 & 817036 & 0 & 0 & 0 & 0 \\ 0 & 0 & 44737.2 & 0 & 0 & 0 \\ 0 & 0 & 0 & 8558450 & 311579 & 0 \\ 0 & 0 & 0 & 311579 & 5608960 & 0 \\ 0 & 0 & 0 & 0 & 0 & 387872 \end{bmatrix}$$

$$[H] = \begin{bmatrix} 37812.8 & 0 \\ 0 & 37964.7 \end{bmatrix}$$

下面的操作步骤代码描述了如何修改 Ex_4.1.a.cae,使用 A-B-D-H 矩阵属性。修改后的文件名为 Ex_4.1.b.cae,可在文献[5]中查询,计算结果见表4.3。

Ex. 4. 1. b using ABDH matrices

Menu: File, Set Work Directory [C:\SIMULIA\user\Ex_4.1], OK

Menu: File, Open [C:\SIMULIA\user\Ex_4.1\Ex_4.1.a.cae], OK

Menu: File, Save As [C:\SIMULIA\user\Ex_4.1\Ex_4.1.b.cae], OK

ⅰ. 定义材料、截面,并将截面赋予部件。

Module: Property

Menu: Material, Delete, Material-1, Yes

Menu: Section, Delete, Section-1, Yes

Menu: Section, Create, Shell, General Shell Stiffness, Cont

  Tab: Stiffness

$$\begin{bmatrix} 817036 & 35937.6 & 0 & 0 & 0 & 0 \\ & 817036 & 0 & 0 & 0 & 0 \\ & & 44737.2 & 0 & 0 & 0 \\ & & & 8.56e6 & 311579 & 0 \\ & & & & 5.61e6 & 0 \\ & & & & & 387872 \end{bmatrix}$$

  Tab: Advanced

  Transverse Shear Stiffnesses: # checkmark: Specify values

  K11 [37812.8], K12 [0], K22 [37964.7], OK

ⅱ. 定义分析步。

Module: Step

Menu: Step, Edit, Step-1

  Maximum number of interactions [70], OK

  # solve the problem using a smaller number for this quantity

  # the following error message may be displayed

  # TOO MANY ITERATIONS NEEDED TO SOLVE THE EIGENVALUE PROBLEM

ⅲ. 求解并可视化结果。

Module: Job

Menu: Job, Manager

  Create, Name [Ex-4-1-b], Cont, OK, Submit, # when completed, Results

  # visualize the results as it is indicated in part (a)

c. 使用 LSS 求解。

  在第三种方法中,需要输入层合板铺层顺序(LSS)和单层板属性(表4.1)。下面的操作步骤代码描述了如何修改 Ex_4.1.a.cae,使用该方法进行求解。修改后的文件名为 Ex_4.1.c.cae,可在文献[5]中查询,计算结果见表4.3。

Ex_4.1.c using LSS

Menu: File, Set Work Directory [C:\SIMULIA\user\Ex_4.1], OK

Menu: File, Open [C:\SIMULIA\user\Ex_4.1\Ex_4.1.a.cae], OK
Menu: File, Save As [C:\SIMULIA\user\Ex_4.1\Ex_4.1.c.cae], OK

表 4.3　分叉载荷　　　　　　　　（单位：N/mm）

模态#	(a)	(b)	(c)
1	253.27	210.1	210.04
2	320.05	319.87	319.75
3	573.36	643.18	642.81
4	998.64	831.65	831.11

ⅰ. 定义材料、截面,并将截面赋予部件。

Module: Property
Menu: Material, Delete, Material-1, Yes
Menu: Section, Delete, Section-1, Yes
Menu: Material, Create
　　Name [Material-1], Mechanical, Elasticity, Elastic, Type: Lamina
　　[1.4588e5, 1.3312e4, 2.63e-1, 4.386e3, 4.386e3, 4.529e3], OK
Menu: Section, Create
　　Shell, Composite, Cont
　　# checkmark: Symmetric layers
　　# right-click: Insert Row After, # until there are six rows
　　Material: Material-1 # for all layers, Thickness [0.85]
　　Orientation Angle: [0,90,0,90,0,90], Integration Points [3]
　　Ply Name [k1, k2, k3, k4, k5, k6], OK

ⅱ. 定义分析步。

Module: Step
Menu: Step, Edit, Step-1
　　Maximum number of interactions [70]
　　# solve the problem using a smaller number for this quantity
　　# the following error message may be displayed
　　# TOO MANY ITERATIONS NEEDED TO SOLVE THE EIGENVALUE PROBLEM

ⅲ. 求解并可视化结果。

Module: Job
Menu: Job, Manager
　　Create, Name [Ex-4-1-c], Cont, OK, Submit,

```
# when Completed, Results
# visualize the results as it is indicated in part (a)
```

4.1.1 缺陷敏感性

为了说明缺陷对结构稳定性的影响,让我们先看一下图4.1中的实线。当外载荷小于分叉临界载荷P_{cr}时,结构横向的变形为0,其P_{cr}为无缺陷结构主路径极限载荷。主路径和次路径的交叉点为分叉点,该点对应的载荷即为P_{cr}。圆柱梁的后临界行为具有一般性并会稍稳一些。具有一般性指的是梁在临界点之后可能发生向左或向右的偏移。稍稳定的后临界路径指的是梁在发生屈曲后,可以进一步承受稍高的载荷。对于圆柱梁而言,这种后临界增强行为非常小,因此很难依靠后临界增强行为承载更高的载荷。事实上,屈曲后,柱体横向偏移很大,从而导致材料强度破坏和结构的最终坍塌。与圆柱梁不同,简支约束的平板在次路径上会明显增强。

4.1.2 非对称分叉

以图4.2中的框架结构为例,两个杆件各采用一个单元模拟(文献[24],5.9节和7.8节)进行屈曲特征值分析,可以得到分叉载荷

$$P_{cr} = 8.932(10^{-6})AE \tag{4.5}$$

但无法给出结构的临界状态的性质,即结构是否稳定、结构的后临界路径是否为对称等。事实上,该框架结构的后临界路径为非对称、不稳定的,如图4.2所示。在力–旋转角的图中,后临界路径为一个斜坡,图中的θ为载荷作用点接头的转角。

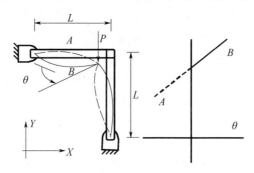

图4.2 两梁框架模型

$$P^{(1)} = 18.73(10^{-9})AE(1/\text{rad}) \tag{4.6}$$

通常,采用特征值方法分析结构稳定性,无法得到结构的后临界路径。如果结构的后临界路径如图4.1中所示是增强的、对称的,那么结构极限承载能力与

分叉点载荷 P_{cr} 相近。但是,如果结构的后临界路径如图 4.2 所示是不稳定的和(或)非对称的,或者存在模态耦合[20,25-29],那么结构的极限承载能力将小于分叉点载荷 P_{cr}。为了更好地评价特征值方法计算的结果,必须理解和分析结构的后屈曲行为。

4.1.3 后临界路径

研究结构后屈曲行为的一种方法是对含缺陷结构进行延续分析,可参见 4.2 节。该方法很适合研究后屈曲行为,但计算复杂、且耗时,本章后续会具体说明。通过软件分析后临界路径的性质(包括对称性、曲率和模态耦合作用等)可以更有效地分析结构的后屈曲行为。如果结构的次路径是稳定的和对称的,那么分叉载荷可认为是结构的极限承载力。后临界路径的曲率提供了一个判断后屈曲刚度变化的指标,并可一定程度上预估后屈曲的变形量。

图 4.1 中后临界路径的分叉载荷、斜率和曲率可以使用 BMI3 程序[26-28]计算,该程序在文献[5]中可下载。后屈曲行为可以通过下式表述

$$\Lambda = \Lambda^{(cr)} + \Lambda^{(1)}s + \frac{1}{2}\Lambda^{(2)}s^2 + \cdots \quad (4.7)$$

式中:s 为摄动参数,被选择作为节点位移的一个分量;$\Lambda^{(cr)}$ 为分叉乘子;$\Lambda^{(1)}$, $\Lambda^{(2)}$ 分别为后临界路径的斜率和曲率(文献[25],(43)),另见文献[20,26-29])。当斜率为 0 时,后临界路径是对称的。因此,屈曲具有一般性,结构向任何一侧屈曲都有可能。除非预先知道平板的缺陷形状,否则无法判断结构向哪侧屈曲。正的曲率代表后屈曲过程中刚度强化,负的曲率代表刚度退化。

例 4.2 以例 4.1 中简支平板为对象,计算分叉乘子 $\Lambda^{(cr)}$,临界载荷 N_{cr},后临界路径的斜率 $\Lambda^{(1)}$ 和曲率 $\Lambda^{(2)}$。计算最大横向位移为平板厚度时的载荷。将第一阶屈曲模态的最大位移分量作为摄动参数。

解 使用 BMI3 程序[5]计算分叉乘子 $\Lambda^{(cr)}$、斜率 $\Lambda^{(1)}$ 和曲率 $\Lambda^{(2)}$。可参考附录 C 中关于软件界面和操作过程的描述。BMI3 程序需要 $A-B-D-H$ 矩阵,可以利用例 4.1(b)建立的.cae 文件(也可在网站[5]中获取 Ex_4.1.b.cae)得到,但为满足 BMI3 的限制,需进行一些更改。

i. 打开 Ex_4.1.b.cae 文件,另存为 Ex_4.2.cae。
Menu: File, Set Work Directory [C:\SIMULIA\user\Ex_4.2], OK
Menu: File, Open [C:\SIMULIA\user\Ex_4.1\Ex_4.1.b.cae], OK
Menu: File, Save As [C:\SIMULIA\user\Ex_4.2\Ex_4.2.cae], OK

ii. 删除已有单元,重新布种子并划分 2×4 个单元,采用 S8R5 单元类型。
Module: Mesh

```
Object: Assembly
Menu: Mesh, Delete Instance Mesh, Yes
Menu: Seed, Edges
    # pick: horizontal lines, Done
    Method: By Number, Number of Elements [4], OK
    # pick: vertical lines, Done
    Method: By Number, Number of Elements [2], OK
Menu: Mesh, Instance, Yes
```

ⅲ. 由于BMI3程序只能施加集中载荷,因此需将模型中的均布边载荷替换为一组集中载荷。对于二次单元S8R5,均布的单元边载荷可以转化为3个作用于节点的集中载荷。对于单位边载荷($N_n = 1.0\text{N/mm}$),单元边的顶点各分得1/6的载荷,中点分得4/6的载荷。对于两个单元共用的节点,那么该节点共分得1/3的载荷。为了进行上述修改,首先要建立用于后续施加集中载荷的节点集。图4.3中给出了通过Set Manager对话框建立的6组所需的节点集。集合的建立可以通过下述过程实现:

```
Module: Load
Menu: Tools, Set, Manager
    Create, Name [x-corner], Type: Node, Cont
    # pick: the two corners of the plate on the edge at x = 500, Done
```

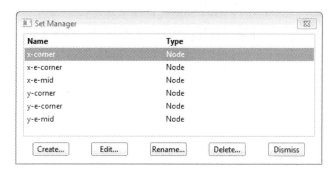

图4.3 划分网格后定义节点集合用于施加集中载荷

ⅳ. 通过相同的方法,建立其余的节点集。x-e-corner节点集为单元共用的节点($x = 500$mm的边上)。x-e-mid节点集为$x = 500$mm边上各单元的中间节点。对于$y = 250$mm边上的单元节点,采用同样的方式进行建立节点集。

ⅴ. 删除所有边载荷,采用集中载荷替代,具体如下:

```
Module: Load
```

```
Menu: Load, Manager
    # delete the existing load
    Create, Step: Step-1, Mechanical, Concentrated force, Cont
    Sets, x-corner, Cont, CF1 [-41.67], OK
    Create, Step: Step-1, Mechanical, Concentrated force, Cont
    Sets, x-e-corner, Cont, CF1 [-83.33], OK
    Create, Step: Step-1, Mechanical, Concentrated force, Cont
    Sets, x-e-mid, Cont, CF1 [-166.67], OK
    # Analogously add compressive loads on the edge y = 250
    # close the Load Manager
```

ⅵ. 运行模型并判断结果的准确性。由于网格大小的不同,会导致计算是特征值与例 4.1 的结果略有不同。

```
Module: Job
Menu: Job, Manager
    Create, Name [Job-1], Cont, OK, Submit, # when Completed, Results
    # visualize the mode shapes as it was indicated in Example 4.1
```

ⅶ. 将 inp2bmi3.exe 程序(可在文献[5]中获取)复制至工作目录,双击执行程序。将基于 Job-1.inp 文件生成 ABQ.inp、BMI3.inp 和 BMI3.dat 三个文件。ABQ.inp 文件是 Abaqus 可以执行的过滤后的输入文件。

ⅷ. 创建新的 Job 并运行 ABQ.inp,确认所有模型能够一起正常运行。本次计算是通过运行 .inp 文件进行计算的,与以往通过模型提交计算不同。具体执行过程如下:

```
Module: Job
Menu: Job, Manager
    Create, Name [ABQ], Source: Input file # not Model
    Select, ABQ.inp, OK, Cont, OK
    Submit, # when Completed, Results
```

ⅸ. 将 bmi3.exe 程序(可在文献[5]中获得)复制至工作目录,双击执行该程序。记录 BMI3.out 文件中输出的结果。

默认情况下,BMI3 程序将结构第一阶模态的最大变形分量作为摄动参数,也可以修改 BMI3.dat 文件改变摄动参数的选取。本例中,默认的摄动参数对应于第一阶屈曲模态、平板中心点垂直平面的变形 $U3$。计算结果,特别是 $\Lambda^{(2)}$,对网格的精细度较为敏感。

BMI3 程序计算时采用的是反力(见附录C)，那么式(4.7)变为

$$-N = \Lambda^{(cr)} + \Lambda^{(1)}s + \frac{1}{2}\Lambda^{(2)}s^2$$

$$N = -\Lambda^{(cr)} - \Lambda^{(1)}s - \frac{1}{2}\Lambda^{(2)}s^2$$

此时，摄动方向为 $s = -\delta$，所以

$$N = -\Lambda^{(cr)} - \Lambda^{(1)}(-\delta) - \frac{1}{2}\Lambda^{(2)}(-\delta)^2$$

$$N = -\Lambda^{(cr)} + \Lambda^{(1)}\delta - \frac{1}{2}\Lambda^{(2)}\delta^2$$

因此，利用更精细网格模型计算的结果，次路径计算为

$$N = -(-209.0) + (0)\delta - (-0.1154)\delta^2 = 209.0 + 0.1544\delta^2$$

由于斜率 $\Lambda^{(1)}$ 为 0，因此后临界路径是对称的。当横向变形(w)与平板厚度($s = Th = 10.2mm$)相等时，得到后屈曲载荷为 221N/mm，如图 4.4 所示。

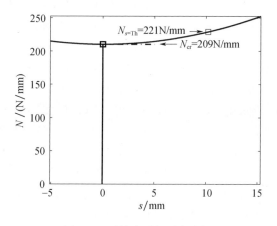

图 4.4　无缺陷平板平衡路径

4.2　延续分析方法

聚合物基复合材料(PMC)的破坏应变一般较高。AS4/3501 复合材料为 1.29%，S-glass/epoxy 复合材料为 2.9%，相比而言，一般钢的破坏应变只有 0.2%，铝只有 0.4%。这就意味着，对于复合材料而言，结构屈曲变形可能已进入后屈曲范围，而材料还处于弹性状态。然而，必须确保材料未出现基体为主的退化损伤，如果出现，那么材料将不再保持弹性(见第 8 章和第 9 章)。只要材料处于弹性状态，采用特征值屈曲分析方法，能够相对简便地进行屈曲分析，因为

可以借助弹性稳定经典理论,如 4.1 节所述。当材料处于非线性状态或者分析缺陷敏感结构时,则需要采用增量分析方法进行求解。

增量分析方法,也称延续分析方法,通常是将载荷逐步增加。每增加一步载荷,计算结构变形以及可能材料性能的变化。增量分析方法必须包含某种初始缺陷,如几何缺陷、材料缺陷或者载荷的偏心等。如果缺乏初始缺陷,那么增量分析将呈现线性结果,无法计算出分叉点或极限值。

延续分析方法是一种几何非线性分析方法。系统必须具有一个非平凡的基本路径,如非对称铺层(LSS)的层合平板受边载荷作用。

如果系统只有平凡的基本路径,如对称铺层的平板受边载荷作用,则必须通引入初始缺陷来制造非平凡的基本路径。比较典型的初始缺陷包括材料的初始缺陷(如非对称 LSS)、几何初始缺陷或者施加偏心载荷。

由于真实的几何缺陷很少被检测到或准确描述,因此计算时首选的方法是通过引入最小分叉载荷对应的分叉模态来人为制造几何初始缺陷。大多数情况下,该方法是有效的;但在某些情况下,应使用对结构更具破坏性的与缺陷相关的第 2 阶模态[30]。如果结构如图 4.2 中那样具有一个非对称后屈曲路径,此时需要注意不要在结构强化的路径上施加载荷。

有限元软件允许用户将初始计算得到的任意阶屈曲模态叠加到无缺陷的有限元模型中,从而制造结构的几何初始缺陷(见例 4.3)。

例 4.3 施加初始几何缺陷 $w_p(x,y) = \delta_0 \phi(x,y)$ 至例 4.1 的模型中,并且绘制缺陷幅值分别为 $\delta_0 =$ Th/10 和 $\delta_0 =$ Th/100 时的载荷-横向变形曲线。其中 Th 为层合板总厚度,$\phi(x,y)$ 为例 4.1 中最小分叉载荷对应的屈曲模态。

解 屈曲模态可以从例 4.1 中获得,同时将使用 Ex_4.1.b.cae 中的模型。首先,打开网站[5]中的 Ex_4.1.b.cae 文件,另存为 Ex_4.3.cae。然后,在 Job Manager 窗口中写出 Job-1.inp 文件。具体的求解过程如下:

ⅰ. 创建工作目录。
Menu: File, Set Work Directory, [C:\SIMULIA\user\Ex_4.3], OK
Menu: File, Open, [C:\SIMULIA\user\Ex_4.1\Ex_4.1.b.cae], OK
Menu: File, Save As, [C:\SIMULIA\user\Ex_4.3\Ex_4.3.cae], OK
Module: Job
Menu: Job, Manager
  Create, Name [Job-1], Cont, OK, Write Input
  # the input file Job-1.inp has been created in the Work Directory

ⅱ. 计算并且保存前几阶模态的形状(位移)。使用 Notepad++(或其他文本编辑器)打开 Job-1.inp 文件,另存为 Job-pert.inp。然后,在 ∗ ∗ BOUNDARY

CONDITIONS 语句之前添加如下语句到 * Buckle 模块中。

```
* * Data lines to specify the reference load,
* NODE FILE, GLOBAL = YES, LAST MODE = 4
U
* *
```

并保存文件。上述语句将前 4 阶模态的模态形状保存至 .fil 文件中。.fil 文件为二进制文件,无法通过 Notepad 文本编辑器打开,仅能通过 Abaqus 软件读取。

ⅲ. 在 Abaqus 的 CAE 模块中,创建新的任务并提交。

Module: Job
Menu: Job, Manager
  Create, Name: [Job-pert], Source: Input file
  Input file: Select, [Job-pert.inp], OK, Cont, OK
  Submit

如果上述操作无错误执行完毕,在工作目录中将自动创建 Job-pert.fil 文件。因为输出的模态变形已进行归一化处理,该文件中模态的最大变形为 1.0。

ⅳ. 建立弧长法(Risk)分析步。

利用如下的步骤,前一个分析得到的位移值将用于更新有限元模型的形状。本例中,第 1 阶模态的乘数因子为 $\delta_1 = 0.05$。第 2 和第 3 阶模态的贡献在 .inp 文件中已被注释掉(以双星 * * 开头的命令行表示)。因此,将中心挠度为 δ_1 的一阶屈曲模态的初始变形被施加至结构中。具体做法是,先打开 Job-1.inp 文件,另存为 Job-risk.inp,然后,将如下 3 行代码

```
* Step, name = Step-1, perturbation
* Buckle
10, , 10, 70
```

使用下述代码替代:

```
* * IMPERFECTION, FILE = Job-pert, STEP = n
* * n : step number where the modes were calculated in the previous run
* * if there was only a Initial Step followed by a Buckle Step, n = 1
* IMPERFECTION, FILE = Job-pert, STEP = 1
* * Multiple data lines to specify the contributions of each mode
* * m, delta_m
* * m: mode number
* * delta_m : scale factor for mode m
```

```
    1,0.05
* * 2,0.02
* * 3,0.01
* STEP, NLGEOM
* * next 2 lines (2nd one is empty) define increment/stopping criteria
* STATIC, RIKS
```

在上述代码后面,是如下相同的多条数据行以指定 Job-pert. inp 中使用的参考载荷条件,

```
* * Data lines to specify the reference load,
* NODE FILE, GLOBAL = YES, LAST MODE = 4
U
* *
```

确保删除所有与 load case = 2 有关的命令行,并保存文件。否则,将出现错误信息。需要删除的命令行与下述命令行类似:

```
* Boundary, op = NEW, load case = 2
_PickedSet5, 3, 3
* Boundary, op = NEW, load case = 2
_PickedSet6, 1, 1
* Boundary, op = NEW, load case = 2
_PickedSet7, 2, 2
```

V. 在 Abaqus 的 CAE 模块中,创建一个新的任务并提交。

Module: Job
Menu: Job, Manager
  Create, name: [Job-risk], Source: Input file
  Input file: Select, [Job-risk. inp], OK, Cont, OK
Submit, # when Completed, Results
  # this finalizes the Risk analysis
Module: Visualization
Menu: Plot, Contours, On Deformed Shape
  Toolbar: Field Output: Primary, U, U3
Menu: Result, Step/Frame, Increment 25, Apply
Menu: Result, Step/Frame, Increment 30, Apply

采用延续分析方法计算含几何初始缺陷的模型,获得延续平衡路径,如

图 4.5 所示。可以看出最终的连续分析解与虚线表示的无初始缺陷结构次路径比较接近。对于初始缺陷较小的情况,连续分析解较接近主路径,然后才是次路径。对于初始缺陷较大的情况,连续分析解与无初始缺陷结构偏差会较大,由图 4.5 中 $\delta_0 = \mathrm{Th}/10$ 的曲线可以看出。

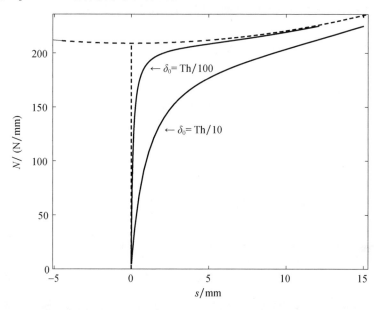

图 4.5 铺层为 $[(0/90)_3]_\mathrm{S}$ 的层合板,$\delta_0 = \mathrm{Th}/10$ 和 $\delta_0 = \mathrm{Th}/100$ 时的平衡路径

例 4.4 计算例 3.11 中复合材料梁的一阶模态和屈曲系数。

解 本例的求解需对 Ex_3.11.cae 文件修改。将静力分析步改为屈曲分析步。下述操作过程代码描述了具体的更改。

ⅰ. 打开 Ex-3.11.cae 文件,另存为 Ex_4.4.cae。
Menu: File, Set Work Directory [C:\SIMULIA\user\Ex_4.4], OK
Menu: File, Open [C:\SIMULIA\user\Ex_3.11\Ex_3.11.cae], OK
Menu: File, Save As [C:\SIMULIA\user\Ex_4.4\Ex_4.4.cae], OK

ⅱ. 在例 3.11 模型中创建屈曲分析步,替代原 dead-weight 分析步。
Module: Step
Menu: Step, Manager
    # select: Step-1, Delete, Yes
    Create, Procedure type: Linear perturbation, Buckle, Cont
    Number of eigenvalues requested [1], OK, # close Step Manager

ⅲ．添加一个集中载荷。

Module: Load

Menu: Load, Create

　　Step: Step-1, Category: Mechanical, Type: Concentrated force, Cont

　　# pick: RP-1 in the WS, Done, CF3 [-1.0], OK

ⅳ．计算模型，可视化结果。

Module: Job

Menu: Job, Manager

　　# select: existing job (Job-1), Delete, Yes

　　Create, Cont, OK, Submit, # when Completed, Results

Module: Visualization

Menu: Plot, Deformed Shape

经计算，屈曲特征值为 -6.22748E +05，屈曲模态如图 4.6 所示。

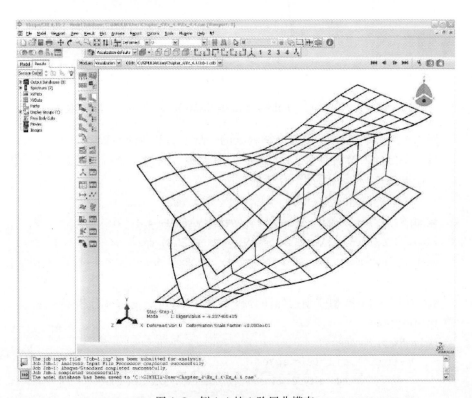

图 4.6　例 4.4 的 1 阶屈曲模态

习题

习题 4.1 计算图 4.2 中双梁框架模型的分叉载荷 P^c,每个梁采用单个二次梁单元模拟。每个梁长度 $L=580\text{mm}$,截面积 $A=41\text{mm}^2$,惯性矩 $I=8.5\text{mm}^4$,高度 $H=10\text{mm}$,弹性模量 $E=200\text{GPa}$。两个梁之间采用刚性连接。

习题 4.2 对习题 4.1 进行收敛性分析,增加每个梁的单元数 N,使分叉载荷收敛在 2% 以内。绘制分叉载荷 P^c-N 曲线。

习题 4.3 将例 4.2 层合板的 LSS 改为 $[(0/90)_6]_T$,重新进行计算。不要引入任何缺陷,仅分析该无缺陷的层合板结构,此时层合板为非对称结构。

习题 4.4 将例 4.2 层合板的 LSS 改为 $[(0/90)_6]_T$,施加载荷 $N_x=1$,$N_y=N_{xy}=0$,重新进行计算。不要引入任何缺陷,仅分析该无缺陷的层合板结构,此时层合板为非对称结构。

习题 4.5 使用有限元商业软件,对于在端部边上施加轴向压力的圆柱壳,计算并绘制如图 4.5 所示的 $\delta_0 = \text{Th}/100$ 时的连续解曲线。圆柱壳的长度 $L=965\text{mm}$,中面半径 $a=242\text{mm}$。LSS 为 $[(0/90)_6]_T$,每层铺层厚度 $t=0.127\text{mm}$。层合板为 E-glass/epoxy 复合材料,$E_1=54\text{GPa}$,$E_2=18\text{GPa}$,$G_{12}=9\text{GPa}$,$\nu_{12}=0.25$ 和 $\nu_{23}=0.38$。

习题 4.6 计算习题 4.5 的结构在分叉载荷 $P=\Lambda^{(\text{cr})}$ 下的最大应力失效因子 I_f。材料强度性能 $F_{1t}=1034\text{MPa}$,$F_{1c}=1034\text{MPa}$,$F_{2t}=31\text{MPa}$,$F_{2c}=138\text{MPa}$ 和 $F_{6c}=41\text{MPa}$。

习题 4.7 绘制习题 4.5 中圆柱壳缺陷在 $(\text{Th}/200)<s<\text{Th}$ 范围的缺陷敏感性曲线。

第 5 章　自由边应力

在面内载荷 N_x、N_y、N_{xy} 作用下,对称铺层层合板只在内部产生面内应力 σ_x、σ_y、σ_{xy}。由于面内应力分量在自由边处不平衡,在自由边附近会产生层间应力 σ_z、σ_{yz}、σ_{xz}。

举例说明,考虑一个长 $2L$,宽 $2b \ll 2L$,厚 $2H < 2b$ 复合材料层合长梁(图 5.1)。该长梁仅受到轴向载荷 N_x 作用。对于平衡、对称层合板,其中面应变和曲率(见式(3.6))是一致的,由下式给出

$$\begin{cases} \epsilon_x^0 = \alpha_{11} N_x \\ \epsilon_y^0 = \alpha_{12} N_x \\ \gamma_{xy}^0 = 0 \\ k_x = k_y = k_{xy} = 0 \end{cases} \quad (5.1)$$

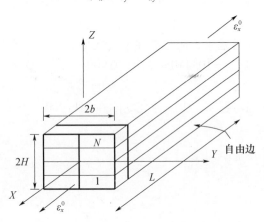

图 5.1　拉伸试样[①]

式中:α_{11},α_{12} 分别为面内层合板柔度分量,可以通过对式(3.8)求逆获得;也可以参考文献[1,式(6.21)]。由本构方程[1,式(6.24)],对于第 k 层单层板,有

---

① 转载自 *Mechanics of Fibrous Composites*,C. T. Herakovich,图 8.1,版权所有(1998),经 John Wiley & Sons,Inc. 许可。

$$\begin{cases} \sigma_x^k = (\overline{Q}_{11}^k \alpha_{11} + \overline{Q}_{12}^k \alpha_{12}) N_x \\ \sigma_y^k = (\overline{Q}_{12}^k \alpha_{11} + \overline{Q}_{22}^k \alpha_{12}) N_x \\ \sigma_{xy}^k = (\overline{Q}_{16}^k \alpha_{11} + \overline{Q}_{26}^k \alpha_{12}) N_x \\ \sigma_z^k = \sigma_{xz}^k = \sigma_{yz}^k = 0 \end{cases} \quad (5.2)$$

从层合板内部取出的一小块层合板,对立面上的 σ_y 和 σ_{xy} 处于平衡状态;不需要任何附加力,其自由体受力图(FBD)处于平衡状态。在这种情况下,应力分量是自平衡的。在图 5.1 所示的自由边上,$\sigma_y = \sigma_{xy} = \sigma_{yz} = 0$。如果 σ_y 和 σ_{xy} 在层合板内部不为 0,而在自由边为 0,那么在自由边一定存在别的应力使其平衡。

5.1 泊松比的不匹配

受到拉伸载荷的单层板,其在垂直于载荷的方向上将会收缩。如果两个或多个泊松比不同的单层板粘合在一起,将会产生层间应力使得所有层在界面处的变形相同(图 5.2)。在整个层合板厚度方向上,这些应力的总和为 0,因为层合板没有横向外载 N_y。换句话说,它们以如下方式自平衡

$$\int_{z_0}^{z_N} \sigma_y \mathrm{d}z = 0 \quad (5.3)$$

式中:z_0,z_N 分别为底面和顶面的 z 向坐标。

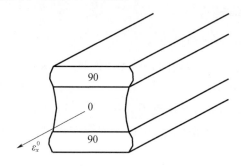

图 5.2 泊松比效应[①]

5.1.1 层间力

如式(5.3)所示,使用经典层合理论(CLT,文献[1],第 6 章),在整个层合板厚度上求和计算得到的面内应力 σ_y 是自平衡的。但在层合板的一小部分上

---

① 转载自 *Mechanics of Fibrous Composites*,C. T. Herakovich,图 8.1,版权所有(1998),经 John Wiley & Sons,Inc. 许可。

(如图 5.3 中的 z_k 上方部分),σ_y 可能不是自平衡的。因此,一层或多层的收缩或膨胀效应必然由层间剪切应力 σ_{yz} 来平衡。由于层合板不存在剪切外载荷,σ_{yz} 在整个试样宽度上的积分必然为 0。但是,在层合板宽度一半的范围内,如果界面上、下的应力 σ_y 不是自平衡的,则必然存在层间剪切力。这些单位长度力的大小可通过将层间剪切应力 σ_{yz} 在层合板宽度一半范围内($0<y<b$)进行积分得到,即

$$F_{yz}(z_k) = \int_0^b \sigma_{yz(z=z_k)} dy = -\int_{z_k}^{z_N} \sigma_y dz \tag{5.4}$$

经典层合理论中,层间剪切应力 σ_{yz} 无法得到,而横向应力 σ_y 可以计算得到。因此,根据已知的横向应力分布 σ_y,可以计算得到层合板的厚度上任意位置层间剪切力的大小。

在纵向拉伸载荷作用下,平衡对称层合板中每个单层的面内应力 σ_y 为常数。因此,当在界面处($z=z_k$)计算层间力时,上述积分简化为

$$F_{yz}(z_k) = -\sum_{i=k}^{N} \sigma_y^i t_i \tag{5.5}$$

式中:t_i 为单层板的厚度。

层间剪切力 F_{yz} 的大小可以用来优化对比不同的层叠顺序,以尽量减小自由边层间剪切应力 σ_{yz}。然而,该层间剪切力不能代表实际应力的大小。因此,它仅可以用于不同的 LSS(层叠顺序)的比较,而不能作为强度失效标准。

5.1.2 层间弯矩

层间剪切应力 σ_{yz} 产生剪切应变 γ_{yz},由于对称性,在试样的中心线上该剪切应变必然消失。因此,中心线处的 $\sigma_{yz}=0$。而且,在自由边缘处 σ_{yz} 也必然为 0,因为 σ_{zy} 在自由边缘处不存在。但是对于 σ_y 不自平衡的任何位置 z_k,σ_{yz} 必然在自由边缘和中心线之间的某些位置不为 0。图 5.5 绘制了 σ_{yz} 的数值解关于自由边距离 y/b 的曲线。结果表明,σ_{yz} 在自由边缘附近迅速增加,然后在层合板内部逐渐减为 0。

非自平衡的应力分布,既产生力 F_{yz}(式(5.5)),也产生弯矩。要计算弯矩 M_z,可使应力 σ_y 对图 5.3 中的 A 点取矩。由 σ_y 产生的力矩只能由层间正应力 σ_z 产生的力矩来平衡。因此,弯矩 M_z 定义为

$$M_z(z_k) = \int_0^b \sigma_{z(z=z_k)} y dy = \int_{z_k}^{z_N} (z-z_k) \sigma_y dz \tag{5.6}$$

式中:z_k 为第 k 层单层板的上表面坐标;z_N 为层合板上表面的坐标(见文献[1],图 6.6 所示的在层合板厚度上定义的坐标系统)。

在拉伸试验中,自由边分层现象证实了 σ_z 的存在,它使层合板的破坏载荷大大降低。同样,弯矩的大小也可以用来比较不同的层叠顺序,以尽量减少层间

第5章 自由边应力

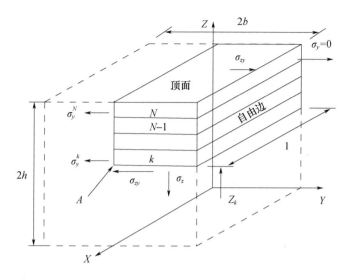

图 5.3 用于计算泊松比不匹配导致的力 F_{yz} 和力矩 M_z 的自由体受力图

正应力 σ_z。然而,弯矩的大小也不能表明实际应力大小,因此也不能作为强度破坏标准。

在纵向拉伸载荷作用下,平衡层合板中每个单层的面内应力 σ_y 均为常数。因此,当在界面处($z=z_k$)计算层间弯矩时,上述积分简化为

$$M_z(z_k) = \sum_{i=k}^{N} \sigma_y^i \left(z_i t_i + \frac{t_i^2}{2} - z_k t_i \right) \tag{5.7}$$

由于 σ_z 是 σ_{yz} 的副产品,σ_{yz} 在 $y=0$ 处时由于对称性而消失,所以 σ_z 同样在试样的中心线上($y=0$)也会消失,但 σ_z 在自由边缘附近却很大。由于试件没有 Z 向的外载,所以 σ_z 的积分必然为零。因此,它在一些区域必然是拉伸的(正值),在另一些区域是压缩的(负值)。数值解表明,σ_z 在自由边附近迅速增大,沿着层合板内部方向下降为负值并逐渐变为零。图 5.5 给出了 σ_z 数值解关于自由边距离 y/b 的曲线。然而,当 $y \to b$ 时,$\sigma_z \to \infty$,这是一个奇异点,不能通过有限元方法很好地处理。因此,即使对于 $y<b$ 情况下,σ_z 的结果也非常依赖于网格密度。另外,因为 $\sigma_z \to \infty$,如果不进一步进行数据处理,就不能将结果用于失效准则。

例 5.1 在载荷 $N_x = 175 \text{kN/m}$ 条件下,计算平衡对称层合板 $[0_2/90_2]_S$(图 5.1)所有界面处的 F_{yz} 和 M_z。其碳纤维/环氧树脂单层板性能 $E_1 = 139 \text{GPa}$,$E_2 = 14.5 \text{GPa}$,$G_{12} = G_{13} = 5.86 \text{GPa}$,$G_{23} = 5.25 \text{GPa}$,$\nu_{12} = \nu_{13} = 0.21$,$\nu_{23} = 0.38$。每层单层板厚 $t_k = 0.127 \text{mm}$。

解 厚度截面上的面内应力分布 σ_y 可通过文献[1],6.2节中所述的计算过程获得,该计算过程可在 CADEC[12] 中执行。计算的应力值如表 5.1 所列。

为计算得到 F_{yz}，利用式(5.5)计算给定界面上方所有层的载荷贡献。在面内载荷作用下的平衡层合板，其面内应力 σ_y 在各层中是常数，因此式(5.5)适用。其他情况下，使用式(5.4)可以进行精确积分计算，因为 σ_y 关于坐标 z 是线性的，或者采用每层中的 σ_y 平均值，使用式(5.5)近似获得 F_{yz}。

由于层合板是平衡的，且仅加载面内荷载，因此可以使用式(5.7)计算 M_z。否则，可以使用式(5.6)或将每层中的 σ_z 平均值代入式(5.7)近似获得 M_z。

计算结果见表5.1和图5.4。

表5.1 泊松比不匹配产生的层间力 F_{yz}

k	位置	σ_y/MPa	t_k/mm	z/mm	F_{yz}/(kN/m)	M_z/(Nm/m)
8	TOP	5.55×10^{-3}		0.508	0.000	
8	BOT	5.55×10^{-3}	0.127	0.381	-0.705	0.045
7	TOP	5.55×10^{-3}		0.381	-0.705	
7	BOT	5.55×10^{-3}	0.127	0.254	-1.410	0.179
6	TOP	-5.55×10^{-3}		0.254	-1.410	
6	BOT	-5.55×10^{-3}	0.127	0.127	-0.705	0.313
5	TOP	-5.55×10^{-3}		0.127	-0.705	
5	BOT	-5.55×10^{-3}	0.127	0.000	0.000	0.358
4	TOP	-5.55×10^{-3}		0.000	0.000	
4	BOT	-5.55×10^{-3}	0.127	-0.127	0.705	0.313
3	TOP	-5.55×10^{-3}		-0.127	0.705	
3	BOT	-5.55×10^{-3}	0.127	-0.254	1.410	0.179
2	TOP	5.55×10^{-3}		-0.254	1.410	
2	BOT	5.55×10^{-3}	0.127	-0.381	0.705	0.045
1	TOP	5.55×10^{-3}		-0.381	0.705	
1	BOT	5.55×10^{-3}	0.127	-0.508	0.000	0.000

例5.2 绘制由泊松比不匹配导致的应力 σ_z 和 σ_{yz} 关于 $y(0<y<b)$（图5.3）的函数曲线。研究对象为 $[0/90]_S$ 层合板，所计算的应力位于中间对称面上方的90/0界面处。其材料性能为 $E_1=139\text{GPa}$，$E_2=14.5\text{GPa}$，$\nu_{12}=\nu_{13}=0.21$，$\nu_{23}=0.38$，$G_{12}=G_{13}=5.86\text{GPa}$，$G_{23}=5.25\text{GPa}$。取 $2b=20\text{mm}$，试样长度 $2L=80\text{mm}$，每层厚度 $t_k=1.25\text{mm}$。通过在 $x=L$ 处施加均匀位移，以施加均匀应变 $\epsilon=0.01$。对每单层使用正交各向异性实体单元进行有限元划分，自由边区域细化网格。对于每层单层板在厚度方向上至少使用两个二次单元，并自由边附近单元横纵比为1。

解 注意到，本例题不需要为整个几何模型建模。可利用对称性，建立 $x>0$、$y>0$、$z>0$ 象限处的模型（平板的1/8部分，见图5.3）。由于任何 $y-z$ 横截面都具有相同的力学行为,因此只需对 $x=0$ 和 $x=L^*$ 之间的一小段进行建模。因为自由边效应也发生在 $x=0$ 和 $x=L^*$ 处,取 $L^*=8h=32t_k$,$t_k=2.5\text{mm}$,并在 $x=L^*/2$ 处显示结果以避免模型由于两端加载产生的自由边效应。计算结果如图 5.5 所示。参见下面的建模计算过程和图 5.5。

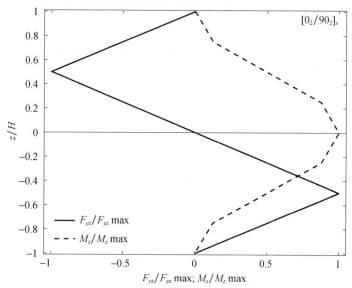

图 5.4 泊松比效应产生的层间力 F_{yz} 和弯矩 M_z

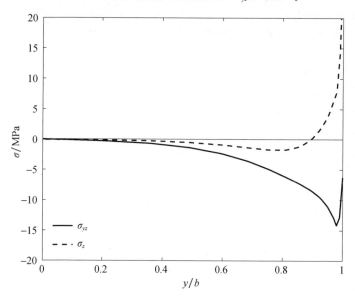

图 5.5 $[0/90]_s$ 铺层的碳纤维/环氧树脂复合材料板在 90/0 界面处的层间应力 σ_{yz} 和 σ_z(FEA)

i. 设置工作目录。

Menu: File, Set Work Directory, [C:\SIMULIA\user\Ex_5.2]

Menu: File, Save As, [C:\SIMULIA\user\Ex_5.2\Ex_5.2.cae]

ii. 创建部件。

Module: Part

    # Look at Fig. 5.3, layer thickness is tk = 1.25 mm along Z-global

    # b = 10 mm, length along x is L = 40 mm (shown as 1 in Fig. 5.3)

    # Draw layers 3 and 4 above the middle surface

Menu: Part, Create

    3D, Deformable, Solid, Extrusion, Approx size[100], Cont

Menu: Add, Line, Rectangle, [0,0], [40,10], X, Done, Depth[2.5], OK

    # We need to cut the volume in 2 layers

Menu: Tools, Datum, Plane, Offset from ppal plane, XY Plane, [1.25]

    X, # close the Create Datum pop-up window

Menu: Tools, Partition, Cell, Use datum plane

    # pick the plane parallel to the XY plane, Create partition, Done

    # close the Create Partition pop-up window

iii. 创建装配体。

Module: Assembly

Menu: Instance, Create, Independent, OK

iv. 定义材料、截面属性并将截面属性赋予部件。

Module: Property

Menu: Material, Create

    Mechanical, Elasticity, Elastic, Type: Engineering Constants

    [1.39E5 1.45E4 1.45E4 0.21 0.21

    0.38 5.86E3 5.86E3 5.2536E3], OK

Menu: Section, Create, Solid, Homogeneous, Cont, OK

Menu: Assign, Section, # pick: layer 3, Done, Section-1, OK

    # pick: layer 4, Done, Section-1, OK, Done

    # assign material orientation for 90-deg layer (near mid surface)

Menu: Assign, Material Orientation, # pick: layer 3, Done

    Use Default Orientation, Definition: Coordinate system

    Additional Rotation Direction: Axis 3

    Additional Rotation: Angle[90], OK

```
# assign material orientation for 0-degree layer (top surface)
Menu: Assign, Material Orientation, #pick layer 4, Done,
    Use Default Orientation, Definition: Coordinate system
    Additional Rotation Direction: Axis 3
    Additional Rotation: None, OK
    # to see the orientations do this:
    # on the left menu tree,
    # Models, Model-1, Parts, Part-1, Orientations
    # you should have only 2 SYSTEM: < Global > entries, one per layer
```

v. 定义分析步。

```
Module: Step
Menu: Step, Create
    Procedure type: General, Static/General, Cont, OK
```

vi. 施加载荷和 BC 信息。

```
Module: Load
Menu: BC, Manager
    Create, Name[DISP], Step: Step-1, Disp/Rota, Cont
    # pick: surface x = 40, Done, # checkmark: U1[0.4], OK
    # you may need to rotate the model to pick the surfaces
    Create, Name[XSYMM], Step: Initial, Symm/Anti/Enca, Cont
    # pick: surface x = 0, Done, XSYMM, OK
    Create, Name[YSYMM], Step: Initial, Symm/Anti/Enca, Cont
    # pick: surface y = 0, Done, YSYMM, OK
    Create, Name[ZSYMM], Step: Initial, Symm/Anti/Enca, Cont
    # pick surface z = 0, Done, ZSYMM, OK
    # close BC Manager pop-up window
```

vii. 网格划分。

```
Module: Mesh
    # Bias towards the edge y = 10 is needed
Menu: Seed, Edges
    # pick: 6 edges parallel to Y-global, Done
Method: By number, Bias: Single
    Number of elements[10], Bias ratio[20], Apply, OK
    Flip bias, # if required adjust bias direction, point toward y = 10
```

```
# pick 6 edges parallel to X-global, Done
Method: By number, Bias: Double
    Flip bias, # pick: lines, # direct the bias toward the center
    Number of elements:[20], Bias ratio:[10], Apply, OK
    # pick 8 edges parallel to Z-global, Done
Method: By number, Bias: None, Number of elements[6], OK
Menu: Mesh, Element Type, # select all regions, Done
    Geometric Order: Quadratic # element C3D20R is assigned, OK
Menu: Mesh, Instance, Yes
```

ⅷ. 求解。
```
Module: Job
Menu: Job, Manager
    Create, Cont, OK, Submit, # when Completed, Results
```

ⅸ. 在单层板坐标系下可视化应力和应变。默认情况下，Abaqus 显示的应力和应变结果均在单层板坐标系下。
```
Module: Visualization
Menu: Plot, Contours, On Deformed Shape
    Toolbox: Field Output, Primary, U, U1 # global coord system
    Toolbox: Field Output, Primary, E, E11 # lamina coord system
    Toolbox: Field Output, Primary, S, S33 # lamina coord system
```

ⅹ. 在全局(global)坐标系中显示应力和应变。该方法可用于在任何用户定义坐标系下显示应力和应变。在本例中，基于这样一个事实，即其中一个单层的坐标方向与全局坐标系方向相同，因此所需的坐标系已存在。
```
Menu: Result, Options
    Tab: Transformation, Transformation Type: User-specified
    # pick: ASSEMBLY_PART-1-1_ORI-2, Apply, # displays CS, OK
    Toolbox: Field Output, Primary, E, E11 # global coord system
    Toolbox: Field Output, Primary, S, S33 # global coord system
Menu: Results, Options
    Tab: Transformation, Transformation Type: Default, OK
```

ⅺ. 在稍低于 0/90 界面层处创建一条绘制 S33 的路径。
```
Menu: Tools, Path, Create
    Name[Path-1], Type: Point List, Cont
```

```
[20,0.0,1.24
20,1.0,1.24
20,2.0,1.24
20,3.0,1.24
20,4.0,1.24
20,5.0,1.24
20,6.0,1.24
20,7.0,1.24
20,8.0,1.24
20,9.0,1.24
20,9.5,1.24
20,9.8,1.24
20,9.9,1.24
20,10.,1.24]
# visualize the path created on the WS, OK
```

ⅻ. 创建 XY 数据集,并对其可视化和(或)保存到一个文件中。

```
Menu: Tools, XY Data, Create
    Source: Path, Cont, Path: Path-1, Model shape: Undeformed
    X Values: True distance, Y Values: Field output: S, S33, OK
    Save As[path1-S33], OK, Plot, # visualize on screen
    # close XY Data pop-up window
```

ⅹⅲ. 将路径上的结果保存在工作目录中的报表文件(.rpt)中。

```
Menu: Report, XY
    # pick: path1-S33, Tab: Setup, File name[S33path1.rpt], OK
```

ⅹⅳ. 在稍高于上述的界面处重新定义路径并可视化(路径定义如下)。

```
[20,0.0,1.26
20,1.0,1.26
20,2.0,1.26
20,3.0,1.26
20,4.0,1.26
20,5.0,1.26
20,6.0,1.26
20,7.0,1.26
20,8.0,1.26
```

```
20,9.0,1.26
20,9.5,1.26
20,9.8,1.26
20,9.9,1.26
20,10.,1.26]
```

然后,将结果保存到文件[S33path2.rpt]中。

沿上述路径的应力值如图5.6所示。注意到,界面上下的应力结果几乎相同。

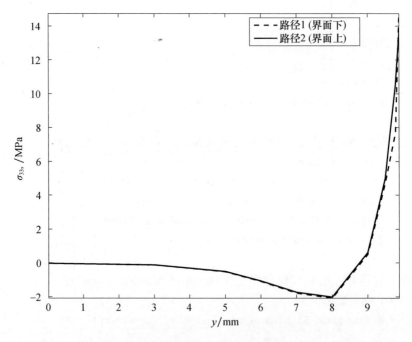

图5.6　90/0界面上、下位置处的应力 σ_{33}

5.2　相互影响系数

在经典层合理论中,假设分析的层合板部位远离层合板的边缘。那么应力合力 N 和 M 施加在层合板的一部分上,由此在每层单层板上产生面内应力 σ_x、σ_y、σ_{xy}。在层合板内部,只有施加相应剪切载荷时才能产生层间应力 σ_{xz}、σ_{yz}。

在单轴加载 N_x 情况下,每一层中由于泊松比效应产生的横向应力必然相互抵消,以达到层合板合力 N_y 为0。此外,偏轴层的面内剪应力也必然与其他层的面内剪应力相互抵消,这样才能使层合板的剪切力 N_{xy} 为0。在自由边附近应力状态会复杂得多,因为面内应力的各组成部分在界面上不会相互抵消。下面重新讨论层合板工程参数的概念。在材料轴上,平面应力柔度方程为

$$\begin{Bmatrix} \epsilon_1 \\ \epsilon_2 \\ \gamma_6 \end{Bmatrix} = \begin{bmatrix} S_{11} & S_{12} & 0 \\ S_{12} & S_{11} & 0 \\ 0 & 0 & S_{66} \end{bmatrix} \begin{Bmatrix} \sigma_1 \\ \sigma_2 \\ \sigma_6 \end{Bmatrix} \qquad (5.8)$$

柔度系数可以用工程参数表示为

$$[S] = \begin{bmatrix} 1/E_1 & -\nu_{12}/E_1 & 0 \\ -\nu_{12}/E_1 & 1/E_2 & 0 \\ 0 & 0 & 1/G_{12} \end{bmatrix} \qquad (5.9)$$

对于偏轴单层板(相对于全局坐标轴,方向任意),有

$$\begin{Bmatrix} \epsilon_x \\ \epsilon_y \\ \gamma_{xy} \end{Bmatrix} = \begin{bmatrix} \overline{S}_{11} & \overline{S}_{12} & \overline{S}_{16} \\ \overline{S}_{12} & \overline{S}_{22} & \overline{S}_{26} \\ \overline{S}_{16} & \overline{S}_{26} & \overline{S}_{66} \end{bmatrix} \begin{Bmatrix} \sigma_x \\ \sigma_y \\ \sigma_{xy} \end{Bmatrix} \qquad (5.10)$$

从这里可以看出,单轴载荷($\sigma_y = \sigma_{xy} = 0$)产生的剪切应变是剪切 – 拉伸耦合的结果。

$$\gamma_{xy} = \overline{S}_{16} \sigma_x \qquad (5.11)$$

式中

$$\begin{aligned} \overline{S}_{16} &= (S_{11} - 2S_{12} - S_{66}) \sin\theta \cos^3\theta \\ &\quad - (2S_{22} - 2S_{12} - S_{66}) \sin^3\theta \cos\theta \end{aligned} \qquad (5.12)$$

现在,对于偏轴的单层板,$[\overline{S}]$ 矩阵的系数可以用工程参数定义为

$$\begin{cases} \overline{S}_{11} = 1/E_x; & \overline{S}_{12} = -\nu_{xy}/E_x = -\nu_{yx}/E_y \\ \overline{S}_{22} = 1/E_y; & \overline{S}_{66} = 1/G_{xy} \end{cases} \qquad (5.13)$$

为完善定义式(5.10)中的$[\overline{S}]$,需要引入两个新的工程参数用来描述剪切 – 拉伸耦合效应,即 $\eta_{xy,x}$ 和 $\eta_{xy,y}$,定义如下:

$$\overline{S}_{16} = \frac{\eta_{xy,x}}{E_x}; \overline{S}_{26} = \frac{\eta_{xy,y}}{E_x} \qquad (5.14)$$

式中:$\eta_{xy,x}$,$\eta_{xy,y}$ 为相互影响系数,它们代表了由拉伸引起的剪切。

通过施加轴向应力并测量由此产生的剪切应变来定义这两个相互影响系数,即

$$\eta_{ij,i} = \frac{\gamma_{ij}}{\epsilon_i} \qquad (5.15)$$

另外一对相互影响系数可以定义为剪切引起的拉伸,即

$$\overline{S}_{16} = \frac{\eta_{x,xy}}{G_{xy}}; \overline{S}_{26} = \frac{\eta_{y,xy}}{G_{xy}} \tag{5.16}$$

它们是通过施加剪应力并测量轴向应变来定义的,即

$$\eta_{i,ij} = \frac{\epsilon_i}{\gamma_{ij}} \tag{5.17}$$

当施加轴向载荷时,偏轴的单层板会产生面内剪切应力,这是因为在一个孤立单层板(图5.7)上发生的自然剪切变形会受到其他单层板的约束。在层合板的整个厚度上,这些应力相互会抵消,但在不平衡的子层合板(如图5.7的顶部单层板)中,这些应力相当于净剪切应力。

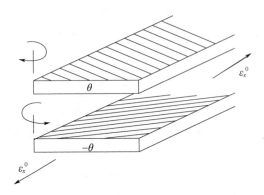

图5.7 相互影响产生的变形[①]

上述剪切力只能通过子层合板底部的层间应力 σ_{zx} 来平衡(图5.8)。那么,沿 x 方向上的剪力求和后产生一个净剪力为

$$F_{xz}(z_k) = \int_0^b \sigma_{zx(z=z_k)} \mathrm{d}y = -\int_{z_k}^{z_N} \sigma_{xy} \mathrm{d}z \tag{5.18}$$

使用经典层合理论(CLT)(文献[1],第6章)计算的面内剪切应力可以用来计算单位长度上的层间剪力 F_{xz}。在面内加载情况下,使用经典层合理论进行计算,每个单层中会得到恒定的剪切应力。当在界面处($z = z_k$ 位置)计算层间剪力时,上述积分简化为

$$F_{xz(z_k)} = -\sum_{i=k}^{N} \sigma_{xy}^i t_i \tag{5.19}$$

根据层间应力最小的原则,层间剪力 F_{xz} 和相互影响系数的值可以定性地用

[①] 转载自 *Mechanics of Fibrous Composites*,C. T. Herakovich,图8.14,版权所有(1998),经 John Wiley & Sons,Inc. 许可。

来选择合适的叠层顺序(LSS)。真实的层间应力值可以通过数值分析计算得到。图 5.10 给出了 $[\pm 45]_S$ 层合板的层间剪切应力近似数值 σ_{xz} 关于距自由边 y 的曲线。

例 5.3　在载荷 $N_x = 175 \text{kN/m}$ 条件下,计算 $[30_2/-30_2]_S$ 平衡对称层合板(图 5.1)的所有界面处的 F_{xz}。材料属性参见例 5.1。单层板厚度 $t_k = 0.127 \text{mm}$。

解　使用与例 5.1 中计算获得 σ_y 相同的过程,计算获得层合板所有单层的面内剪应力 σ_{xy}。

对于在面内载荷条件下的对称平衡层合板,使用式(5.19)。对于一般载荷条件下的普通层合板,使用式(5.18)或对每层单层板内的 σ_{xy} 取平均值,再使用式(5.19)近似得到 F_{xz}。

计算结果见表 5.2 和图 5.9。

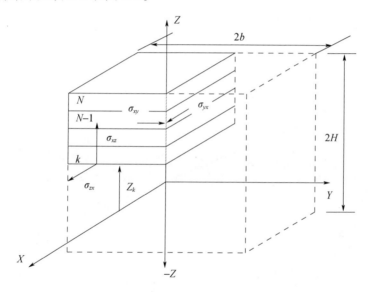

图 5.8　计算由相互影响产生的层间力 F_{xz} 的子层合板自由体受力图

表 5.2　由相互影响产生的层间剪力 F_{xz}

k	位置	σ_{xy}/MPa	t_k/mm	z/mm	$F_{xz}/(\text{kN/m})$
8	TOP	78.6×10^{-3}		0.508	0.000
8	BOT	78.6×10^{-3}	0.127	0.381	-9.982
7	TOP	78.6×10^{-3}		0.381	-9.982
7	BOT	78.6×10^{-3}	0.127	0.254	-19.964
6	TOP	-78.6×10^{-3}		0.254	-19.964
6	BOT	-78.6×10^{-3}	0.127	0.127	-9.982

(续)

k	位置	σ_{xy}/MPa	t_k/mm	z/mm	F_{xz}/(kN/m)
5	TOP	-78.6×10^{-3}		0.127	-9.982
5	BOT	-78.6×10^{-3}	0.127	0.000	0.000
4	TOP	-78.6×10^{-3}		0.000	0.000
4	BOT	-78.6×10^{-3}	0.127	-0.127	9.982
3	TOP	-78.6×10^{-3}		-0.127	9.982
3	BOT	-78.6×10^{-3}	0.127	-0.254	19.964
2	TOP	78.6×10^{-3}		-0.254	19.964
2	BOT	78.6×10^{-3}	0.127	-0.381	9.982
1	TOP	78.6×10^{-3}		-0.381	9.982
1	BOT	78.6×10^{-3}	0.127	-0.508	0.000

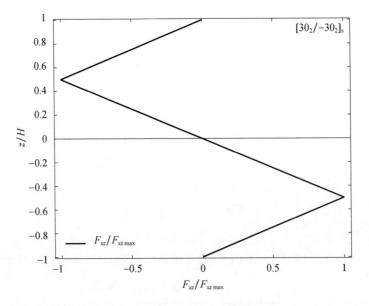

图 5.9 由相互影响产生的层间剪力 F_{xz}

例 5.4 绘制 $[\pm 45]_s$ 层合板中间对称界面上方位置的 σ_{xz} 曲线,材料属性、几何参数和载荷参见例 5.2。

解 由于 LSS 对称,可以对层合板的 $1/2(z > 0)$ 进行简化建模。但不能使用例 5.2 中采用的对称条件,因为材料(例如,30°层)关于 $x-z$ 和 $y-z$ 平面不对称(见图 5.8)。$x = 0$ 处的 $y-z$ 平面不是对称面,而是 $\epsilon_x = 0$ 的平面。此外,模型端部 $x = 0$ 和 $x = L^*$ 处的自由边效应现在很重要,因此必须在 $x = L^*/2$ 处绘制应力结果曲线,以避免加载端部的自由边缘效应。计算结果如图 5.10 所示。

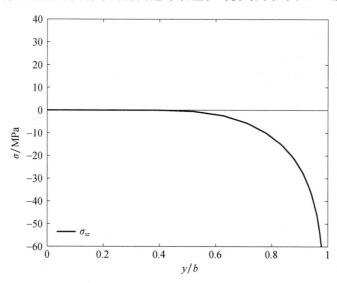

图 5.10　碳纤维/环氧树脂[±45]$_s$铺层层合板在对称界面上方处的层间剪切应力 σ_{xz}(FEA)。

ⅰ. 设置工作目录。

Menu: File, Set Work Directory,[C:\SIMULIA\user\Ex_5.4]
Menu: File, Save As,[C:\SIMULIA\user\Ex_5.4\Ex_5.4.cae]

ⅱ. 创建部件。

Module: Part
Menu: Part, Create
    3D, Deformable, Solid, Extrusion, Approx size[100], Cont
      Menu: Add, Line, Rectangle,[0,0],[80,20], X, Done, Depth[2.5], OK
    # We need to cut the volume in 2 layers
Menu: Tools, Datum, Plane,
    Offset from principal plane, XY Plane,[1.25]
    X, # close the pop-up window
Menu: Tools, Partition, Cell, Use datum plane
    # pick the plane parallel to the XY plane, Create partition, Done
    # close pop-up window

ⅲ. 创建装配体。

Module: Assembly

Menu: Instance, Create, Independent, OK

ⅳ. 定义材料、截面属性并赋予部件。

Module: Property

Menu: Material, Create

  Mechanical, Elasticity, Elastic, Type: Engineering Constants

  [1.39E5 1.45E4 1.45E4 0.21 0.21

  0.38 5.86E3 5.86E3 5.2536E3], OK

Menu: Section, Create, Solid, Homogeneous, Cont, OK

Menu: Assign, Section, # pick: layer 3, Done, Section-1, OK

  # pick: layer 4, Done, Section-1, OK, Done

  # assign material orientation -45 layer (near middle surface)

Menu: Assign, Material Orientation, # pick: layer 3, Done

  Use Default Orientation, Definition: Coordinate system

  Additional Rotation Direction: Axis 3

  Additional Rotation: Angle[-45], OK

  # assign material orientation for +45-degree layer (top surface)

Menu: Assign, Material Orientation, #pick layer 4, Done,

  Use Default Orientation, Definition: Coordinate system

  Additional Rotation Direction: Axis 3

  Additional Rotation: Angle[45], OK

  # to see the orientations do this:

  # on the left menu tree,

  # Models, Model-1, Parts, Part-1, Orientations

  # you should have only 2 SYSTEM: < Global > entries, one per layer

ⅴ. 定义分析步。

Module: Step

Menu: Step, Create

  Procedure type: General, Static/General, Cont, OK

ⅵ. 施加载荷和 BC。

Module: Load

Menu: BC, Manager

  Create, Name[DISP], Step: Step-1, Disp/Rota, Cont

  # pick: surface x = 80, Done, # checkmark: U1[0.8], OK

  # you may need to rotate the model to pick the surfaces

Create, Name[XEND], Step: Initial, Rota/Disp, Cont
   # pick: surface x = 0, Done, # checkmark: U1, OK
Create, Name[ZSYMM], Step: Initial, Symm/Anti/Enca, Cont
   # pick: surface z = 0, Done, # checkmark: U3, OK
   # close BC Manager pop-up window

ⅶ. 划分网格。

Module: Mesh
Menu: Seed, Edges
   # pick: 6 edges parallel to Y-global, Done
   Method: By number, Bias: Double
   # make sure the bias is directed toward the end of the lines
   Number of elements[20], Bias ratio[10], Apply, OK
   # pick 6 edges parallel to X-global, Done
   Method: By number, Bias: Double
   Flip bias, # all the lines toward the center
   Number of elements:[20], Bias ratio:[10], Apply, OK
   # pick 8 edges parallel to Z-global, Done
   Method: By number, Bias: None, Number of elements[6], Apply, OK
Menu: Mesh, Element Type, # select all regions, Done
   Geometric Order: Quadratic, # C3D20R is assigned, OK
Menu: Mesh, Instance, Yes

ⅷ. 求解并可视化结果。

Module: Job
Menu: Job, Manager
   Create, Cont, OK, Submit, # when Completed, Results
Module: Visualization
Menu: Plot, Contours, On Deformed Shape
   # create a path for plotting S13 slightly above the -45/+45 interface
Menu: Tools, Path, Create
   Name[Path-1], Type: Point List, Cont
   [40, 0.0, 1.26
   40, 0.5, 1.26
   40, 1.0, 1.26
   40, 1.5, 1.26
   40, 2.0, 1.26
   40, 2.5, 1.26

```
        40, 3.0, 1.26
        40, 3.5, 1.26
        40, 4.0, 1.26
        40, 4.5, 1.26
        40, 5.0, 1.26
        40, 6.0, 1.26
        40, 7.0, 1.26
        40, 8.0, 1.26
        40, 9.0, 1.26
        40, 10.0, 1.26
        40, 11.0, 1.26
        40, 12.0, 1.26
        40, 13.0, 1.26
        40, 14.0, 1.26
        40, 15.0, 1.26
        40, 15.5, 1.26
        40, 16.0, 1.26
        40, 16.5, 1.26
        40, 17.0, 1.26
        40, 17.5, 1.26
        40, 18.0, 1.26
        40, 18.5, 1.26
        40, 19.0, 1.26
        40, 19.5, 1.26
        40, 20.0, 1.26]
        # visualize the path created on the WS, OK
Menu: Tools, XY Data, Create
        Source: Path, Cont, Path: Path-1, Model shape: Undeformed
        X Values: True distance, Y Values: Field output: S, S13
        Save As[S13plot], OK, Plot, # visualize result on screen
        # close XY Data pop-up window
Menu: Report, XY
        # pick: S13plot, Tab: Setup, File name[S13plot.rpt], OK
        # the file S13plot.rpt must have been saved in the Work Directory
```

$-45/+45$ 界面上下处的 σ_{13} 应力如图 5.11 所示。

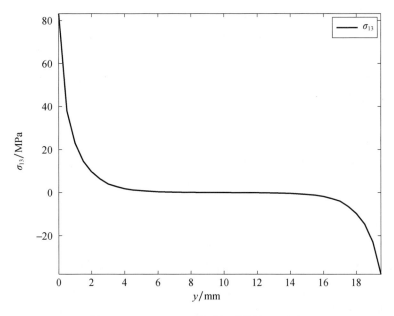

图 5.11　-45/+45 界面上下处的 σ_{13} 应力

习题

习题 5.1 编写一个计算程序,使用 σ_y 和 σ_{xy} 的列表数据(在每个单层板的上下处)计算具有任意层数的层合板厚度内所有界面位置的 F_{yz}、F_{xz} 和 M_z。使用这个程序,在厚度范围内 $-4t < z < 4t$ 绘制 F_{yz}、F_{xz} 和 M_z。层合板的铺层顺序为 $[\pm 45/0/90]_s$,单层厚度 $t = 0.125\text{mm}$,施加的载荷 $N_x = 100\text{kN/m}$。采用的碳纤维/环氧预浸料性能为 $E_1 = 139\text{GPa}$,$E_2 = 14.5\text{GPa}$,$G_{12} = G_{13} = 5.86\text{GPa}$,$G_{23} = 5.25\text{GPa}$,$\nu_{12} = \nu_{13} = 0.21$,$\nu_{23} = 0.38$。

习题 5.2 重新施加载荷 $M_x = 1\text{Nm/m}$,重新求解习题 5.1。

习题 5.3 在 $[0/90/90/0]$ 层合板的中面上方的第一个交界面上,绘制 $x = L/2$ 时的 σ_z/σ_{x0} 和 σ_{yz}/σ_{x0} 与 $y/b(0 < y/b < 1)$ 关系曲线。层合板中单层厚度 $t = 0.512\text{mm}$,施加的载荷为 $\epsilon_x = 0.01$。根据施加的应变计算远场均匀应力 σ_{x0}。使用二次实体单元和偏向自由边加密(偏移 0.1)的网格来建立 1/8 的拉伸试样模型(见例 5.2),模型宽度为 $2b = 25.4\text{mm}$,长度为 $2L = 20\text{mm}$。碳纤维/环氧力学性能 $E_1 = 139\text{GPa}$,$E_2 = 14.5\text{GPa}$,$G_{12} = G_{13} = 5.86\text{GPa}$,$G_{23} = 5.25\text{GPa}$,$\nu_{12} = \nu_{13} = 0.21$,$\nu_{23} = 0.38$。尝试使自由边附近单元的横纵比保持接近 1。

习题 5.4 对于习题 5.3 中所述的层合板和载荷,绘制中面上方、距离自由

边 $0.1t_k$、$x = L/2$ 处的 σ_z/σ_{x0} 和 σ_{yz}/σ_{x0} 与 z/t_k ($0 < z/t_k < 2$) 关系曲线。沿 z - 方向提供 4 种不同的单元划分数目来研究网格细化带来的效果。尝试使自由边附近的单元的长宽比保持接近 1。

习题 5.5 针对 $[\pm 10_2]_S$ 层合板的中面以上的所有界面,按照习题 5.3,绘制 σ_{xz}/σ_{x0} 曲线。

习题 5.6 针对 $[\pm 10_2]_S$ 层合板,按照习题 5.4,绘制 σ_{xz}/σ_{x0} 曲线。

习题 5.7 使用实体单元和偏移的网格划分建立 1/8 拉伸试样模型(见例 5.2),其宽度 $2b = 24$mm,长度 $2L = 20$mm。层合板铺层顺序为 $[\pm 45/0/90]_S$,单层厚度为 $t = 0.125$mm,载荷为 $N_x = 175$kN/m。采用的碳纤维/环氧力学性能 $E_1 = 139$GPa,$E_2 = 14.5$GPa,$G_{12} = G_{13} = 5.86$GPa,$G_{23} = 5.25$GPa,$\nu_{12} = \nu_{13} = 0.21$,$\nu_{23} = 0.38$。在每单层的中面,绘制从试样边缘到中心线的 3 个层间应力分量曲线。把同一应力下的 4 个图集合成一个图。

习题 5.8 对于角度为 θ 的单向单层板,在同一个图中绘制 E_x/E_2、G_{xy}/G_{12}、$10V_{xy}$、$-\eta_{xy,x}$ 和 $-\eta_{x,xy}$ 与 θ 的关系曲线,θ 的范围为 $-\pi/2 < \theta < \pi/2$。材料为 S - 玻璃纤维/环氧树脂,性能参见文献[1,表1.3~表1.4]。

习题 5.9 利用习题 5.8 的图,并考虑一个 $[\theta_1/\theta_2]_S$ 层合板,在(a)泊松比不匹配和(b)剪切不匹配两种情况下,最差的 θ_1,θ_2 值组合是什么?

习题 5.10 在一个图中,在 $-\pi/2 < \theta < \pi/2$(文献[1],表1.3~表1.4)范围内,比较 E - 玻璃纤维/环氧树脂、Kevlar49/环氧树脂和 T800/3900 - 2 的 $-\eta_{xy,x}$。

习题 5.11 绘制 $[\pm 45]_S$ 层合板上表面处的 3 个变形 u_x、u_y、u_z(独立)云图。使用习题 5.7 中的试件尺寸、载荷和材料属性。解释发现的现象。

习题 5.12 针对 $[0/90]_S$ 层合板重复求解习题 5.11。解释发现的现象。

习题 5.13 使用实体单元和偏移网格对总宽度 $2b = 12$mm 和长度 $2L = 24$mm 的拉伸试样(图 5.1)建立 1/4 模型。针对 50% 纤维体积含量的 SCS-6/铝材料,当采用 $[\pm 15/\pm 45]_S$ 和 $[\pm(15/45)]_S$ 两种铺层顺序时,在同一曲线中比较 σ_z 与 z/H 的关系曲线。使用微观力学方法,式(6.8),来预测单向复合材料属性。单层板厚度 $t_k = 0.25$mm。载荷 $\epsilon_x = 0.01$。

参数	Al-2014-T6(文献[31],附录 B)	SCS-6(文献[1],表2.1 和表2.2)
E/GPa	75.0	427.0
G/GPa	27.0	177.9

习题 5.14 使用与习题 5.13 相似的有限元模型进行计算,对于一个受到载荷为 $\epsilon_x = 0.01$ 的 $[\pm\theta]_S$SCS-6/Al 层合板,绘制 $\sigma_{xz}/\sigma_{zx\max}$ 与 θ($0 < \theta < \pi/2$)关系图。

习题 5.15 使用习题 5.13 的有限元模型进行计算,绘制 $[\pm 15/\pm 45]_S$ 层合板中面处的 σ_z 和 $y/b(0<y<0.95b)$ 关系图。注意在接近 $y=b$ 时 $\sigma_z\to\infty$,因此 $y=b$ 处的有限元分析结果取决于网格密度。使用不同的网格密度分别计算并结果列表,研究 $y=0.95b$ 处的网格依赖性。

习题 5.16 采用类似于习题 5.13 的有限元分析模型进行计算,对于一个受到轴向 1% 应变载荷($\epsilon_x=0.01$)的 $[\pm\theta]_S$ SCS-6/A1 层合板,绘制 σ_x、σ_{xy} 和 σ_{xz} 与 $y/b(0<y<b)$ 的关系曲线。

习题 5.17 图 5.1 所示的 $[0/90]_S$ 层合板性能为 $E_1=139\text{GPa}$,$E_2=14.5\text{GPa}$,$G_{12}=G_{13}=5.86\text{GPa}$,$G_{23}=5.25\text{GPa}$,$\nu_{12}=\nu_{13}=0.21$,$\nu_{23}=0.38$。单层板预浸料的强度性能为 $F_{1t}=1550\text{MPa}$,$F_{1c}=1090\text{MPa}$,$F_{2t}=F_{2c}=59\text{MPa}$ 和 $F_6=75\text{MPa}$。取 $2b=20\text{mm}$,试样长度 $2L=20\text{mm}$,每层厚度 $t_k=1.25\text{mm}$。通过施加均匀位移,对试件加载均匀应变 $\epsilon_x=0.01$。使用对称性仅对 $x>0,y>0,y>0$ 象限部分进行建模。每层均使用正交各向异性实体单元,且每层厚度上至少有两个二次单元。对实体单元使用 UMAT 子程序,计算 3D Tsai-Wu 失效指数 I_F。获得各层中 I_F 云图(不要对结果进行平均)。

第6章 细观力学计算

第1章中,复合材料的宏观弹性性能以弹性模量E、剪切模量G、泊松比ν等形式给出。对于复合材料等非均质材料,需要大量的材料性能进行表征,而这些性能的实验测定是一个繁琐且昂贵的过程。此外,这些性能参数的值随增强相的体积分数等因素的变化而变化。另外,还有一种方法,至少是对实验的补充。该方法根据组分(基体和增强项)的弹性性能,利用材料均匀化技术来预测复合材料的弹性性能。由于均匀化模型基本都是建立在细观结构模型的基础上,所以均匀化模型也被称为细观力学模型,用于计算获得复合材料性能近似值的技术称为细观力学方法或技术[1]。细观力学模型可分为经验模型、半经验模型、解析模型和数值模型。精确的半经验模型在文献[1]中有详细的描述。

本书只涉及不需要经验修正系数的严格的解析或数值模型,所以不需要进行实验。本书的大部分内容围绕3D问题进行分析,所以重点放在仅利用一个单独的模型来估计所有弹性性能集合的细观力学模型上,而不是利用基于不同假设的多个不相交的模型来组合得到所需的性能集合。许多均匀化的分析技术均基于等效本征应变法[32,33],该方法考虑了无限大弹性介质中的单椭球夹杂问题。文献[34]中利用Eshelby解来发展一种近似考虑夹杂物之间相互作用的方法。一种更常用的均匀化技术是自洽方法[35],它考虑了无限大介质中夹杂物的随机分布,而假设该无限大介质的材料属性与所寻求的材料属性相同。因此,采用迭代法来获得全部参数。具有周期性细观结构的复合材料均匀化可以使用多种技术,包括Eshelby夹杂问题的推广[32,33]、傅里叶级数技术(见6.1.3节和文献[36,37])和变分原理。周期本征应变法的进一步发展确定了线性黏弹性复合材料的全部松弛弹性模量(见7.6节和文献[38,39])。在文献[40]中针对一个特殊情况,对周期性介质使用胞元法,考虑含有方形夹杂的单胞进行简化计算。

以上提到的分析方法给出了细观力学问题精确解的近似计算。这些近似计算必然介于精确解的下限和上限之间。文献[41]中为了计算得到宏观各向同性非均质材料的均匀化弹性参数上下限,发展了一些变分原理方法。这些上下限仅取决于组分材料的体积分数和物理参数。

为了研究具有周期性细观结构复合材料的非线性材料行为,本书主要采用基于有限元法的数值计算方法。文献[42]通过观察指定载荷历程中细观结构的行为,对金属基复合材料的非线性有限元分析进行研究。文献[43]通过有限元法导出了复合材料整体瞬时弹塑性性能的上下限。

6.1 均匀化的分析方法

现有的分析模型在复杂性和准确性方面差异都很大。较为简单的解析模型得出复合材料的刚度 C 和柔度 S 张量的公式(文献[40],式(2.9)和式(2.12)),如下

$$C = \sum V_i C^i A^i \quad ; \quad \sum V_i A^i = I$$
$$S = \sum V_i S^i B^i \quad ; \quad \sum V_i B^i = I \tag{6.1}$$

式中:V_i,C^i,S^i 分别为复合材料中第 i 相的体积分数、刚度和柔度张量(指标符号缩写形式)①;I 为 6×6 单位矩阵;A^i,B^i 分别为第 i 相的应变和应力集中张量(指标符号缩写形式)[40]。对于纤维增强复合材料,i = f,m 分别代表纤维相和基体相。

6.1.1 Reuss 模型

Reuss 模型(也称混合规则)假定纤维、基体和整体复合材料中的应变张量②相同,即 $\varepsilon = \varepsilon^f = \varepsilon^m$。因此,应变集中张量均等于 6×6 单位矩阵,即 $A^i = I$。基于此,推导和计算混合规则(ROM)的 E_1 和 ν_{12} 公式。

6.1.2 Voigt 模型

Voigt 模型(也称混合逆规则)假定纤维、基体和整体复合材料中的应力张量是相同的,即 $\sigma = \sigma^f = \sigma^m$。因此,应力集中张量均等于 6×6 单位矩阵,即 $B^i = I$。基于此,导出和计算混合逆规则(IROM)的 E_2 和 G_{12} 公式。更真实的应力集中张量在文献[8,附录 B]中给出。

6.1.3 周期性细观结构模型

如果复合材料具有周期性细观结构,或者近似具有这种细观结构(见6.1.4节),则可以使用傅里叶级数估算复合材料刚度张量的所有分量。文献[37]中给出了由周期性方阵排列的各向同性圆柱形纤维(图6.1)增强的复合材料性能显式计算公式。纤维方向与 x_1 轴一致,且分布间距相等($2a_2 = 2a_3$)。如果纤维

---

① 利用缩写形式的优势,使用 6×6 矩阵表示次对称的四阶张量。
② 张量采用黑体字表示,或使用带下标的分量表示。

在横截面上随机分布,则该复合材料具有横观各向同性特性,6.1.4 节中会进行详细阐释。文献[39]中考虑了横观各向同性的纤维增强复合材料情况,并在文献[12]中应用了相应的推导公式。

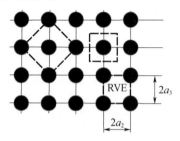

图 6.1 具有周期性方形纤维阵列复合材料的 3 种 RVE 形式

由于细观结构具有方形对称特点,刚度张量只有 6 个独立系数,如下:

$$C_{11}^* = \lambda_m + 2\mu_m - \frac{V_f}{D}\left[\frac{S_3^2}{\mu_m^2} - \frac{2S_6 S_3}{\mu_m^2 g} - \frac{aS_3}{\mu_m c} + \frac{S_6^2 - S_7^2}{\mu_m^2 g^2} + \frac{aS_6 + bS_7}{\mu_m gc} + \frac{a^2 - b^2}{4c^2}\right]$$

$$C_{12}^* = \lambda_m + \frac{V_f}{D}b\left[\frac{S_3}{2c\mu_m} - \frac{S_6 - S_7}{2c\mu_m g} - \frac{a+b}{4c^2}\right]$$

$$C_{23}^* = \lambda_m + \frac{V_f}{D}\left[\frac{aS_7}{2\mu_m gc} - \frac{ba + b^2}{4c^2}\right]$$

$$C_{22}^* = \lambda_m + 2\mu_m - \frac{V_f}{D}\left[-\frac{aS_3}{2\mu_m c} + \frac{aS_6}{2\mu_m gc} + \frac{a^2 - b^2}{4c^2}\right]$$

$$C_{44}^* = \mu_m - V_f\left[-\frac{2S_3}{\mu_m} + (\mu_m - \mu_f)^{-1} + \frac{4S_7}{\mu_m(2-2\nu_m)}\right]^{-1}$$

$$C_{66}^* = \mu_m - V_f\left[-\frac{S_3}{\mu_m} + (\mu_m - \mu_f)^{-1}\right]^{-1}$$

(6.2)

其中

$$D = \frac{aS_3^2}{2\mu_m^2 c} - \frac{aS_6 S_3}{\mu_m^2 g c} + \frac{a(S_6^2 - S_7^2)}{2\mu_m^2 g^2 c} + \frac{S_3(b^2 - a^2)}{2\mu_m c^2} + \frac{S_6(a^2 - b^2)}{2\mu_m gc^2} + \frac{(a^3 - 2b^3 - 3ab^2)}{8c^3}$$

(6.3)

和

$$a = \mu_f - \mu_m - 2\mu_f \nu_m + 2\mu_m \nu_f$$
$$b = -\mu_m \nu_m + \mu_f \nu_f + 2\mu_m \nu_m \nu_f - 2\mu_f \nu_m \nu_f$$
$$c = (\mu_m - \mu_f)(\mu_f - \mu_m + \mu_f \nu_f - \mu_m \nu_m + 2\mu_m \nu_f - 2\mu_f \nu_m + 2\mu_m \nu_m \nu_f - 2\mu_f \nu_m \nu_f)$$
$$g = (2 - 2\nu_m)$$

(6.4)

式中：下标$(\ )_m$，$(\ )_f$分别表示基体和纤维。假设纤维和基体都是各向同性的(1.12.5节)，利用式(1.75)计算得到由弹性模量E、泊松比ν和剪切模量$G=\mu$组成的这两种材料的拉梅常数。

对于由长圆柱形纤维、周期性排列成方阵形式(图6.1)增强的复合材料，纤维方向与x_1轴一致，$a_2 = a_3$，常数S_3，S_6，S_7计算如下[37]：

$$\begin{cases} S_3 = 0.49247 - 0.47603 V_f - 0.02748 V_f^2 \\ S_6 = 0.36844 - 0.14944 V_f - 0.27152 V_f^2 \\ S_7 = 0.12346 - 0.32035 V_f - 0.23517 V_f^2 \end{cases} \quad (6.5)$$

由于周期性细观结构呈方阵形式排列，所以产生的张量$\boldsymbol{C}^*$具有方形对称性。因此，张量$\boldsymbol{C}^*$需要6个常量进行描述。然而，大多数复合材料的纤维均为随机分布(见图1.12)，产生的是横观各向同性刚度张量。横观各向同性材料的一般化形式见6.1.4节。

6.1.4 横观各向同性平均化

为了得到横观各向同性刚度张量(1.12.4节)，采用以下平均化步骤对方形对称刚度张量在平均意义上进行等效。张量$\boldsymbol{C}^*$关于x_1轴旋转θ角度时，有

$$\boldsymbol{B}(\theta) = \overline{\boldsymbol{T}}^T(\theta) \boldsymbol{C}^* \overline{\boldsymbol{T}}(\theta) \quad (6.6)$$

式中：$\overline{\boldsymbol{T}}(\theta)$是坐标变换矩阵(见式(1.50))。然后，平均化得到等效横观各向同性张量为

$$\overline{\boldsymbol{B}} = \frac{1}{\pi} \int_0^\pi \boldsymbol{B}(\theta) \mathrm{d}\theta \quad (6.7)$$

然后，利用工程常数与$\overline{\boldsymbol{B}}$张量分量之间的关系，得到由$\boldsymbol{C}^*$张量的分量(式(6.2)~式(6.5))组成的下列表达式：

$$\begin{aligned} E_1 &= C_{11}^* - \frac{2 C_{11}^{*\,2}}{C_{22}^* + C_{23}^*} \\ E_2 &= \frac{(2 C_{11}^* C_{22}^* + 2 C_{11}^* C_{23}^* - 4 C_{12}^{*\,2})(C_{22}^* - C_{23}^* + 2 C_{44}^*)}{3 C_{11}^* C_{22}^* + C_{11}^* C_{23}^* + 2 C_{11}^* C_{44}^* - 4 C_{12}^{*\,2}} \\ G_{12} &= G_{13} = C_{66}^* \\ \nu_{12} &= \nu_{13} = \frac{C_{12}^*}{C_{22}^* + C_{23}^*} \\ \nu_{23} &= \frac{C_{11}^* C_{22}^* + 3 C_{11}^* C_{23}^* - 2 C_{11}^* C_{44}^* - 4 C_{12}^{*\,2}}{3 C_{11}^* C_{22}^* + C_{11}^* C_{23}^* + 2 C_{11}^* C_{44}^* - 4 C_{12}^{*\,2}} \end{aligned} \quad (6.8)$$

请注意，横向剪切模量G_{23}可以由其他工程常数写为

$$G_{23} = \frac{C_{22}^*}{4} - \frac{C_{23}^*}{4} + \frac{C_{44}^*}{4} = \frac{E_2}{2(1+\nu_{23})}$$

或者直接由 μ_m, μ_f 表达为

$$\begin{aligned}G_{23} = \mu_m - \frac{f}{4D}\Bigg[&\left(-\frac{aS_3}{2\mu_m c} + \frac{a(S_7+S_6)}{2\mu_m gc} - \frac{ba+2b^2-a^2}{4c^2} \right) \\ &+ 2\left(-\frac{2S_3}{\mu_m} + (\mu_m-\mu_f)^{-1} + \frac{4S_7}{\mu_m(2-2\nu_m)} \right)^{-1} \Bigg]\end{aligned} \quad (6.9)$$

式中:D 由式(6.3)给出;a,b,c,g 由式(6.4)给出;S_3,S_6,S_7 可由式(6.5)计算。这些方程可使用网站[5]中的 PMMIE.m 和 PMMIE.xls 程序实现计算。对于横观各向同性情况,文献[12]中应用了上述计算公式。

例 6.1 计算平行圆柱形纤维增强复合材料的弹性性能,其横截面上纤维随机分布。组分材料性能为 $E_f = 241\text{GPa}, \nu_f = 0.2, E_m = 3.12\text{GPa}, \nu_m = 0.38$,纤维体积含量 $V_f = 0.4$。

解 文献[12]中针对横观各向同性纤维情况,实现 PMM 方程编程计算。利用文献[12]中的程序,本例的计算结果如表 6.1 所列。

表 6.1 单层板的弹性性能($V_f = 0.4$)

弹性模量	泊松比	剪切模量
$E_1 = 98306\text{MPa}$	$\nu_{12} = \nu_{13} = 0.298$	$G_{12} = G_{13} = 2594\text{MPa}$
$E_2 = E_3 = 6552\text{MPa}$	$\nu_{23} = 0.6$	

6.2 均匀化的数值计算方法

本节考虑的复合材料具有无限长的圆柱形纤维,嵌在弹性基体中,如图 6.2 所示。与纤维轴正交的平面作为复合材料的横截面,如图 6.3 所示,横截面清楚地显示了复合材料具有周期性的细观结构。由于周期性特点,图 6.4 所示的三维代表性体积单元(RVE)可用于有限元分析中。

一般来说,平行纤维增强的复合材料在中尺度(单层板层面)上表现出正交各向异性的材料性能(见1.12.3节)。在特殊情况下,像图 6.2 和图 6.3 所示的纤维呈六边形阵列形式分布,其性能变为横观各向同性(见1.12.4节)。在大多数商业制备的复合材料中,不可能如此精确地控制纤维的位置,而且大多数情况下细观结构中纤维是随机分布的,如图 1.12 所示。而纤维随机分布的细观结构在中尺度上也会产生横观各向同性性能。因此,对于具有纤维随机分布的细观结构复合材料,仍然可以使用图 6.1 所示的虚拟的周期性细观结构进

行分析，然后对 6.1.4 节中介绍的刚度张量 C 进行平均化，从而得到横观各向同性复合材料的刚度张量。另外一个更简单的方法是假设随机分布的细观结构近似等价于图 6.3 所示的六边形分布的细观结构。对这种形式的细观结构分析可以直接得到由式(1.70)表示的横观各向同性刚度张量。为了方便阅读，重写如下：

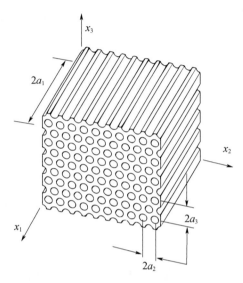

图 6.2　纤维正六边形阵列的复合材料

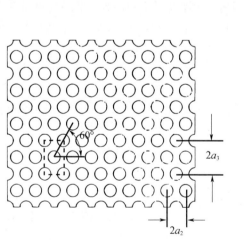

图 6.3　复合材料的横截面

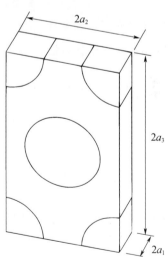

图 6.4　代表性体积单元(RVE)

$$\begin{Bmatrix} \bar{\sigma}_1 \\ \bar{\sigma}_2 \\ \bar{\sigma}_3 \\ \bar{\sigma}_4 \\ \bar{\sigma}_5 \\ \bar{\sigma}_6 \end{Bmatrix} = \begin{bmatrix} C_{11} & C_{12} & C_{13} & 0 & 0 & 0 \\ C_{12} & C_{22} & C_{23} & 0 & 0 & 0 \\ C_{13} & C_{23} & C_{33} & 0 & 0 & 0 \\ 0 & 0 & 0 & \frac{1}{2}(C_{22}-C_{23}) & 0 & 0 \\ 0 & 0 & 0 & 0 & C_{66} & 0 \\ 0 & 0 & 0 & 0 & 0 & C_{66} \end{bmatrix} \begin{Bmatrix} \bar{\epsilon}_1 \\ \bar{\epsilon}_2 \\ \bar{\epsilon}_3 \\ \bar{\gamma}_4 \\ \bar{\gamma}_5 \\ \bar{\gamma}_6 \end{Bmatrix} \quad (6.10)$$

式中：1-轴为纤维方向，而符号上的横线表示在 RVE 体积上计算的平均值。一旦横观各向同性张量 C 的分量已知，则可通过式(6.11)计算均匀化后材料的 5 个弹性参数，即纵向和横向弹性模量 E_1 和 E_2、纵向和横向泊松 ν_{12} 和 ν_{23}、剪切模量 G_{12}，有

$$\begin{cases} E_1 = C_{11} - 2C_{12}^2/(C_{22}+C_{23}) \\ \nu_{12} = C_{12}/(C_{22}+C_{23}) \\ E_2 = [C_{11}(C_{22}+C_{23}) - 2C_{12}^2](C_{22}-C_{23})/(C_{11}C_{22}-C_{12}^2) \\ \nu_{23} = [C_{11}C_{23} - C_{12}^2] \\ G_{12} = C_{66} \end{cases} \quad (6.11)$$

通过经典关系式(1.74)或直接按照下式可得到横向剪切模量 G_{23}：

$$G_{23} = C_{44} = \frac{1}{2}(C_{22}-C_{23}) = \frac{E_2}{2(1+\nu_{23})} \quad (6.12)$$

为计算得到复合材料整体刚度矩阵 C，可以对 RVE 施加平均应变 $\bar{\epsilon}_\beta$[44]。通过设置如下的位移分量形式边界条件，对 RVE 施加应变 ε_{ij}^0 的 6 个分量。

$$u_i(a_1,x_2,x_3) - u_i(-a_1,x_2,x_3) = 2a_1\varepsilon_{i1}^0 \quad \begin{matrix} -a_2 \leqslant x_2 \leqslant a_2 \\ -a_3 \leqslant x_3 \leqslant a_3 \end{matrix} \quad (6.13)$$

$$u_i(x_1,a_2,x_3) - u_i(x_1,-a_2,x_3) = 2a_2\varepsilon_{i2}^0 \quad \begin{matrix} -a_1 \leqslant x_1 \leqslant a_1 \\ -a_3 \leqslant x_3 \leqslant a_3 \end{matrix} \quad (6.14)$$

$$u_i(x_1,x_2,a_3) - u_i(x_1,x_2,-a_3) = 2a_3\varepsilon_{i3}^0 \quad \begin{matrix} -a_1 \leqslant x_1 \leqslant a_1 \\ -a_2 \leqslant x_2 \leqslant a_2 \end{matrix} \quad (6.15)$$

请注意，式(6.13)~式(6.15)中使用了式(1.5)中定义的应变张量分量。另外，上标( )^0 表示施加的应变载荷，而符号上方的横线表示体积上的平均值。此外，$2a_j\varepsilon_{ij}^0$ 是在 $2a_j$ 长度上强制施加应变 ε_{ij}^0 所必需的位移量（见图6.4）。

利用式(6.13)~式(6.15)在边界上施加应变 ε_{ij}^0 后，RVE 内部会产生复杂的应变状态。然而，RVE 中应变的体积平均值等于施加的应变，即

$$\bar{\varepsilon}_{ij} = \frac{1}{V}\int_V \varepsilon_{ij}\mathrm{d}V = \varepsilon_{ij}^0 \qquad (6.16)$$

对于均匀化后的复合材料,平均应力与应变的关系为

$$\bar{\sigma}_\alpha = C_{\alpha\beta}\bar{\epsilon}_\beta \qquad (6.17)$$

式中:$i,j = 1,\cdots,3$ 和 $\beta = 1,\cdots,6$ 之间的关系由式(1.9)中的指标缩写的定义给出。由此,通过求解在边界条件(式(6.13)~式(6.15))下的 6 个 RVE 弹性模型,可以确定张量 C 的所有分量。对于这 6 个模型中的每一个,每次只需施加一个非 0 的 ϵ_β^0 应变分量。

通过施加单位应变,一旦由边界条件(式(6.13)~式(6.15))定义的模型得到求解,就可以进一步计算出应力场 σ_α,其平均值即为所需的弹性矩阵分量。施加一次单位应变可以计算弹性矩阵的一列分量,即

$$C_{\alpha\beta} = \bar{\sigma}_\alpha = \frac{1}{V}\int_V \sigma_\alpha(x_1,x_2,x_3)\mathrm{d}V \quad \epsilon_\beta^0 = 1 \qquad (6.18)$$

式中:$\alpha,\beta = 1,\cdots,6$(见 1.5 节)。有限元方法中,每个单元中采用 Gauss-Legendre 积分法对式(6.18)进行积分计算。商业计算软件,如 Abaqus,能够计算每个单元的平均应力和体积。因此,式(6.18)的积分计算是一件比较容易的事情。更多详情见例题 6.2。

通过为式(6.10)中的每一列设置不同的问题(模型),就得到张量 C 中的系数,如下所示。

(1) C 的第 1 列。为了确定 $C_{i1}\quad i = 1,2,3$,可以将如下的应变载荷沿纤维方向(x_1 方向)施加在 RVE 上

$$\epsilon_1^0 = 1 \quad \epsilon_2^0 = \epsilon_3^0 = \gamma_4^0 = \gamma_5^0 = \gamma_6^0 = 0 \qquad (6.19)$$

因此,对于如图 6.4 所示的 RVE,位移边界条件(式(6.13)~式(6.15))变为

$$u_1(+a_1,x_2,x_3) - u_1(-a_1,x_2,x_3) = 2a_1$$
$$u_2(+a_1,x_2,x_3) - u_2(-a_1,x_2,x_3) = 0 \quad -a_2 \leqslant x_2 \leqslant a_2$$
$$u_3(+a_1,x_2,x_3) - u_3(-a_1,x_2,x_3) = 0 \quad -a_3 \leqslant x_3 \leqslant a_3$$
$$u_i(x_1,+a_2,x_3) - u_i(x_1,-a_2,x_3) = 0 \quad \begin{matrix}-a_1 \leqslant x_1 \leqslant a_1 \\ -a_3 \leqslant x_3 \leqslant a_3\end{matrix}$$
$$u_i(x_1,x_2,+a_3) - u_i(x_1,x_2,-a_3) = 0 \quad \begin{matrix}-a_1 \leqslant x_1 \leqslant a_1 \\ -a_2 \leqslant x_2 \leqslant a_2\end{matrix} \qquad (6.20)$$

条件式(6.20)是对 RVE 两个对立面之间相对位移的约束。鉴于 RVE 的对称性和约束条件式(6.20)的对称性,仅需对 RVE 建立 1/8 有限元模型。假设对右上角部分进行建模(图 6.5),则可以使用如下等效的外部边界条件,即关于位移和应力分量的边界条件,如下:

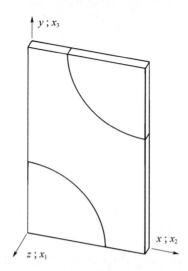

图 6.5 1/8 的 RVE 模型

注意,模型中纤维沿着 Z 轴方向,Z 轴对应于方程中的 x_1 方向。

$$\begin{aligned}
&u_1(a_1,x_2,x_3) = a_1 \\
&u_1(0,x_2,x_3) = 0 \\
&\sigma_{12}(a_1,x_2,x_3) = 0 \quad 0 \leqslant x_2 \leqslant a_2 \\
&\sigma_{12}(0,x_2,x_3) = 0 \quad 0 \leqslant x_3 \leqslant a_3 \\
&\sigma_{13}(a_1,x_2,x_3) = 0 \\
&\sigma_{13}(0,x_2,x_3) = 0 \\
&u_2(x_1,a_2,x_3) = 0 \\
&u_2(x_1,0,x_3) = 0 \\
&\sigma_{21}(x_1,a_2,x_3) = 0 \quad 0 \leqslant x_1 \leqslant a_1 \\
&\sigma_{21}(x_1,0,x_3) = 0 \quad 0 \leqslant x_3 \leqslant a_3 \\
&\sigma_{23}(x_1,a_2,x_3) = 0 \\
&\sigma_{23}(x_1,0,x_3) = 0 \\
&u_3(x_1,x_2,a_3) = 0 \\
&u_3(x_1,x_2,0) = 0 \\
&\sigma_{31}(x_1,x_2,a_3) = 0 \quad 0 \leqslant x_1 \leqslant a_1 \\
&\sigma_{21}(x_1,x_2,0) = 0 \quad 0 \leqslant x_2 \leqslant a_2 \\
&\sigma_{23}(x_1,x_2,a_3) = 0 \\
&\sigma_{23}(x_1,x_2,0) = 0
\end{aligned} \quad (6.21)$$

这些边界条件很容易施加。对称边界条件施加在 $x_1=0, x_2=0, x_3=0$ 平面上。然后,在 $x_1=a_1$ 平面上施加均匀位移。在基于位移的算法中,应力边界条件不需要施加。式(6.21)中的位移分量代表了沿 x_1 方向施加的非 0 应变,沿其他两个方向的应变为 0。式(6.21)中列出的应力边界条件反映了在所使用的坐标系中,复合材料在宏观上是正交异性的,并且其组分材料也是正交各向异性的。因此,拉伸应变与剪切应变之间不存在耦合,式(6.10)中第 4~第 6 列的对角线上方系数为 0 也证实了这一点。

利用式(6.18)求出式(6.10)第 1 列中的系数,即

$$C_{\alpha 1} = \overline{\sigma}_\alpha = \frac{1}{V}\int_V \sigma_{\alpha 1}(x_1, x_2, x_3)\,\mathrm{d}V \tag{6.22}$$

(2) C 的第 2 列。$C_{\alpha 2}$,$\alpha = 1, 2, 3$ 可以通过设置如下条件确定:

$$\epsilon_2^0 = 1 \quad \epsilon_1^0 = \epsilon_3^0 = \gamma_4^0 = \gamma_5^0 = \gamma_6^0 = 0 \tag{6.23}$$

由此,可以使用如下的位移边界条件

$$\begin{cases} u_1(a_1, x_2, x_3) = 0 \\ u_1(0, x_2, x_3) = 0 \\ u_2(x_1, a_2, x_3) = a_2 \\ u_2(x_1, 0, x_3) = 0 \\ u_3(x_1, x_2, a_3) = 0 \\ u_3(x_1, x_2, 0) = 0 \end{cases} \tag{6.24}$$

应力边界条件没有列出,因为它们由基于位移的有限元方法自动施加。利用式(6.18),C 第 2 列系数计算如下:

$$C_{\alpha 2} = \overline{\sigma}_\alpha = \frac{1}{V}\int_V \sigma_{\alpha 2}(x_1, x_2, x_3)\,\mathrm{d}V \tag{6.25}$$

(3) C 的第 3 列。由于材料为横观各向同性,矩阵 C 的第 3 列分量可以由第 1 和第 2 列数据确定,不需要进一步计算。但是,如果确实需要的话,可以通过应用以下应变条件得到 $C_{\alpha 3}$,$\alpha = 1, 2, 3$ 各分量。

$$\epsilon_3^0 = 1 \quad \epsilon_1^0 = \epsilon_2^0 = \gamma_4^0 = \gamma_5^0 = \gamma_6^0 = 0 \tag{6.26}$$

因此,可以使用以下位移边界条件:

$$\begin{cases} u_1(a_1, x_2, x_3) = 0 \\ u_1(0, x_2, x_3) = 0 \\ u_2(x_1, a_2, x_3) = 0 \\ u_2(x_1, 0, x_3) = 0 \\ u_3(x_1, x_2, a_3) = a_3 \\ u_3(x_1, x_2, 0) = 0 \end{cases} \tag{6.27}$$

C 的第 3 列系数通过式(6.18)对应力场平均化得到。

例 6.2 计算单向复合材料的 E_1, E_2, ν_{12} 和 ν_{23}，该单向复合材料由各向同性纤维和各向同性基体组成，其中 $E_f = 241\text{GPa}$, $\nu_f = 0.2$, $E_m = 3.12\text{GPa}$, $\nu_m = 0.38$, 纤维体积分数 $V_f = 0.4$。其纤维直径为 $d_f = 7\mu\text{m}$，如图 6.3 所示，呈六边形阵列分布。

解 如图 6.4 所示，选择 RVE 的尺寸 a_2 和 a_3，以获得六边形阵列的细观结构 $V_f = 0.4$。RVE 纤维体积和总体积为

$$\nu_f = 4a_1\pi\left(\frac{d_f}{2}\right)^2; \quad \nu_t = 2a_1 2a_2 2a_3$$

纤维体积含量即为两者的比值，即

$$V_f = \pi\frac{(d_f/2)^2}{2a_2 a_3} = 0.4$$

此外，由六边形几何关系，a_2 和 a_3 关系为

$$a_3 = a_2 \tan 60°$$

以上两个关系式可得到 a_2 和 a_3，而 a_1 尺寸可以任意选择。该例子中，RVE 尺寸为

$$a_1 = a_2/4 = 1.3175\mu\text{m}; a_2 = 5.270\mu\text{m}; a_3 = 9.128\mu\text{m}$$

依次单个施加单位应变 ϵ_1^0, ϵ_2^0 和 ϵ_3^0，利用式(6.18)计算得到 C_{11}, C_{12}, C_{13}, C_{22}, C_{23} 和 C_{33}。在没有施加剪切变形情况下，其平均变形关于 3 个坐标平面对称。图 6.5 中的 RVE 也关于 3 个坐标平面几何对称。因此，可以对 1/8 的 RVE 进行建模，如图 6.5 所示。下面给出了在 Abaqus 中对 1/8 的 RVE 进行建模的过程。

ⅰ. 创建工作目录。
Menu: File, Set Work Directory, [C:\SIMULIA\user\Ex_6.2]
Menu: File, Save As, [C:\SIMULIA\user\Ex_6.2\Ex_6.2.cae]

ⅱ. 创建部件。
\# Values used below, replace the variables accordingly.
\# values in microns
\# a_1 = 1.3175 # half RVE along x_1 (extrusion direction)
\# a_2 = 5.270 # half RVE along x_2
\# a_3 = 9.138 # half RVE along x_3
\# r_f = 3.5 # fiber radius
Module: Part
Menu: Part, Create

3D,Deformable,Solid,Extrusion,Cont
Menu: Add,Line,Rectangle,[0,0],[a_2,a_3],X,Done
　　Depth [a_1],OK
Menu: Tools,Partition
　　Cell,Sketch planar partition,Sketch Origin: Specify
　　# pick: front face (z = a_1),Sketch origin X,Y,Z: [0,0,a_1]
　　Select and edge: vertical and on the right,# pick: right vertical
Menu: Add,Circle
　　[0,0],[0,r_f]
　　[a_2,a_3],[8.77,a_3],X,Done
　　Extrude/Sweep,# pick: lines forming lower quarter-circle,Done
　　Extrude Along Direction,# pick edge parallel to Z-global,OK
　　Create Partition
　　# pick: top cell (larger cell),Done
　　# pick: the lines forming the upper quarter-circle,Done
　　Extrude Along Direction,# pick edge parallel to Z-global,OK
　　Create Partition,Done,# close Create Partition pop-up window

ⅲ. 定义材料、截面属性并赋予部件。
Module: Property
Menu: Material,Create
　　Name [Fiber],Mechanical,Elasticity,Elastic,Type: Isotropic
　　[241000,0.2],OK
Menu: Material,Create
　　Name [Matrix],Mechanical,Elasticity,Elastic,Type: Isotropic
　　[3120,0.38],OK
Menu: Section,Create
　　Name [Fiber],Solid,Homogeneous,Cont,Material: Fiber,OK
Menu:　Section,Create
　　Name [Matrix],Solid,Homogeneous,Cont,Material: Matrix,OK
Menu: Assign,Section
　　# pick: the matrix,Done,Section: Matrix,OK
　　# pick: both fibers,Done,Section: Fiber,OK,Done

ⅳ. 创建装配体。
Module: Assembly
Menu: Instance,Create,Independent,OK

ⅴ. 定义分析步。这里需要 3 个分析步,每一个分析步对应 6.2 节中的一列。为求解本例题,在输出请求(Output Request)中必须添加输出变量 S 和 IVOL。

Module: Step
Menu: Step,Manager
  Create,Name [Column-1],Insert new step after: Initial
  Procedure type: Linear perturbation,Static,Cont,OK
  Create,Name [Column-2],Insert new step after: Column-1
  Procedure type: Linear perturbation,Static,Cont,OK
  Create,Name [Column-3],Insert new step after: Column-2
  Procedure type: Linear perturbation,Static,Cont,OK
  # close Step Manager pop-up window
Step: Column-1
Menu: Output,Field Output Requests,Edit,F-Output-1
  Edit variables,[S,E,U,IVOL],OK

ⅵ. 施加载荷和 BC。

将所有边界条件分别定义在 3 个载荷步中,然后利用这些载荷步计算获得 $C_{\alpha\beta}$ 第 1 列、第 2 列和第 3 列中的系数。每个分析步在每个单独的坐标方向上施加单位应变。然后使用方程式(6.18)计算获得刚度系数。过程如下:

Module: Load
  # each step has its own set of BC
  # for step: Column-1
Menu: BC,Manager
  Create,Name [xsymm-C1],Step: Column-1,Mechanical,
    Symm/Anti/Enca,Cont
  # pick: faces @ x = 0 and a_2,Done,XSYMM,OK
  Create,Name [ysymm-C1],Step: Column-1,Mechanical,
    Symm/Anti/Enca,Cont
  # pick: faces @ y = 0 and a_3,Done,YSYMM,OK
  Create,Name [zsymm-C1],Step: Column-1,Mechanical,
    Symm/Anti/Enca,Cont
  # pick: face @ z = 0,Done,ZSYMM,OK,# note only one face is picked
  # imposed displacement(unit strain)
  Create,Name [disp-C1],Step: Column-1,Mech. ,Disp/Rota,Cont
  # pick: faces @ z = a_1,Done,# Checkmark: U3 [a_1],OK
  # for step: Column-2
  Create,Name [zsymm-C2],Step: Column-2,Mechanical,

```
        Symm/Anti/Enca,Cont
    # pick: faces @ z = 0,and a_1,Done,ZSYMM,OK
    Create,Name [ysymm-C2],Step: Column-2,Mechanical,
        Symm/Anti/Enca,Cont
    # pick: faces @ y = 0 and a_3,Done,YSYMM,OK
    Create,Name [xsymm-C2],Step: Column-2,Mechanical,
        Symm/Anti/Enca,Cont
    # pick: face @ x = 0,Done,XSYMM,OK,# note only one face is picked
    # imposed displacement(unit strain)
    Create,Name [disp-C2],Step: Column-2,Mech. ,Disp/Rota,Cont
    # pick: face @ x = a_2,Done,# Checkmark: U1 [a_2],OK
    # for step: Column-3
    Create,Name [xsymm-C3],Step: Column-3,Mechanical,
        Symm/Anti/Enca,Cont
    # pick: faces @ x = 0,and a_2,Done,XSYMM,OK
    Create,Name [zsymm-C3],Step: Column-3,Mechanical,
        Symm/Anti/Enca,Cont
    # pick: faces @ z = 0,and a_1,Done,ZSYMM,OK
    Create,Name [ysymm-C3],Step: Column-3,Mechanical,
        Symm/Anti/Enca,Cont
    # pick: face @ y = 0,Done,YSYMM,OK,# note only one face is picked
    # imposed displacement(unit strain)
    Create,Name [disp-C3],Step: Column-3,Mech. ,Disp/Rota,Cont
    # pick: face @ y = a_3,Done,# Checkmark: U2 [a_3],OK
    # close the BC Manager pop-up window
```

ⅶ. 模型划分网格。

```
Module: Mesh
Menu: Seed,Instance,Approximate global size [0.5],OK
Menu: Mesh,Controls,# select all cells,Done
    Element Shape: Hex,Technique: Structured,OK
Menu: Mesh,Element Type,# select all cells,Done
    Geometric Order: Linear,# checkmark: Reduced integration,OK
Menu: Mesh,Instance,Yes
```

ⅷ. 求解并可视化结果。

```
Module: Job
Menu: Job,Manager
```

Create,Cont,OK,Submit,# when Completed,Results

系数 $C_{\alpha\beta}$ 和层合板弹性常数由下面给出的 Python 脚本计算,该脚本可以通过在 Python shell 中键入或粘贴来执行。或者,可以通过 Menu:file,run script 来执行 Python 脚本。本例中使用的 srecover.py 脚本也可在文献[5]中获得。现在,在 Abaqus 中单击 >>>,信息窗口切换到 Python shell。然后,键入或复制下面的命令到 python shell。

```python
# srecover.py
from visualization import *
# Open the Output Data Base for the current Job
odb = openOdb(path = 'Job-1.odb');
myAssembly = odb.rootAssembly;
#
# Creating a temporary variable to hold the frame repository
# provides the same functionality and speeds up the process
# Column-1
frameRepository = odb.steps['Column-1'].frames;
frameS = [];
frameIVOL = [];
# Get only the last frame [-1]
frameS.insert(0,frameRepository[-1].fieldOutputs['S']
    .getSubset(position = INTEGRATION_POINT));
frameIVOL.insert(0,frameRepository[-1].fieldOutputs['IVOL']
    .getSubset(position = INTEGRATION_POINT));
# Total Volume
Tot_Vol = 0;
# Stress Sum
Tot_Stress = 0;
#
for II in range(0,len(frameS[-1].values)):
    Tot_Vol = Tot_Vol + frameIVOL[0].values[II].data;
    Tot_Stress = Tot_Stress + frameS[0].values[II].data * frameIVOL[0]\
        .values[II].data;

# Calculate Average
Avg_Stress = Tot_Stress/Tot_Vol;
print 'Abaqus/Standard Stress Tensor Order:'
```

```
# from Abaqus Analysis User's Manual-1.2.2 Conventions -
# Convention used for stress and strain components
print 'Average stresses Global CSYS: 11-22-33-12-13-23';
print Avg_Stress;
C11 = Avg_Stress[2]#z-component,1-direction
C21 = Avg_Stress[0]#x-component,2-direction
C31 = Avg_Stress[1]#y-component,3-direction in Fig. 6.5

# Column-2
frameRepository = odb.steps['Column-2'].frames;
frameS = [];
frameIVOL = [];
# Get only the last frame [-1]
frameS.insert(0,frameRepository[-1].fieldOutputs['S']
    .getSubset(position = INTEGRATION_POINT));
frameIVOL.insert(0,frameRepository[-1].fieldOutputs['IVOL']
    .getSubset(position = INTEGRATION_POINT));
# Total Volume
Tot_Vol = 0;
# Stress Sum
Tot_Stress = 0;
#
for II in range(0,len(frameS[-1].values)):
    Tot_Vol = Tot_Vol + frameIVOL[0].values[II].data;
    Tot_Stress = Tot_Stress + frameS[0].values[II].data * frameIVOL[0] \
        .values[II].data;

# Calculate Average
Avg_Stress = Tot_Stress/Tot_Vol;
print 'Abaqus/Standard Stress Tensor Order:'
print 'Average stresses Global CSYS: 11-22-33-12-13-23';
print Avg_Stress;
C12 = Avg_Stress[2]#z-component,1-direction
C22 = Avg_Stress[0]#x-component,2-direction
C32 = Avg_Stress[1]#y-component,3-direction in Fig. 6.5

# Column-3
frameRepository = odb.steps['Column-3'].frames;
```

```python
frameS = [];
frameIVOL = [];
# Get only the last frame [-1]
frameS.insert(0,frameRepository[-1].fieldOutputs['S']
    .getSubset(position = INTEGRATION_POINT));
frameIVOL.insert(0,frameRepository[-1].fieldOutputs['IVOL']
    .getSubset(position = INTEGRATION_POINT));
# Total Volume
Tot_Vol = 0;
# Stress Sum
Tot_Stress = 0;
#
for II in range(0,len(frameS[-1].values)):
    Tot_Vol = Tot_Vol + frameIVOL[0].values[II].data;
    Tot_Stress = Tot_Stress + frameS[0].values[II].data * frameIVOL[0] \
        .values[II].data;

# Calculate Average
Avg_Stress = Tot_Stress/Tot_Vol;
print 'Abaqus/Standard Stress Tensor Order:'
print 'Average stresses in Global CSYS: 11-22-33-12-13-23';
print Avg_Stress;
C13 = Avg_Stress[2] #z-component,1-direction
C23 = Avg_Stress[0] #x-component,2-direction
C33 = Avg_Stress[1] #y-component,3-direction in Fig. 6.5
#
EL = C11-2* C12* C21/(C22 + C23) # Longitudinal E1 modulus
nuL = C12/(C22 + C23) # 12 Poisson coefficient
ET = (C11* (C22 + C23)-2* C12* C12)* (C22-C23)/(C11* C22-C12* C21) #E2
nuT = (C11* C23-C12* C21)/(C11* C22-C12* C21) # 23 Poisson coefficient
GT = (C22-C23)/2 # or GT = ET/2/(1 + nuT) # 23 Shear stiffness
#
print "If Moduli are in TPa and dimensions in microns,results are in TPa"
print "E1 = ",EL
print "E2 = ",ET
print "PR12 = ",nuL
print "PR23 = ",nuT
# end srecover.py
```

计算结果汇总于表6.2。如果读者计算过程中出现任何错误,请检查在每个for循环结束后的空白行是否完全为空。在Python中,这些空白行可以终结for循环。否则,请尝试按如下操作运行脚本:Menu:File,Run Script:[srecover.py],OK。

表6.2 单向层板弹性性能计算结果

性能	PMM	FEA
E_1/MPa	98306	98197
E_2/MPa	6552	7472
ν_{12}	0.298	0.299
ν_{23}	0.6	0.540
G_{12}/MPa	2594	(*)

注:(*)为本例中使用的边界条件无法获得。

例6.2中的建模过程较为繁琐。如果进行参数化研究,使用CAE图形用户界面(GUI)在交互式会话中重复该过程将非常耗时并且容易出错。读者可以在交互会话过程中获得由CAE生成的Python脚本,并使用它来自动执行该过程。

(1)C矩阵的第4列。对于横观各向同性材料,根据式(6.10),只有C_{44}项可能非0,C_{44}可以由其他分量的函数形式表示,因此不需要单独进行分析。C_{44}可以由下式计算:

$$C_{44} = \frac{1}{2}(C_{22} - C_{23}) \qquad (6.28)$$

如果材料是正交各向异性的,则必须按照类似于求解第6列系数的过程进行计算得到。

(2)C矩阵的第5列。对于横观各向同性材料,根据式(6.10),只有C_{55}项可能非0,且等于C_{66},所以可以从第6列系数计算中得到。同样,如果材料是正交各向异性的,则必须按照类似于求解第6列系数的过程进行计算得到。

(3)C矩阵的第6列。由于荷载不对称,这种情况下,不能像计算前3列系数那样使用对称边界条件。因此,边界条件必须使用耦合约束方程施加(大多数FEA商业软件中称为CE)。

根据式(6.10),只有C_{66}项非0。分量$C_{\alpha 6}$通过设置如下条件计算得到

$$\gamma_6^0 = \varepsilon_{12}^0 + \varepsilon_{21}^0 = 1.0 \quad \epsilon_1^0 = \epsilon_2^0 = \epsilon_3^0 = \gamma_4^0 = \gamma_5^0 = 0 \qquad (6.29)$$

注意,$\varepsilon_{12}^0 = 1/2$施加在$x_1 = \pm a_1$之间,另一半施加在$x_2 = \pm a_2$之间。这种情况下,应用于两个周期面(边和顶点中的点除外)之间的CE作为式(6.13)~

式(6.15)的一个特殊情况给出,如下:

$$u_1(a_1,x_2,x_3) - u_1(-a_1,x_2,x_3) = 0$$
$$u_2(a_1,x_2,x_3) - u_2(-a_1,x_2,x_3) = a_1 \quad -a_2 \leq x_2 \leq a_2$$
$$u_3(a_1,x_2,x_3) - u_3(-a_1,x_2,x_3) = 0 \quad -a_3 \leq x_3 \leq a_3$$
$$u_1(x_1,a_2,x_3) - u_1(x_1,-a_2,x_3) = a_2$$
$$u_2(x_1,a_2,x_3) - u_2(x_1,-a_2,x_3) = 0 \quad -a_1 \leq x_1 \leq a_1 \quad (6.30)$$
$$u_3(x_1,a_2,x_3) - u_3(x_1,-a_2,x_3) = 0 \quad -a_3 \leq x_3 \leq a_3$$
$$u_1(x,x_2,a_3) - u_1(x_1,x_2,-a_3) = 0$$
$$u_2(x,x_2,a_3) - u_2(x_1,x_2,-a_3) = 0 \quad -a_1 \leq x_1 \leq a_1$$
$$u_3(x,x_2,a_3) - u_3(x_1,x_2,-a_3) = 0 \quad -a_2 \leq x_2 \leq a_2$$

注意,式(6.30)条件应用于 RVE 面上的对立点之间,边上的点和顶点除外。在 FEA 中,CE 施加在自由度(DOF)之间。一旦 CE 条件中使用了某个 DOF,该自由度就不能在其他 CE 中使用。例如,当 $x_2 = a_2$ 时,式(6.30)中的第一个式子变成

$$u_1(a_1,a_2,x_3) - u_1(-a_1,a_2,x_3) = 0 \quad (6.31)$$

按照式(6.31)的约束需求并施加一个 CE 条件,其与 $u_1(a_1,a_2,x_3)$($-a_3 < x_3 < a_3$)相关的 DOF 被消除掉,因为它们与 $u_1(-a_1,a_2,x_3)$ 的 DOF 相同。一旦 DOF 消除掉,它们就不能再被使用到其他 CE 条件中。例如,式(6.30)的第四个式子,当 $x_1 = a_1$ 时

$$u_1(a_1,a_2,x_3) - u_1(a_1,-a_2,x_3) = a_2 \quad (6.32)$$

该 CE 条件不能施加,因为与 $u_1(a_1,a_2,x_3)$ 关联的 DOF 已被与式(6.31)关联的 CE 条件消除掉。作为推论,RVE 的边和顶点上的约束方程必须与式(6.30)分开定义。此外,只有 3 个方程(分别对应位移 u_i 的每个分量)可以定义在一对边或一对顶点之间。简单地说,只有 3 种位移约束可以用来施加周期性边界条件。

对于成对的边,首要任务是将式(6.30)的前 6 个方程简化为 3 个方程,简化的 3 个方程施加于成对边之间($-a_3 < x_3 < a_3$)。请注意,简化的方程不能施加在 $x_3 = \pm a_3$ 点处,因为这些是顶点,将单独处理。因此,式(6.30)的最后 3 个方程不需要施加。

从 6 个独立的 DOF 角度考虑,将 6 个方程简化为 3 个方程的唯一方法是增加对角边位置的方程。图 6.6 是从正 x_3 轴方向看过去的 RVE 俯视图。图 6.6 中的 A 点表示由平面 $x_1 = a_1$ 和 $x_2 = a_2$ 相交形成的边。该位置由式(6.30)的第一个方程进行约束,即式(6.31)。图 6.6 中的点 C 表示由平面 $x_1 = -a_1$ 和 $x_2 = -a_2$ 相交形成的边。该位置由式(6.30)的第四个方程进行约束,简化如下:

$$u_1(-a_1,a_2,x_3) - u_1(-a_1,-a_2,x_3) = a_2 \quad (6.33)$$

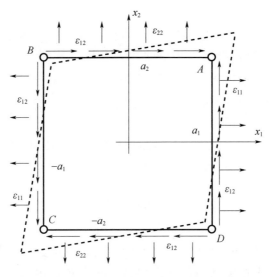

图 6.6 RVE 的俯视图,其表明必须在边缘处施加两个位移(垂直和水平)以形成剪切应变(如图中 A、B、C 和 D 点所示)

根据式(6.31),式(6.33)变为

$$u_1(a_1,a_2,x_3) - u_1(-a_1,-a_2,x_3) = a_2 \quad (6.34)$$

对分量 u_2 和 u_3 的重复以上过程,并与式(6.34)进行组合,得

$$\begin{aligned} u_1(a_1,a_2,x_3) - u_1(-a_1,-a_2,x_3) &= a_2 \\ u_2(a_1,a_2,x_3) - u_2(-a_1,-a_2,x_3) &= a_1 \quad -a_3 < x_3 < a_3 \\ u_3(a_1,a_2,x_3) - u_3(-a_1,-a_2,x_3) &= 0 \end{aligned} \quad (6.35)$$

在图 6.6 中 B 边和 D 边之间应用式(6.30),得

$$\begin{aligned} u_1(a_1,-a_2,x_3) - u_1(-a_1,a_2,x_3) &= -a_2 \\ u_2(a_1,-a_2,x_3) - u_2(-a_1,a_2,x_3) &= a_1 \quad -a_3 < x_3 < a_3 \\ u_3(a_1,-a_2,x_3) - u_3(-a_1,a_2,x_3) &= 0 \end{aligned} \quad (6.36)$$

平面 $x_1 = \pm a_1$ 和 $x_3 = \pm a_3$ 定义了两对边,由以下 6 个 CE 条件约束。

$$\begin{aligned} u_1(+a_1,x_2,+a_3) - u_1(-a_1,x_2,-a_3) &= 0 \\ u_2(+a_1,x_2,+a_3) - u_2(-a_1,x_2,-a_3) &= a_1 \quad -a_2 < x_2 < a_2 \\ u_3(+a_1,x_2,+a_3) - u_3(-a_1,x_2,-a_3) &= 0 \\ u_1(+a_1,x_2,-a_3) - u_1(-a_1,x_2,+a_3) &= 0 \\ u_2(+a_1,x_2,-a_3) - u_2(-a_1,x_2,+a_3) &= a_1 \quad -a_2 < x_2 < a_2 \\ u_3(+a_1,x_2,-a_3) - u_3(-a_1,x_2,+a_3) &= 0 \end{aligned} \quad (6.37)$$

平面 $x_2 = \pm a_2$ 和 $x_3 = \pm a_3$ 定义了两对边,由以下 6 个 CE 条件约束。

$$u_1(x_1, +a_2, +a_3) - u_1(x_1, -a_2, -a_3) = a_2$$
$$u_2(x_1, +a_2, +a_3) - u_2(x_1, -a_2, -a_3) = 0 \quad -a_1 < x_1 < a_1$$
$$u_3(x_1, +a_2, +a_3) - u_3(x_1, -a_2, -a_3) = 0$$
$$u_1(x_1, +a_2, -a_3) - u_1(x_1, -a_2, +a_3) = a_2 \quad (6.38)$$
$$u_2(x_1, +a_2, -a_3) - u_2(x_1, -a_2, +a_3) = 0 \quad -a_1 < x_1 < a_1$$
$$u_3(x_1, +a_2, -a_3) - u_3(x_1, -a_2, +a_3) = 0$$

请注意,式(6.35)~式(6.38)未施加在顶点位置处,因为关于 RVE 体心对称的成对顶点将会出现冗余的 CE。因此,4 对顶点中的每一对均需要在 3 个方向上约束一次。产生的 CE 条件如下:

$$u_1(+a_1, +a_2, +a_3) - u_1(-a_1, -a_2, -a_3) = a_2$$
$$u_2(+a_1, +a_2, +a_3) - u_2(-a_1, -a_2, -a_3) = a_1$$
$$u_3(+a_1, +a_2, +a_3) - u_3(-a_1, -a_2, -a_3) = 0$$
$$u_1(+a_1, +a_2, -a_3) - u_1(-a_1, -a_2, +a_3) = a_2$$
$$u_2(+a_1, +a_2, -a_3) - u_2(-a_1, -a_2, +a_3) = a_1$$
$$u_3(+a_1, +a_2, -a_3) - u_3(-a_1, -a_2, +a_3) = 0$$
$$u_1(-a_1, +a_2, +a_3) - u_1(+a_1, -a_2, -a_3) = a_2$$
$$u_2(-a_1, +a_2, +a_3) - u_2(+a_1, -a_2, -a_3) = -a_1$$
$$u_3(-a_1, +a_2, +a_3) - u_3(+a_1, -a_2, -a_3) = 0$$
$$u_1(+a_1, -a_2, +a_3) - u_1(-a_1, +a_2, -a_3) = -a_2$$
$$u_2(+a_1, -a_2, +a_3) - u_2(-a_1, +a_2, -a_3) = a_1$$
$$u_3(+a_1, -a_2, +a_3) - u_3(-a_1, +a_2, -a_3) = 0 \quad (6.39)$$

式(6.30)~式(6.39)按照式(6.29)给出的单位应变对 RVE 的体积进行了约束。该模型的有限元分析可以得到应力的所有分量。如前所述,这些应力分量在每个单元上的平均值可从有限元计算中获得(见例 6.2 中的宏指令 srecover),或者,它们也可以很容易地通过后处理程序计算获得。因此,利用式(6.18),系数 C_{66} 可以写为

$$C_{66} = \bar{\sigma}_6 = \frac{1}{V}\int_V \sigma_6(x_1, x_2, x_3)\mathrm{d}V \quad \gamma_6^0 = 1 \quad (6.40)$$

最后,使用式(6.11)确定复合材料的弹性特性。

例 6.3 计算例 6.2 中复合材料的 G_{12}。

解 首先注意,对于纤维六边形阵列增强的复合材料,$G_{12} = G_{13}$。因此,计算 $G_{13} = C_{55}$,可以施加应变 $\gamma_5^0 = 2\epsilon_{13}^0 = 1.0$。这种平均变形状态相对于 (x_1, x_2) 坐标平面(图 6.7 中 Abaqus 模型的 X, Z 平面)是反对称的,相对于 (x_1, x_3) 平面(模

型中的 Y,Z 平面)是对称的。因此,需要建立完整的 RVE 模型。应变 γ_{13}(模型中的 E23)的变形形状和云图如图 6.7 所示。该云图对应的平均变形为 $\epsilon_{13}=1.0$ 和平均应力为 $\sigma_{13}=2579\mathrm{MPa}$,因此根据如下等式得到 $G_{13}=2579\mathrm{MPa}$。

$$C_{55}=\overline{\sigma_5}=\frac{1}{V}\int_V \sigma_5(x_1,x_2,x_3)\mathrm{d}V \quad \gamma_5^0=1 \tag{6.41}$$

纤维直径 $d_\mathrm{f}=7\mu\mathrm{m}$,RVE 尺寸为

$$a_1=a_2/4=1.3175\mu\mathrm{m}; a_2=5.270\mu\mathrm{m}; a_3=9.128\mu\mathrm{m}$$

创建几何模型(Module:Part),定义材料和截面属性(Module:Property),创建装配体(Module:Assembly),定义分析步(Module:Step)和划分网格(Module:Mesh)等步骤与例 6.2 一致。

本章节中使用模型坐标 X、Y、Z(图 6.7)描述边界条件。RVE 的底面,即 $Y=-a_3$,固支约束。模型顶面施加位移 $U_3=2\times a_3$,以代表施加剪切应变 $\epsilon_{13}=1.0$。在 RVE 的 $X=\pm a_2$ 面施加对称 BC,这样模拟了周期性边界条件,因为变形相对模型中的 Y,Z 平面对称。下面的代码描述了如何施加边界条件。

```
Module: Load
Menu: BC,Manager
    Create,Name [ysupp],Step: Step-1,Mechanical,Disp/Rota,Cont
    # pick the face @ y = -a_3,Done,# Checkmark: U2 [0] and U3 [0],OK
    Create,Name [ydisp],Step: Step-1,Mechanical,Disp/Rota,Cont
    # pick the face @ y = a_3,Done,# Checkmark: U3 [18.276] # 2* a_3,OK
    Create,Name [xsymm],Step: Step-1,Mechanical,Symm/Anti/Enca,Cont
    # pick the faces @ x = -a_2 and x = a_2,Done,XSYMM,OK
    # close BC Manager pop-up window
```

在 $Z=\pm a_1$ 面上施加周期性条件更为困难。这两个面不能处于自由状态,因为会产生正应变且施加的边界条件 $U_3=2a_3$ 不会对 RVE 产生一个单位平均应变。虽然中尺度下(复合材料)的平均变形为纯剪切 γ_{13},但细观尺度下(纤维和基体)的变形包括 E11、E22 和 E33。因此,必须在 $Z=\pm a_1$ 面上施加周期性条件。由于此处没有施加 E33 应变,我们可以直接使用 Tie 约束。

Tie 约束在相互作用模块下设置,如下所示:

```
Module: Interaction
Menu: Constraint,Create
    Type: Tie,Cont
        Choose the master type: Surface
        # pick: front face (surfaces z = 2* a_1),Done
        Choose the slave type: Surface
```

```
# pick: back face (surfaces z = 0),Done
Position Tolerance: Specify distance: [2.635] # 2* a_1
# Uncheckmark: Adjust slave surface initial position
# Uncheckmark: Tie rotational DOF if applicable
```

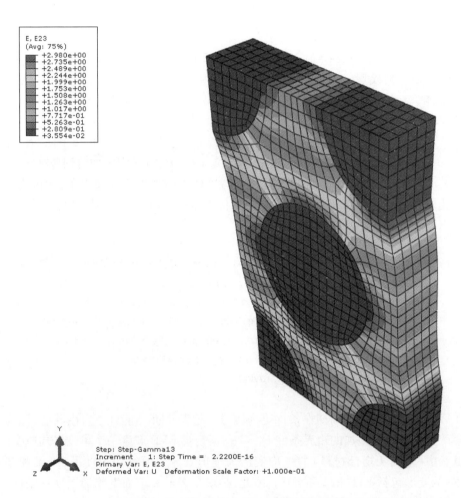

图 6.7　例 6.3 中应变 E23 的变形形状及云图,注意坐标系 X,Y,Z

指定面面之间的距离(位置公差)非常重要,这样 CAE 就可以找到两个曲面上共用的节点。

通过查看 .rec 文件,可以得到与上文操作过程对应的 python 脚本。该例子的完整脚本可以在网站[5,Ex_6.3-TC.py]中找到。该脚本可以在 Abaqus/CAE 中执行,方法是执行 menu：file,Run script：Ex 6.3-TC.py。脚本中包含[5,srecover.py]计算平均应变和平均应力程序,由此得到表 6.3 中所示的计算结果。

表6.3 单向单层板面内剪切模量

性能	PMM	FEA
G_{12}/MPa	2594	2579

6.3 局部–整体分析

在局部–整体(local-global)分析中(图6.8),先在整体模型(global model)中的每个高斯积分点上计算应变,再根据计算出的应变在 RVE 中计算得到应力。假设整体模型材料均质,使用整体模型计算位移和产生的应变。局部模型(loacl model)通过使用 RVE 模型进行模拟来考虑不均匀性。

6.2 节中使用式(6.13)~式(6.15)每次对模型施加一个应变分量,目的是求出材料的等效弹性参数。对于一般的应变状态施加到 RVE 中,式(6.13)~式(6.15)仍然有效,但必须注意在边和顶点处周期性边界条件的指定。式(6.13)~式(6.15)包含9个约束方程,施加在 RVE 表面上所有成对的周期性节点上,边和顶点位置的节点除外。

在面 $x_1 = \pm a_1$ 上,u_1 方向施加 ε_{11}^0,u_2 方向施加 $\varepsilon_{21}^0 = \gamma_6/2$,$u_3$ 方向施加 $\varepsilon_{31}^0 = \gamma_5/2$。为此,将式(6.13)扩展为3个组成部分,使用指标符号形式表示应变张量分量,如下所示:

$$\begin{cases} u_1(a_1,x_2,x_3) - u_1(-a_1,x_2,x_3) = 2a_1\varepsilon_{11}^0 \\ u_2(a_1,x_2,x_3) - u_2(-a_1,x_2,x_3) = 2a_1\varepsilon_{21}^0 \\ u_3(a_1,x_2,x_3) - u_3(-a_1,x_2,x_3) = 2a_1\varepsilon_{31}^0 \end{cases} \quad (6.42)$$

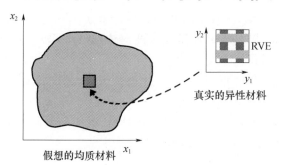

图6.8 利用 RVE 进行局部–整体分析

在面 $x_2 = \pm a_2$ 上,u_1 方向施加 $\varepsilon_{12}^0 = \gamma_6/2$,$u_2$ 方向施加 $\varepsilon_{22}^0 = \varepsilon_{22}^0$,$u_3$ 方向施加 $\varepsilon_{32}^0 = \gamma_4/2$。为此,将式(6.14)扩展为3个组成部分,使用指标符号形式表示应变张量分量,如下所示:

$$\begin{cases} u_1(x_1,a_2,x_3) - u_1(x_1,-a_2,x_3) = 2a_2\varepsilon_{12}^0 \\ u_2(x_1,a_2,x_3) - u_2(x_1,-a_2,x_3) = 2a_2\varepsilon_{22}^0 \\ u_3(x_1,a_2,x_3) - u_3(x_1,-a_2,x_3) = 2a_2\varepsilon_{32}^0 \end{cases} \quad (6.43)$$

在面 $x_3 = \pm a_3$ 上，u_1 方向施加 $\varepsilon_{13}^0 = \gamma_5/2$，$u_2$ 方向施加 $\varepsilon_{23}^0 = \gamma_4/2$，$u_3$ 方向施加 ε_{33}^0。为此，将式(6.15)扩展为 3 个组成部分，使用指标符号形式表示应变张量分量，如下所示：

$$\begin{cases} u_1(x_1,x_2,a_3) - u_1(x_1,x_2,-a_3) = 2a_3\varepsilon_{13}^0 \\ u_2(x_1,x_2,a_3) - u_2(x_1,x_2,-a_3) = 2a_3\varepsilon_{23}^0 \\ u_3(x_1,x_2,a_3) - u_3(x_1,x_2,-a_3) = 2a_3\varepsilon_{33}^0 \end{cases} \quad (6.44)$$

由于两个面共用一条边，在每条边上，每个位移分量都被施加了两个如式(6.42)～式(6.44)所示的 CE 条件。然而，求解 C 矩阵的第 6 列中所述，只能为位移的每个分量施加一个 CE 条件。因此，边上节点必须分开处理。同样，由于 3 个面共用一个顶点，每个面上的该处顶点分配一个位移分量，然后对该顶点的 3 个位移分量施加 3 个周期性的 CE 条件。与求解 C 矩阵第 6 列中的推导过程类似，可以得到以下结果：

平面 $x_1 = \pm a_1$ 和 $x_2 = \pm a_2$ 定义了两对边，式(6.42)～式(6.44)简化为以下 6 个方程（$i=1,2,3$）：

$$\begin{cases} u_i(+a_1,+a_2,x_3) - u_i(-a_1,-a_2,x_3) - 2a_1\varepsilon_{i1} - 2a_2\varepsilon_{i2} = 0 \\ u_i(+a_1,-a_2,x_3) - u_i(-a_1,+a_2,x_3) - 2a_1\varepsilon_{i1} + 2a_2\varepsilon_{i2} = 0 \end{cases} \quad (6.45)$$

平面 $x_1 = \pm a_1$ 和 $x_3 = \pm a_3$ 定义了两对边，式(6.42)～式(6.44)简化为以下 6 个方程（$i=1,2,3$）：

$$\begin{cases} u_i(+a_1,x_2,+a_3) - u_i(-a_1,x_2,-a_3) - 2a_1\varepsilon_{i1} - 2a_3\varepsilon_{i3} = 0 \\ u_i(+a_1,x_2,-a_3) - u_i(-a_1,x_2,+a_3) - 2a_1\varepsilon_{i1} + 2a_3\varepsilon_{i3} = 0 \end{cases} \quad (6.46)$$

平面 $x_2 = \pm a_2$ 和 $x_3 = \pm a_3$ 定义了两对边，式(6.42)～式(6.44)简化为以下 6 个方程（$i=1,2,3$）：

$$\begin{cases} u_i(x_1,+a_2,+a_3) - u_i(x_1,-a_2,-a_3) - 2a_2\varepsilon_{i2} - 2a_3\varepsilon_{i3} = 0 \\ u_i(x_1,+a_2,-a_3) - u_i(x_1,-a_2,+a_3) - 2a_2\varepsilon_{i2} + 2a_3\varepsilon_{i3} = 0 \end{cases} \quad (6.47)$$

对于 4 对顶点，每次对其中一对进行分析，其中每对均相对于 RVE 的中心坐标为 $(0,0,0)$ 处对称。得到的 CE 条件如下：

$$\begin{cases} u_i(+a_1,+a_2,+a_3) - u_i(-a_1,-a_2,-a_3) - 2a_1\varepsilon_{i1} - 2a_2\varepsilon_{i2} - 2a_3\varepsilon_{i3} = 0 \\ u_i(+a_1,+a_2,-a_3) - u_i(-a_1,-a_2,+a_3) - 2a_1\varepsilon_{i1} - 2a_2\varepsilon_{i2} + 2a_3\varepsilon_{i3} = 0 \\ u_i(-a_1,+a_2,+a_3) - u_i(+a_1,-a_2,-a_3) + 2a_1\varepsilon_{i1} - 2a_2\varepsilon_{i2} - 2a_3\varepsilon_{i3} = 0 \\ u_i(+a_1,-a_2,+a_3) - u_i(-a_1,+a_2,-a_3) - 2a_1\varepsilon_{i1} + 2a_2\varepsilon_{i2} - 2a_3\varepsilon_{i3} = 0 \end{cases}$$

$$(6.48)$$

例 6.4 对例 6.2 中的复合材料同时施加 $\epsilon_2^0 = 0.2\%$ 和 $\gamma_4^0 = 0.1\%$。计算 RVE 中的平均应力 $\overline{\sigma}_2$、$\overline{\sigma}_{12}$ 和最大应力 σ_2、σ_{12}。

解 当两个面之间要施加的应变不为 0 时,不能使用 Tie 约束代替周期边界条件。因为施加的变形 $\epsilon_2^0 = 0.2\%$ 和 $\gamma_4^0 = 0.1\%$ 相对于坐标平面 (x_1, x_2) 和 (x_1, x_3)(图 6.9 模型中的 Z,X 和 Z,Y)不对称,所以也不能应用对称边界条件。相反,必须建立如图 6.4 所示的完整 RVE 模型,并施加式(6.30)~式(6.39)中的 CE 条件。然而,可以采用 2D 模型进行分析,因为几何是棱柱形的,即,所有与 x_1-轴垂直的横截面相同且变形一致。在 $\pm a_1$ 面处,$\epsilon_1 = \gamma_{12} = \gamma_{13} = 0$,在 2D 模型中,其周期性条件可以由平面应变条件表示,即可以使用平面应变单元。

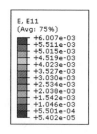
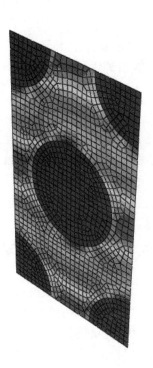

图 6.9 例 6.4 中 ϵ_{22}(模型 X,Y,Z 坐标下 E11)云图

纤维直径 $d_f = 7\mu m$，RVE 尺寸为

$$a_1 = a_2/4 = 1.3175\mu m; a_2 = 5.270\mu m; a_3 = 9.128\mu m$$

几何模型，材料和截面属性，装配体，分析步和网格划分的创建过程与例6.2一致。但建立的RVE模型为2D模型，其中包含1根全纤维和4个1/4纤维截面，如图6.4所示。在这种情况下，不能使用对称边界条件。必须在module:load模块中按照式(6.13)~式(6.15)施加周期边界条件(PBC)。这些PBC可以通过约束方程(CE)施加。模型中本身并没有外载荷，而是由式(6.13)~式(6.15)右侧(RHS)的常数项(由施加的应变$\epsilon_2^0,\epsilon_3^0,\epsilon_{23}^0$产生)对模型实施加载效应。

要编写CE条件，需要定义一个参考点以及主节点和从节点。主节点和从节点通过式(6.13)~式(6.15)表示的约束方程进行连接。如前所述，顶点处的约束方程与面的约束方程分开处理。这部分工作可以由文献[5,pbc 2d.py]文件自动完成。

本例的完整求解脚本可以在文献[5,Ex 6.4-CE.py]中获得。该脚本生成网格和在CE中使用节点集。该脚本可以在Abaqus/CAE中按如下方式执行，Menu:File,Run script,[Ex 6.4-CE.py],OK。脚本调用了文献[5,pbc 2d.py]来编写约束方程，调用了文献[5,srecover2d.py]来计算平均应力。计算结果见表6.4。

表6.4　例6.4的RVE中平均应力和基体中的最大应力(单位:MPa)

应力分量	模型中标记	方向	平均值	基体中最大值
σ_{11}	S33	纤维方向	9.920	16.710
σ_{22}	S11	横向	21.534	28.710
σ_{33}	S22	横向	11.841	16.230
σ_{23}	S12	横向	2.425	5.640

6.4　层合板的RVE

在细观尺度上获取RVE的方法也可用于在中尺度上对层合板性能进行分析。在这种情况下，RVE代表一个层合板，因此，厚度方向应该可以自由膨胀变形。例如，各单层板与$x-y$平面平行，不施加$\sigma_z = 0$和式(6.15)条件，那么厚度方向可以自由变形(图6.10)。一般来说，层合板的RVE必须包括层合板整个厚度。对于面内载荷作用下的对称层合板，可以利用对称边界条件将RVE定义为层合板厚度的1/2(见例6.5)。

针对层合板RVE的CE条件比较简单。仅式(6.13)和式(6.14)约束条件必须施加。在六面体RVE中，如图6.10所示，只需要考虑4个面($x_1 = \pm a_1$和$x_2 = \pm a_2$)和由这些面定义的4条边。

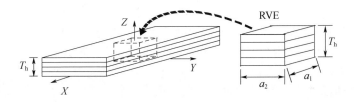

图 6.10 层合板 RVE

因此,在层合板 RVE 中,约束方程式(6.13)和式(6.14)发生改变。在周期面 $x_1 = \pm a_1$ 上,由式(6.13)导出的 CE 条件为

$$\begin{cases} u_1(a_1,x_2,x_3) - u_1(-a_1,x_2,x_3) - 2a_1\varepsilon_{11} = 0 \\ u_2(a_1,x_2,x_3) - u_2(-a_1,x_2,x_3) - 2a_1\varepsilon_{21} = 0 \\ u_3(a_1,x_2,x_3) - u_3(-a_1,x_2,x_3) - 2a_1\varepsilon_{31} = 0 \end{cases} \quad (6.49)$$

在 $x_2 = \pm a_2$ 面上,由式(6.14)导出的 CE 条件为

$$\begin{cases} u_1(x_1,a_2,x_3) - u_1(x_1,-a_2,x_3) - 2a_2\varepsilon_{12} = 0 \\ u_2(x_1,a_2,x_3) - u_2(x_1,-a_2,x_3) - 2a_2\varepsilon_{22} = 0 \\ u_3(x_1,a_2,x_3) - u_3(x_1,-a_2,x_3) - 2a_2\varepsilon_{32} = 0 \end{cases} \quad (6.50)$$

面 $x_1 = \pm a_1$ 和 $x_2 = \pm a_2$ 定义了两对周期性边,此处的条件式(6.13)~式(6.14)简化为以下方程

$$\begin{cases} u_1(+a_1,+a_2,x_3) - u_1(-a_1,-a_2,x_3) - 2a_1\varepsilon_{11} - 2a_2\varepsilon_{12} = 0 \\ u_2(+a_1,+a_2,x_3) - u_2(-a_1,-a_2,x_3) - 2a_1\varepsilon_{21} - 2a_2\varepsilon_{22} = 0 \\ u_3(+a_1,+a_2,x_3) - u_3(-a_1,-a_2,x_3) - 2a_3\varepsilon_{32} = 0 \end{cases} \quad (6.51)$$

和

$$\begin{cases} u_1(+a_1,-a_2,x_3) - u_1(-a_1,+a_2,x_3) - 2a_1\varepsilon_{11} + 2a_2\varepsilon_{12} = 0 \\ u_2(+a_1,-a_2,x_3) - u_2(-a_1,+a_2,x_3) - 2a_1\varepsilon_{21} + 2a_2\varepsilon_{22} = 0 \\ u_3(+a_1,-a_2,x_3) - u_3(-a_1,+a_2,x_3) - 2a_3\varepsilon_{32} = 0 \end{cases} \quad (6.52)$$

对于面内问题分析,$\varepsilon_{31} = \varepsilon_{32} = 0$,式(6.49)~式(6.52)中的第三个方程自动满足。

例 6.5 复合材料铺层信息为 $[0/90/-45/45]_s$,单层板材料性能为 $E_1 = 139\text{GPa}$,$E_2 = 14.5\text{GPa}$,$G_{12} = G_{13} = 5.86\text{GPa}$,$G_{23} = 2.93\text{GPa}$,$\nu_{12} = \nu_{13} = 0.21$,$\nu_{23} = 0.38$,其单层厚度为 $t_k = 1.25\text{mm}$,试计算 G_{xy}。

解 由层合板的对称性,可以使用层合板厚度的 1/2 建立代表性体积元(RVE),在 $z=0$ 处设置对称边界条件。使用用 3D 实体单元离散的 RVE 模型。在 Abaqus CAE GUI 中构建模型的步骤如下:

ⅰ. 设置工作目录。

Menu: File,Set Work Directory,[C:\Simulia\User\Ex_6.5],OK
Menu: File,Save As,[C:\Simulia\User\Ex_6.5\Ex_6.5.cae],OK

ⅱ. 创建部件。

Module: Part
Menu: Part,Create
    3D,Deformable,Solid,Extrusion,Approx. size [10],Cont
Menu: Add,Line,Rectangle,[0,0],[1,1],X,Done,Depth [5],OK
Menu: Tools,Datum
    Type: Plane,Method: Offset from principal plane
    XY Plane,Offset [1.25]
    XY Plane,Offset [2.5]
    XY Plane,Offset [3.75],X,# close Create Datum pop-up window
Menu: Tools,Partition
    Cell,Use datum plane,# pick: Datum plane z = 1.25,
    Create Partition
    # pick: larger cell,Done,# pick: Datum plane z = 2.5,
    Create Partition
    # pick: larger cell,Done,# pick: Datum plane z = 3.75,
    Create Partition
    Done,# close Create partition pop-up window

ⅲ. 定义材料、截面属性并赋予部件。

Module: Property
Menu: Material,Create
    Mechanical,Elasticity,Elastic,Type: Engineering Constants
    [139000,14500,14500,0.21,0.21,0.38,5860,5860,2930],OK
Menu: Assign,Material Orientation
    # pick: layer 1 on Z = 4* 1.25 (free surface),Done
    Use Default Orientation or Other Method
    Definition: Coordinate system,
    Additional Rotation Direction: Axis 3
    Additional Rotation: Angle [0],OK
    # pick: layer 2 on Z = 3* 1.25,Done
    Use Default Orientation or Other Method
    Definition: Coordinate system,

```
Additional Rotation Direction: Axis 3
Additional Rotation: Angle [90],OK
# pick: layer 3 on Z = 2* 1.25,Done
Use Default Orientation or Other Method
Definition: Coordinate system,
Additional Rotation Direction: Axis 3
Additional Rotation: Angle [-45],OK
# pick: layer 4 on Z = 1* 1.25,Done
Use Default Orientation or Other Method
Definition: Coordinate system,
Additional Rotation Direction: Axis 3
Additional Rotation: Angle [45],OK,Done
Menu: Section,Create
    Solid,Homogeneous,Cont,Material: Material-1,OK
Menu: Assign,Section,# select all,Done,OK,Done
```

ⅳ. 创建装配体。

```
Module: Assembly
Menu: Instance,Create,Independent,OK
```

ⅴ. 定义分析步。

```
Module: Step
Menu: Step,Create
    Procedure type: Linear perturbation,Static,Cont,OK
Menu: Output,Field Output Requests,Edit,F-Output-1
    Output Variables: [S,E,U,IVOL],OK
```

ⅵ. 施加 BC 和载荷。

对 RVE 施加面内剪切应变 $\gamma_{xy}^0 = 1$，并求解模型。层合板的剪切刚度 G_{xy} 可以直接由 RVE 中的平均应力得到。边界条件式(6.49)~式(6.52)按照如下施加：

```
# Apply simple shear
Module: Load
Menu: BC,Manager
    Create,Name [yfix],Step: Initial,Disp/Rota,Cont
    # pick: surface @ y=0,Done,# Checkmark: U1 and U2,OK
    Create,Name [disp],Step: Step-1,Disp/Rota,Cont
```

```
# pick: surface @ Y = 1,Done
# checkmark: U1 [1],# checkmark: U2 [0],OK
Create,Name [zsymm],Step: Initial,Symm/Anti/Enca,Cont
# pick: surfaces @ Z = 0,Done,ZSYMM,OK
# close BC Manager pop-up window
```

ⅶ. 定义约束。

在 $x=0/1$ 的面上施加 Tie 约束以获得 $\epsilon_{xx}=0$。

```
Module: Interaction
Menu: Constraint,Create,Tie,Cont
    Choose the master type: Surface,# pick: surface x = 0,Done
    Choose the slave type: Surface,# pick: surface x = 1,Done
    Position Tolerance: Specify distance [1]
    # Uncheckmark: Adjust slave surface initial position # !!!
    # Uncheckmark: Tie rotational DOFs if applicable,OK
```

ⅷ. 模型划分网格。

```
Module: Mesh
Menu: Seed,Instance
    Approx global size [1.25],OK
    # one element/layer-thickness enough for pure shear
Menu: Mesh,Instance,Yes
```

ⅸ. 求解并可视化结果。

```
Module: Job
Menu: Job,Manager
    Create,Cont,OK,Submit,# when Completed,Results
Menu: File,Save
Menu: File,Run Script,[Ex_6.5-post.py],OK
```

脚本[5,Ex_6.5-post.py]可用于计算平均剪切应力,从而计算剪切模量。由于施加的应变为单位应变,计算得到的平均应力 σ_{12} 等于 C_{66}。因此,利用该脚本的最后一行可以得到 $G_{12}=21441\text{MPa}$。

```
""" Post-processing Ex. 6.5 """
# make sure the Work Directory is OK
import os
os.chdir(r'C:\Simulia\User\Chapter_6\Ex_6.5')
```

```python
# Open the Output Database for the current Job
from visualization import *
odb = openOdb(path = 'Job-1.odb');
myAssembly = odb.rootAssembly;

# Create temporary variable to hold frame repository speeds up the process
frameRepository = odb.steps['Step-1'].frames;
frameS = [];
frameIVOL = [];

# Create a Coordinate System in the Laminate direction (Global)
coordSys = odb.rootAssembly.DatumCsysByThreePoints(name = 'CSYSLAMINATE',
    coordSysType = CARTESIAN, origin = (0,0,0),
    point1 = (1.0,0.0,0), point2 = (0.0,1.0,0.0) )

# Transform stresses from Lamina Coordinates
# to Laminate Coordinate System defined in CSYSLAMINATE
# stressTrans = odb.steps['Step-1'].frames[-1].fieldOutputs['S'] \
#   .getTransformedField(datumCsys = coordSys)
stressTrans = frameRepository[-1].fieldOutputs['S'] \
    .getTransformedField(datumCsys = coordSys)

# Insert transformed stresses into frameS
frameS.insert(0,stressTrans.getSubset(position = INTEGRATION_POINT));
frameIVOL.insert(0,frameRepository[-1].fieldOutputs['IVOL'] \
    .getSubset(position = INTEGRATION_POINT));

Tot_Vol = 0; # Total Volume
Tot_Stress = 0; # Stress Sum

for II in range(0,len(frameS[-1].values)):
    Tot_Vol = Tot_Vol + frameIVOL[0].values[II].data;
    Tot_Stress = Tot_Stress + frameS[0].values[II].data * frameIVOL[0] \
        .values[II].data;

#Calculate Average
Avg_Stress = Tot_Stress/Tot_Vol;
print 'Abaqus/Standard Stress Tensor Order:'
```

```
# From Abaqus Analysis User's Manual-1.2.2 Conventions
print ' 11-22-33-12-13-23';
print Avg_Stress;

print 'G12 = ',Avg_Stress[3]
```

习题

习题 6.1 考虑一个单向复合材料,其各向同性材料纤维 $E_f = 241\text{GPa}$, $\nu_f = 0.2$,各向同性材料基体 $E_m = 3.12\text{GPa}$, $\nu_m = 0.38$,纤维体积含量 $V_f = 0.4$。纤维直径 $d_f = 7\mu\text{m}$,按图 6.1 所示方形排列分布。选择一个 RVE 形式,其中心包含一根完整纤维截面,垂直间距为 $2a_2$,水平间距为 $2a_3$,试计算(1)式(6.10)中刚度矩阵的前 3 列;(2) C_{66}。

习题 6.2 考虑一个性能和纤维分布与习题 6.1 相同的复合材料,但是在图 6.1 中选择一个沿垂直和水平方向均旋转 45°的 RVE。因此,RVE 尺寸变为垂直间距为 $2\sqrt{2}a_2$,水平间距为 $2\sqrt{2}a_3$,其中包含两个纤维(1 个完整的截面和 4 个 1/4 截面)。注意选择一个正确的周期性 RVE。

(1)试计算式(6.10)中刚度矩阵的前 3 列;

(2) C_{66}。

习题 6.3 利用习题 6.1 中得到的刚度阵计算 E_1, E_2, ν_{12}, ν_{23}, G_{12} 和 G_{23}。

习题 6.4 利用习题 6.2 中得到的刚度阵计算 E_1, E_2, ν_{12}, ν_{23}, G_{12} 和 G_{23}。

习题 6.5 说明从例 6.3 中移除 Tie 约束后会导致非 0 正应变,并导致错误的 G_{13} 结果。

习题 6.6 如例 6.3 所示计算 G_{12},但使用对称边界条件,建立 1/4 的 RVE 并离散。

习题 6.7 通过修改例 6.3 中的边界条件和 Tie 约束条件来计算 G_{23}。

习题 6.8 以例 6.4 为参考计算 G_{23}。

第7章 黏弹性

聚合物基复合材料(PMC)的蠕变行为引起了我们对其黏弹性的兴趣,而蠕变就是黏弹性的一种表现。基于材料随时间的响应特征,可以把材料分为弹性材料、黏性材料和黏弹性材料。对于弹性材料,突然施加一个载荷,然后保持恒定,材料会产生瞬时变形。在一维应力状态下,弹性应变为 $\varepsilon = D\sigma$,其中 $D = 1/E$ 为弹性模量 E 的倒数。材料的变形会保持恒定,卸载后,材料的弹性应变恢复至初始值。因此,所有的弹性变形都是可恢复的。

黏性材料会以恒定应变率 $\dot{\varepsilon} = \sigma/\eta$ 流动,其中 $\eta = \tau E_0$ 为牛顿黏度,E_0 为初始模量,τ 为材料时间常数。材料累积应变 $\varepsilon = \int \dot{\varepsilon} dt$ 卸载后无法恢复。

黏弹性材料综合了弹性材料和黏性材料的特性,但其响应远比仅将弹性应变和黏性应变叠加更为复杂。令 H 表示阶跃函数,有

$$H(t - t_0) = \begin{cases} 0 & (t < t_0) \\ 1 & (t \geq t_0) \end{cases} \quad (7.1)$$

在脉冲荷载 $\sigma = H(t - t_0)\sigma_0$ 作用下,其中 σ_0 为恒定,黏弹性材料会像弹性材料一样,出现瞬时弹性变形。随后,黏弹性材料的变形增长由可恢复变形和不可恢复黏性流共同驱动。

简单的黏性流和弹性应变串联模型(Maxwell 模型,图 7.1(a),其中 $\eta = \tau E_0$)表达为总的不可恢复黏性流与可恢复弹性变形之和,即

$$\dot{\varepsilon}(t) = \frac{\sigma(t)}{\tau E_0} + \frac{\dot{\sigma}(t)}{E_0} \quad (7.2)$$

简单的弹性应变和黏性流并联模型(Kelvin 模型,图 7.1(b),其中 $\eta = \tau E$)表达为总的不包含黏性流的可恢复变形,但该变形不能像弹性材料一样立即恢复,即

$$\sigma(t) = \tau E \dot{\varepsilon}(t) + E \varepsilon(t) \quad (7.3)$$

如式(7.2)所示,具有不可恢复黏性流特性的材料,称为流体,即使材料的黏性流动过程可能非常缓慢。玻璃就是一种流体材料,其黏性流动可能需要跨越上百年的时间。中世纪大教堂的窗户玻璃底部会比顶部更厚,这个现象揭示了玻璃在重力作用下的流体性质。如式(7.3)所示,如果材料的黏性变形可完全恢复,则称为固体。可以看出采用固体性材料进行结构设计,比流体性材料更容易。

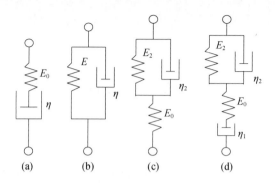

图 7.1 黏弹性模型
（a）Maxwell 模型；（b）Kelvin 模型；（c）标准固体模型；（d）Maxwell-Kelvin 模型。

早期材料力学课程中有个普遍的误解，就是认为大部分结构材料都是弹性的，事实上只有理想晶体材料才是弹性的。如果用足够长的时间进行观察，或者在足够高的温度条件下，绝大部分材料都是黏弹性的。或者说，大多数真实的材料性质都是黏弹性的。

对于弹性材料，柔量（度）D 是弹性模量 E 的倒数，两者都是常数，它们之间的关系是

$$DE = 1 \tag{7.4}$$

考虑时域内的黏弹性材料，柔量表示为 $D(t)$，与松弛模量 $E(t)$ 类似，均与时间相关但更复杂，其在 7.3 节中将详细说明。注意到松弛模量 $E(t)$ 代替了模量 E。为了在 7.1 节中方便介绍黏弹性模型，接下来给出柔量和松弛模量之间的主要推导过程。当柔量 D 和松弛模量 E 都是时间的函数时，式（7.4）表示为

$$D(t)E(t) = 1 \tag{7.5}$$

$D(t)$ 和 $E(t)$ 都是时间的函数，因此无法对式（7.5）进行代数运算求解得到其中任一显示函数表达式。为了由其中一个函数表达另一个函数，需要对其进行拉普拉斯变换

$$s^2 D(s)E(s) = 1 \tag{7.6}$$

式中：$D(s)$，$E(s)$ 都是 s 的代数函数，不涉及时间 t，因此两个函数间可以进行代数运算，有

$$E(s) = \frac{1}{s^2 D(s)} \tag{7.7}$$

最后，时域内的松弛模量由式（7.7）进行拉普拉斯逆变换得到

$$E(t) = L^{-1}[E(s)] \tag{7.8}$$

同理，柔量 $D(t)$ 可由松弛模量 $E(t)$ 表示为

$$D(t) = L^{-1}\left[\frac{1}{s^2 L[E(t)]}\right] \tag{7.9}$$

式中:$L[\ ]$表示拉普拉斯变换;$L^{-1}[\ ]$表示拉普拉斯逆变换。

7.1 黏弹性模型

本节介绍的黏弹性材料模型均便于对实验数据的曲线拟合。时域内,常用的测试试验为蠕变试验和应力松弛实验。蠕变实验时,施加恒定的应力 σ_0,然后持续测量产生的应变。实测应变与施加的应力比率即为柔量 $D(t) = \varepsilon(t)/\sigma_0$。应力松弛实验时,施加恒定应变 ε_0,然后测量维持该应变所需的应力。测得的应力与施加的应变之比即为松弛模量 $E(t) = \sigma(t)/\varepsilon_0$。

7.1.1 Maxwell 模型

为了得出 Maxwell 模型的柔量[45],在图 7.1(a)所示的模型末端施加恒定应力 σ_0,并进行蠕变测试。该模型的应变率由式(7.2)给出,对时间进行积分,得

$$\varepsilon(t) = \frac{1}{\tau E_0}\int_0^t \sigma_0 \mathrm{d}t + \frac{\sigma_0}{E_0} \quad (7.10)$$

式中:E_0 为弹簧的弹性常数;τ 为材料的时间常数。

图 7.1(a)中 $\eta_0 = \tau E_0$。弹簧和阻尼受相同的载荷、相同的恒定应力 σ_0,因此对时间积分后可得

$$\varepsilon(t) = \frac{\sigma_0 t}{\tau E_0} + \frac{\sigma_0}{E_0} \quad (7.11)$$

则柔量为

$$D(t) = \frac{1}{E_0} + \frac{t}{\tau E_0} \quad (7.12)$$

要求解 Maxwell 模型的松弛模量,可对式(7.12)进行拉普拉斯变换(使用表 7.1 或 MATLAB),得

表 7.1 常用的拉普拉斯变换

函数	$f(t) = L^{-1}\{f(s)\}$	$f(t) = L\{f(s)\}$
常数	a	a/s
线性	at	a/s^2
导数	$\mathrm{d}f/\mathrm{d}t$	$sf(s) - f(0)$
幂指数	$\exp(at)$	$1/(s-a)$
卷积积分	$\int_0^t f(t-\tau)g(\tau)\mathrm{d}\tau$	$L\{f\}L\{g\}$

$$D(s) = \frac{1}{sE_0} + \frac{1}{s^2\tau E_0} = \frac{s\tau+1}{s^2\tau E_0} \tag{7.13}$$

$t=0$ 时,阻尼位移为 0,因此 E_0 是材料的初始弹性模量。现在,拉普拉斯域的松弛模量为

$$E(s) = \frac{1}{s^2 D(s)} = \frac{\tau E_0}{s\tau+1} \tag{7.14}$$

进行拉普拉斯逆变换后,可得到时域的松弛模量(使用表 7.1 或 MATLAB)

$$E(t) = E_0 \exp(-t/\tau) \tag{7.15}$$

注意:$t=\tau$ 时,松弛模量衰减至初始值的 36.8%,因此 τ 称为材料的时间常数。

7.1.2 Kelvin 模型

对于 Kelvin 模型,只可进行蠕变试验,因为应力松弛试验需要无限大的应力才能将图 7.1(b) 中的阻尼瞬时拉伸到常数。对于蠕变试验,施加恒定应力 $\sigma = \sigma_0$。那么,式(7.3)为 $\varepsilon(t)$ 的常微分方程(ODE),并且其满足 $\varepsilon(t) = (\sigma_0/E)[1-\exp(-t/\tau)]$。因此,柔量 $D(t) = \varepsilon(t)/\sigma_0$ 可表示为

$$D(t) = 1/E_0[1-\exp(-t/\tau)] \tag{7.16}$$

利用式(7.8),松弛模量函数可由单位阶跃函数 $H(t)$ 和 Dirac 函数 $\delta(t)$ 表示为

$$E(t) = EH(t) + E\tau\delta(t) \tag{7.17}$$

其中当 $t=t_0$ 时,$\delta(t-t_0) = \infty$;当 $t \neq t_0$ 时,$\delta(t-t_0) = 0$。以下为运算得到式(7.17)的 MATLAB 代码:

```
syms s complex; syms Dt Et t E tau real;
Dt = expand((1-exp(-t/tau))/E)
Ds = laplace(Dt)
Es = 1/Ds/s^2
Et = ilaplace(Es)
```

7.1.3 标准线性固体模型

为了拥有一个初始柔量 $1/E_0$,在 Kelvin 模型中添加一个弹簧模型(图 7.1(c))。那么,其模型柔量可表示为

$$D(t) = 1/E_0 + 1/E_2\left[1-\exp\left(\frac{-t}{\tau_2}\right)\right] \tag{7.18}$$

松弛模量可表示为

$$E(t) = E_\infty + (E_0 - E_\infty)\exp\left[\frac{-t(E_0 + E_2)}{\tau_2 E_2}\right] \quad (7.19)$$

式中:$E_\infty = (E_0^{-1} + E_2^{-1})^{-1}$ 为时间达到无穷时的平衡模量。

为了获得更好的关联性,需要添加更多的弹簧-阻尼单元,式(7.18)变为

$$D(t) = D_0 + \sum_{j=1}^{n} D_j[1 - \exp(-t/\tau_j)] \quad (7.20)$$

式中:τ_j 为延迟时间[17]。

当 $n \to \infty$ 时,有

$$D(t) = \int_0^\infty \Delta(\tau)[1 - \exp(-t/\tau)\mathrm{d}\tau] \quad (7.21)$$

式中 $\Delta(\tau)$ 为柔量谱[17]。

7.1.4 Maxwell-Kelvin 模型

Maxwell-Kelvin 模型可作为流体材料的粗略近似模型,也称为四参数模型,如图 7.1(d)所示。此模型中 Maxwell 单元和 Kelvin 单元被串联放置,因此柔量可以分别将两种单元的柔量进行叠加得到:

$$D(t) = \frac{1}{E_0} + \frac{t}{\tau_1 E_0} + \frac{1}{E_2}[1 - \exp(-t/\tau_2)] \quad (7.22)$$

其中:E_0 是弹性模量,τ_1 代替式(7.12)中的 τ,E_2 和 τ_2 代替式(7.16)中的 E 和 τ。松弛模量可参考文献[45]给出:

$$E(t) = (P_1^2 - 4P_2)^{-1/2}\left[\left(q_1 - \frac{q_2}{T_1}\right)\exp(-t/T_1) - \left(q_1 - \frac{q_2}{T_2}\right)\exp(-t/T_2)\right]$$

$$\eta_1 = E_0\tau_1;\ \eta_2 = E_0\tau_2$$

$$q_1 = \eta_1;\ q_2 = \frac{\eta_1\eta_2}{E_2}$$

$$T_1 = \frac{1}{2P_2}[P_1 + \sqrt{P_1^2 - 4P_2}];\ T_2 = \frac{1}{2P_2}[P_1 - \sqrt{P_1^2 - 4P_2}]$$

$$P_1 = \frac{\eta_1}{E_0} + \frac{\eta_1}{E_2} + \frac{\eta_2}{E_2};\ P_2 = \frac{\eta_1\eta_2}{E_0 E_2}$$

(7.23)

另一种判断材料是流体材料还是固体材料的方法是看材料的长时形变,如果材料可以无限变形,则为流体材料,如果材料的变形是可停止的,则为固体材料。

7.1.5 幂指数模型

另一种经常用来表示聚合物相对短时变形的模型称为幂指数模型

$$E(t) = At^{-n} \tag{7.24}$$

参数 A 和 n 可由试验数据进行修正。幂指数模型之所以受欢迎,是因为它与聚合物的短时变形行为吻合较好,并且容易进行数据拟合;只需对式(7.24)两边取对数,这样式(7.24)就变成了直线方程,然后再使用线性回归方法进行参数拟合。通过式(7.9)可得柔量为

$$\begin{cases} D(t) = D_0 + D_c(t) \\ D_c(t) = [A\Gamma(1-n)\Gamma(1+n)]^{-1}t^n \end{cases} \tag{7.25}$$

式中:Γ 为 Gamma 函数[46];$D_0 = 1/E_0$ 为弹性柔量;下标$()_c$ 为应力松弛函数和柔量函数的蠕变分量。

7.1.6 Prony 级数模型

尽管聚合物的短时蠕变和应力松弛可以用幂指数模型较为准确地描述,但是随着时间区间变长,就需要更精细化的模型进行描述。Prony 级数模型就是这样的一种模型,它由 n 个衰减指数组成。

$$E(t) = E_\infty + \sum_{i=1}^{n} E_i \exp(-t/\tau_i) \tag{7.26}$$

式中:τ_i 为松弛时间;E_i 为松弛模量;E_∞ 为平衡模量(如果存在)。例如,Maxwell 材料具有流体属性,因此 $E_\infty = 0$。τ_i 越大,衰减越慢。注意,当 $t=0$ 时,$E_0 = E_\infty + \sum E_i$。式(7.26)改写为

$$E_0 = E_\infty + \sum_{1}^{n} m_i E_0 \exp(-t/\tau_i) \tag{7.27}$$

式中:$m_i = E_i/E_0$。

Prony 级数模型可以用剪切模量和体积模量来表示(G, K,见 1.12.5 节),如下:

$$\begin{cases} G(t) = G_\infty + \sum_{1}^{n} g_i G_0 \exp(-t/\tau_i) \\ k(t) = k_\infty + \sum_{1}^{n} k_i K_0 \exp(-t/\tau_i) \end{cases} \tag{7.28}$$

其中:$g_i = G_i/G_0$ 和 $k_i = K_i/K_0$。G_0, K_0 分别为剪切模量和体积模量的初始值。当 $t=0$ 时,$G_\infty = G_0(1 - \sum_{1}^{n} g_i)$ 和 $k_\infty = K_0(1 - \sum_{1}^{n} k_i)$,Prony 级数模型可

以改写为

$$\begin{cases} G(t) = G_0\left(1 - \sum_1^n g_i\right) + \sum_1^n g_i G_0 \exp(-t/\tau_i) \\ k(t) = K_0\left(1 - \sum_1^n k_i\right) + \sum_1^n k_i K_0 \exp(-t/\tau_i) \end{cases} \quad (7.29)$$

在 Abaqus/CAE 中，其数据可以通过 Eidt Material 对话框输入，包括弹性属性（E_0 和 ν_0），Prony 级数中第 i 个单元的黏性属性（g_i, k_i, τ_i）。Abaqus 通过式（1.74）和式（1.76），以及输入的参数 E_0 和 ν_0，自动计算出 G_0 和 K_0。对于大多数聚合物材料和复合材料，通常假设泊松比不随时间变化，其可以通过设置 $k_i = g_i$ 满足（根据式（1.74）和式（1.76）的定义）。同样，如果 ν 不随时间变化，则式（7.27）中的 $m_i = g_i$。

注意，如果 Moduli time scale (for viscoelasticity) 设置为 Instantaneous，那么在 Eidt Material 对话框中输入的弹性模量代表 E_0。否则，如果将 Moduli time scale (for viscoelasticity) 设置为 Long term，那么弹性模量代表 E_∞。当然，两者也可以通过关系式 $E_0 = E_\infty / (1 - \sum m_i)$ 进行相互计算。

例如，Maxwell 材料（式（7.15））可以通过设置 $g_i = k_i = 1$ 的一级 Prony 级数模型模拟。但是，Abaqus 中无法对 $E_\infty = 0$ 的材料进行计算，因此必须使 $\sum g_i < 1$，要模拟 Maxwell 材料可令 $g_i = k_i = 0.999$。

7.1.7 标准非线性固体模型

虽然采用足够多的级数项，Prony 级数模型可以描述任何材料的行为，但对于难以用数学方法拟合的材料，采用其他类型的模型可能更有效。例如，标准非线性固体模型能够很好地描述聚合物蠕变 α 时域内的长时柔量[47]，即

$$D(t) = D_0 + D_1'[1 - \exp(-t/\tau)^m] \quad (7.30)$$

在室温下，这也是结构工程师比较关心的时域，因为它涵盖了从秒到年的跨度。而振动、噪声工程师更关心 β 时域[47]，因为它在亚秒级的时间范围内。换句话说，对于长时域模型模拟，所有 β 时域内的柔量可集中到 D_0 项中，而 D_1' 项代表全部 α 时域柔量的累积。式（7.30）中有 4 个参数，当数据在短时域内时，无法确定这 4 个参数，因为从 3 参数幂指数模型式（7.31）中无法区分材料行为。这很容易理解，如果将式（7.30）按幂级数展开，并且舍去第一项之后的部分[47]，有

$$D(t) = D_0 + D_1'(t/\tau)^m[1 - (t/\tau)^m + \cdots] \approx D_0 + D_1 t^m; D_1 = D_1'/\tau \quad (7.31)$$

对于短时问题，所有 t 的高阶幂级数都可以忽略。剩下只有 3 个参数的修正幂指数模型。此时，参数 D_1 由 τ 和 D_1' 表示，短时域内的数据拟合无法同时调

整式(7.31)中的 τ 和 D_1';事实上,只有将 τ 和 D_1' 进行组合后才能数据拟合。这意味着短时间问题数据必须由较少的参数进行模拟,本模型中采用3个参数。

7.1.8 非线性幂指数模型

上述所介绍的所有模型均用来描述线性黏弹性材料,在黏弹性理论中,线性意味着模型中的参数不是应力函数(见7.2.1节)。这也意味着在任何确定时间时的形变都可以根据应力的增加成比例的增大。如果模型中任何参数是应力的函数,那么该模型表征的材料即是非线性黏弹性的。例如非线性幂指数模型,形式为

$$\dot{\varepsilon} = At^B\sigma^D \quad (7.32)$$

对式(7.32)两侧取对数,得到两个变量组成的线性方程

$$y = \overline{A} + BX_1 + DX_2; \overline{A} = \log(A), X_1 = \log(t), X_2 = \log(\sigma) \quad (7.33)$$

其可以通过MATLAB中多重线性回归算法拟合。

尽管大多数材料并不是线性黏弹性的,但对于结构应力变化不大的情况,可以近似看作线性黏弹性材料。

例7.1 分别用(a)Maxwell模型(式(7.12)),(b)幂指数模型(式(7.31))和(c)标准非线性固体模型(式(7.30))拟合表7.2中的蠕变数据。

表7.2 例7.1中的蠕变数据

时间/s	1	21	42	62	82	102	123	143	163	184	204
$D(t)/\text{GPa}^{-1}$	1.49	1.99	2.21	2.35	2.56	2.66	2.75	2.85	2.92	2.96	3.01

解 使用Maxwell模型,将蠕变数据拟合成一条直线,即忽略短时蠕变数据中的弯曲部分,可得 $E_0 = 0.460\text{GPa}, \tau = 495\text{s}$。

使用幂指数模型,将式(7.25)写为

$$D(t) - D_0 = D_1 t^m$$

式中:$D_0 = 1.49\text{GPa}^{-1}$ 为表7.2中的第一个数据(另见式(7.31))。将上述方程的两边取对数,并使用线性回归方法拟合出一条曲线,得到 $D_0 = 1.49\text{GPa}^{-1}$,$D_1 = 0.1117(\text{GPa} \cdot \text{s})^{-1}, m = 0.5$。

使用标准非线性固体模型,必须使用非线性方法对预测值(期望值)e_i 和试验值 o_i 之间的误差进行最小化求解。误差定义为所有数据点偏差之和:$\chi^2 = \sum(e_i - o_i)^2/o_i^2$。通过此方法,可以得到:$D_0 = 1.657\text{GPa}^{-1}, D_1' = 1.617\text{GPa}^{-1}$,$\tau = 0.273\text{s}, m = 0.0026$。

实验数据和拟合函数如图7.2所示。

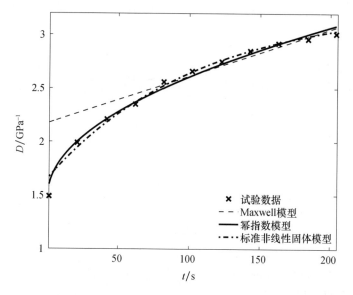

图 7.2　黏弹性数据拟合:Maxwell 模型、幂指数模型、标准非线性固体模型

7.2　Boltzmann 叠加原理

7.2.1　线性黏弹性材料

如果应用叠加原理,那么黏弹性材料认为是线性的,即给定一个应力加载历史

$$\sigma(t) = \sigma_1(t) + \sigma_2(t) \tag{7.34}$$

那么应变表示为

$$\varepsilon(t) = \varepsilon_1(t) + \varepsilon_2(t) \tag{7.35}$$

式中:$\varepsilon_1(t)$,$\varepsilon_2(t)$ 分别为 $\sigma_1(t)$ 和 $\sigma_2(t)$ 对应的应变历史。

对于线性黏弹性材料,蠕变柔量和松弛模量均独立于应力,有

$$\begin{cases} D(t) = \dfrac{\varepsilon(t)}{\sigma_0} \\ E(t) = \dfrac{\sigma(t)}{\varepsilon_0} \end{cases} \tag{7.36}$$

对于非线性黏弹性材料,$D(t,\sigma)$ 为应力的函数,$E(t,\varepsilon)$ 为应变的函数。

对于线性黏弹性材料,在 $t = \theta_0$ 时施加一个应力 σ_0(图 7.3),有

$$\varepsilon(t) = \sigma_0 D(t,\theta_0); t > \theta_0 \tag{7.37}$$

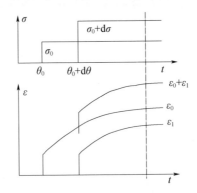

图 7.3 应变的 Boltzmann 叠加假设

在 $\theta_0 + \mathrm{d}\theta$ 时,增加一个无穷小载荷步 $\mathrm{d}\sigma$,得

$$\varepsilon(t) = \sigma_0 D(t,\theta_0) + \mathrm{d}\sigma D(t,\theta_0 + \mathrm{d}\theta) \quad (t > \theta_0 + \mathrm{d}\theta) \tag{7.38}$$

如果在时间间隔 $\mathrm{d}\theta$ 内,应力连续变化 $\mathrm{d}\sigma$,则可以通过积分对式(7.38)求和得到累积应变,即

$$\varepsilon(t) = \sigma_0 D(t,\theta_0) + \int_{\theta_0}^{t} D(t,\theta)\mathrm{d}\sigma = \sigma_0 D(t,\theta_0) + \int_{\theta_0}^{t} D(t,\theta)\frac{\mathrm{d}\sigma}{\mathrm{d}\theta}\mathrm{d}\theta \tag{7.39}$$

其中离散的时间 $\theta_0, \theta_0 + \mathrm{d}\theta$ 等由连续函数 θ 代替。尽管在每个无穷小 $\mathrm{d}\theta$ 上,材料老化效应可以忽略,但随着时间的增加,老化效应开始作用。因此,柔量 $D(t,\theta)$ 是当前时间 t 的函数和代表所有时间历史的 θ 函数。

7.2.2 耐老化黏弹性材料

如果 $\varepsilon_1(t,\theta)$ 曲线与 $\varepsilon_1(t,\theta_0)$ 曲线的形状相同,仅是时间轴水平平移,那么任何曲线都可以平移至原点(图 7.3)

$$D(t,\theta) = D(t-\theta) \tag{7.40}$$

式(7.40)是对耐老化材料的定义,其更详细的讨论可参见文献[45,48]。式(7.40)表示所有曲线都具有相同形状,与时域区间 θ 无关,只是在时间轴上平移。注意,式(7.40)中的 θ 为 $\theta<t$ 时域内的一个连续函数,表示每个载荷(σ_0,$\mathrm{d}\sigma$ 等)作用的时间。

在一个固定时间 t 时的材料响应 $\varepsilon(t,\theta)$ 是所有 $\theta<t$ 时域内响应的函数。因此,可以说该响应具有遗传性。如果材料是可老化的,则 t 和 θ 都是 $D(t,\theta)$ 中的独立变量。对于耐老化材料,只有 $t-\theta$ 一个独立变量,所以不需要关心目前材料(t)有多久,只需关注材料在施加的载荷 $\mathrm{d}\sigma(\theta)$ 下经历多长($t-\theta$)。

蠕变柔量是材料在应力下的响应,总是在施加应力载荷时开始出现。如果

蠕变的发生是渐进的,则从式(7.39)可得

$$\varepsilon(t) = \int_0^t D(t-\theta)\dot{\sigma}(\theta)\mathrm{d}\theta \tag{7.41}$$

松弛应力表示为

$$\sigma(t) = \int_0^t E(t-\theta)\dot{\varepsilon}(\theta)\mathrm{d}\theta \tag{7.42}$$

线性黏弹性材料的依时性行为一般都具有遗传性,这意味着材料在时间 t 时的行为取决于自 $t=0$ 开始以来材料被作用的历史。

例 7.2 考虑由 $D(t-\theta) = 1/E + (t-\theta)/\eta$ 表示的耐老化材料,加载(a) $\sigma_0 H(\theta)$ 和(b) $\sigma_0 H(\theta-1)$。在两种载荷下,试计算 $\varepsilon(t)$。

解

加载(a):

$$\sigma = \sigma_0 H(\theta) \Rightarrow \mathrm{d}\sigma/\mathrm{d}t = \sigma_0 \delta(0)$$

$$\varepsilon(t) = \int_0^t \left[\frac{1}{E} + \frac{(t-\theta)}{\eta}\right]\sigma_0 \delta(0)\mathrm{d}\theta$$

$$\varepsilon(t) = \left[\frac{1}{E} + \frac{t}{\eta}\right]\sigma_0 \quad (t > 0)$$

加载(b):

$$\sigma = \sigma_0 H(\theta-1) \Rightarrow \mathrm{d}\sigma/\mathrm{d}t = \sigma_0 \delta(1)$$

$$\varepsilon(t) = \int_0^t \left[\frac{1}{E} + \frac{(t-\theta)}{\eta}\right]\sigma_0 \delta(1)\mathrm{d}\theta$$

$$\varepsilon(t) = \left[\frac{1}{E} + \frac{(t-1)}{\eta}\right]\sigma_0 \quad (t > 1)$$

可以看出,(b)与(a)相同,只是在时间轴上移位了,意味着材料没有老化。

7.3 对应原理

对一个时域(t-域)函数 $f(t)$ 进行拉普拉斯变换,得到对应拉普拉斯域(s-域)函数 $f(s)$。拉普拉斯变换定义为

$$L[f(t)] = f(s) = \int_0^\infty \exp(-st)f(t)\mathrm{d}t \tag{7.43}$$

在大多数情况下,拉普拉斯变换仅需要使用转换表(见表 7.1)解析获得。对式(7.41)、式(7.42)进行拉普拉斯变换,得

$$\varepsilon(s) = L[D(t)]L[\dot{\sigma}(t)] = sD(s)\sigma(s) \tag{7.44}$$

$$\sigma(s) = L[E(t)]L[\dot{\varepsilon}(t)] = sE(s)\varepsilon(s) \tag{7.45}$$

将式(7.44)乘以式(7.45),得

$$s^2 D(s) E(s) = 1 \tag{7.46}$$

或者

$$sD(s) = [sE(s)]^{-1} \tag{7.47}$$

可以看出 $sD(s)$ 是 $sE(s)$ 的倒数。这一特点类似于弹性材料的式(7.4)。

对应原理表明,所有适用于弹性材料的弹性方程,在拉普拉斯域中对线性黏弹性材料都有效。对应原理是利用标准细观方法研究由纤维和基体组成的聚合物基复合材料蠕变和松弛的基础,详见7.6节。

从拉普拉斯域到时域的逆变换更难计算

$$f(t) = L^{-1}(f(s)) \tag{7.48}$$

对此,部分分解方法[49]是一种有用的工具,可以将 $f(s)$ 分解为多个简化的函数,这样可以解析地进行拉普拉斯逆变换。另一种求解方法是通过卷积理论,参见表7.1。另外,采用极限值理论

$$\begin{cases} f(0) = \lim_{s \to \infty} [sF(s)] \\ f(\infty) = \lim_{s \to 0} [sF(s)] \end{cases} \tag{7.49}$$

可以在拉普拉斯域中直接计算材料时域内的初始和最终响应。除此之外,文献[50]中介绍了拉普拉斯逆变换的数值求解,或可采用文献[8,附录D]中所述的方法。

Carson 变换定义为

$$\hat{f}(s) = sf(s) \tag{7.50}$$

在 Carson 域中,本构式(7.41)、式(7.42)变为

$$\begin{cases} \varepsilon(s) = \hat{D}(s)\sigma(s) \\ \sigma(t) = \hat{E}(s)\varepsilon(s) \end{cases} \tag{7.51}$$

在 Carson 域中,上式类似于时域内弹性材料的应力-应变方程。此外,柔量和松弛模量的关系变为

$$\hat{D}(s) = 1/\hat{E}(s) \tag{7.52}$$

7.4 频域

傅里叶变换将时域映射到频域,定义为

$$F[f(t)] = f(\omega) = \int_{-\infty}^{\infty} \exp(-\mathrm{i}\omega t) f(t) \mathrm{d}t \tag{7.53}$$

傅里叶逆变换定义为

$$f(t) = \frac{1}{\sqrt{2\pi}} \int_{-\infty}^{\infty} \exp(-\mathrm{i}\omega t) f(\omega) \mathrm{d}\omega \tag{7.54}$$

将傅里叶变换应用于式(7.41)、式(7.42),得

$$\begin{cases} \varepsilon(\omega) = D(\omega)\dot{\sigma}(\omega) \\ \sigma(\omega) = E(\omega)\dot{\varepsilon}(\omega) \end{cases} \quad (7.55)$$

$$D(\omega) = -\frac{1}{\omega^2 E(\omega)} \quad (7.56)$$

其中:$D(\omega) = D' + iD''$ 和 $E(\omega) = E' + iE''$ 是复数。在这里 D', D'' 分别为储能柔量和耗散柔量,E', E'' 分别为储能模量和耗散模量。

使用标准复数分析方法,可得

$$\begin{cases} D' = \dfrac{E'}{E'^2 + E''^2} \\ D'' = \dfrac{E''}{E'^2 + E''^2} \end{cases} \quad (7.57)$$

频域具有明确的物理意义。正弦应力 $\sigma(\omega,t) = \sigma_0 \exp(-i\omega t)$ 作用于黏弹性材料,其响应为异相位的正弦应变 $\varepsilon(\omega,t) = \varepsilon_0 \exp(-i\omega t + \phi)$。此外,复数柔量为 $D(\omega) = \varepsilon(\omega,t)/\sigma(\omega,t)$,复数松弛模量为 $E(\omega) = \sigma(\omega,t)/\varepsilon(\omega,t)$,它们互为倒数。

7.5 频谱表示

Prony 级数模型式(7.26)对聚合物的特性进行了物理解释,其表示为系列 Maxwell 模型串联,每个 Maxwell 模型都有自己的衰减时间。极限情况下,真实的聚合物特性可以由无限数量的此类模型表示[51],因此,有

$$E_t - E_\infty = \int_{-\infty}^{\infty} H(\theta) \exp(-t/\theta) \mathrm{d}\ln\theta = \int_{-\infty}^{\infty} \frac{H(\theta)}{\theta} \exp(-t/\theta) \mathrm{d}\theta \quad (7.58)$$

式中:$H(\theta)$ 为松弛谱[52]。

对于柔量,有

$$D(t) - D_0 = \frac{t}{\eta} + \int_{-\infty}^{\infty} \frac{L(\theta)}{\theta}[1 - \exp(-t/\theta)] \mathrm{d}\theta \quad (7.59)$$

式中:$L(\theta)$ 为延迟谱[52];D_0 为弹性柔量;η 为流体材料的渐进黏度,当 $\eta \to \infty$ 时,材料为固体(另见文献[53])。

7.6 黏弹性复合材料的细观力学

7.6.1 一维情况

回顾 Carson 域中的本构方程式(7.51)。由对应原理,弹性材料的所有细观

力学方程都适用于 Carson 域中的线性黏弹性材料。例如,Reuss 细观力学模型假设基体和纤维应变是相同的。因此,复合材料的刚度 C 是各组分(纤维和基质)刚度与它们体积分数 V_m,V_f 加权后线性组合而成

$$C = V_m C^m + V_f C^f \tag{7.60}$$

其中:由式(6.1),$A^m = A^f = I$,考虑到黏弹性材料的对应原理(见第 7.3 节),通过类比式(7.60)形式,可将 Carson 域的刚度张量简化为

$$\hat{C} = V_m \hat{C}^m + V_f \hat{C}^f \tag{7.61}$$

从上式可得拉普拉斯域的刚度张量(见式(7.47))为

$$C(s) = \frac{1}{s}\hat{C} \tag{7.62}$$

最后,通过拉普拉斯逆变换式(7.48),可以获得时域的刚度张量为

$$C(t) = L^{-1}[C(s)] \tag{7.63}$$

例 7.3 在时域内推导复合材料单向板的横向柔量 $D_2(t)$,其纤维是弹性的,基体是黏弹性的,基体模型为 $D_m = 1/E_m + t/\eta_m$。取 $0 < t < 0.1$,$E_f = 10$,$V_f = 0.5$,$E_m = 5$,$\eta_m = 0.05$,绘制 D_f,$D_m(t)$ 和 $D_2(t)$。使用 Reuss 模型并讨论结果。

解 纤维的弹性行为和基体的黏弹性行为定义如下:

纤维(弹性):$E_f = $ 常数 $\to D_f = \dfrac{1}{E_f}$

基体(Maxwell 模型式(7.12),$E_m = E_0$,$\eta_m = \tau E_0$):$D_m = \dfrac{1}{E_m} + \dfrac{t}{\eta_m}$

进行拉普拉斯变换,有

$D_f(s) = \dfrac{1}{sE_f}$,因为 $\dfrac{1}{E_f}$ 是常数。

$$D_m(s) = \frac{1}{sE_m} + \frac{1}{s^2\eta_m}$$

然后,进行 Carson 变换,

$$\hat{D}_f(s) = sD_f(s) = \frac{1}{E_f}$$

$$\hat{D}_m(s) = sD_m(s) = \frac{1}{E_m} + \frac{t}{s\eta_m}$$

使用 Reuss 模型计算复合材料属性,有

$$\hat{D}_2 = V_f \hat{D}_f + V_m \hat{D}_m$$

$$\hat{D}_2 = V_f \frac{1}{E_f} + V_m \left(\frac{1}{E_m} + \frac{1}{s\eta_m}\right)$$

返回拉普拉斯域,有

$$D_2(s) = \frac{V_f}{sE_f} + \frac{V_m}{sE_m} + \frac{V_m}{s^2\eta_m}$$

返回时域(拉普拉斯逆变换),有

$$D_2(t) = L^{-1}(D_2(s)) = \frac{V_f}{E_f} + \frac{V_m(E_m t + \eta_m)}{E_m \eta_m}$$

为绘制曲线,取 $E_f = 10, V_f = 0.5, E_m = 5, \eta_m = 0.05$,得

$$\begin{cases} D_f = 0.1 = 1/10 \\ D_m(t) = 0.2 + 20t \\ D_2(t) = 0.15 + 10t \end{cases}$$

由于 $V_f = 0.5$,因此初始柔量介于纤维柔量和基体柔量之间。弹性纤维具有恒定的柔量。复合材料的蠕变率 $1/\eta_c$ 是基体蠕变率 $1/\eta_m$ 的 $1/2$。

7.6.2 三维情况

弹性的各向同性材料的本构方程式(1.78)可以由两个材料参数 λ 和 $\mu = G$ 表示,即

$$\boldsymbol{\sigma} = (\lambda \boldsymbol{I}^{(2)} \otimes \boldsymbol{I}^{(2)} + 2\mu \boldsymbol{I}^{(4)}) : \boldsymbol{\varepsilon} \tag{7.64}$$

式中:$\boldsymbol{I}^{(2)}$、$\boldsymbol{I}^{(4)}$ 为二阶和四阶等同张量①(见附录 A)。各向同性黏弹性材料的本构方程可以由黏弹性拉梅常数 $\lambda(s)$ 和 $\mu(s)$ 表示为[53]

$$\boldsymbol{\sigma}(t) = \int_0^t \lambda(t-\theta)\boldsymbol{I}^{(2)} \otimes \boldsymbol{I}^{(2)} : \dot{\boldsymbol{\varepsilon}}(\theta)\mathrm{d}\theta + \int_0^t 2\mu(t-\theta)\boldsymbol{I}^{(4)} : \dot{\boldsymbol{\varepsilon}}(\theta)\mathrm{d}\theta$$

$$\tag{7.65}$$

使用卷积定理(表 7.1),对式(7.65)进行拉普拉斯变换,有

$$\boldsymbol{\sigma}(s) = s\lambda(s)\boldsymbol{I}^{(2)} \otimes \boldsymbol{I}^{(2)} : \boldsymbol{\varepsilon}(s) + s2\mu(s)\boldsymbol{I}^{(4)} : \boldsymbol{\varepsilon}(s) \tag{7.66}$$

或进行 Carson 变换,有

$$\hat{\boldsymbol{\sigma}}(s) = \hat{\boldsymbol{C}}(s) : \hat{\boldsymbol{\varepsilon}}(s) \tag{7.67}$$

假设纤维是弹性的,而基体是黏弹性的,后者用 Maxwell 模型表示为

$$D_m(t) = 1/E_m + t/\eta_m \tag{7.68}$$

对上式进行 Carson 变换,有

$$\hat{D}_m = 1/E_m + t/s\eta_m = \frac{E_m + s\eta_m}{s\eta_m E_m} \tag{7.69}$$

使用对应原理,得

$$\hat{E}_m = 1/\hat{D}_m = \frac{s\eta_m E_m}{E_m + s\eta_m} = \frac{sE_m}{E_m/\eta_m + s} \tag{7.70}$$

---

① 张量使用黑体字符表示,或者使用其分量形式表示。

假设基体泊松比 ν_m 为常数并利用式(1.75)，Carson 域中基体的拉梅常数为

$$\hat{\lambda}_m = \frac{\hat{E}_m \nu_m}{(1+\nu_m)(1-2\nu_m)} \tag{7.71}$$

基体的剪切模量为

$$\hat{\mu}_m = \frac{\hat{E}_m}{2(1+\nu_m)} \tag{7.72}$$

Barbero 和 Luciano[39] 使用傅里叶级数展开法得到 Carson 域中复合材料松弛张量的各个分量，此复合材料含圆柱形纤维并按正方形排布，纤维体积含量为 V_f。弹性的、横观各向同性纤维由式(1.70)和式(1.92)定义的横观各向同性张量 C' 表示，张量 C' 由纤维的轴向和径向材料参数 E_A, E_T, G_A, G_T 和 ν_T 表示。在拉普拉斯域($\tilde{\ }$)和 Carson 域($\hat{\ }$)中定义基体属性为 $\hat{\lambda}_m = s\tilde{\lambda}_m(s)$ 和 $\hat{\mu}_m = s\tilde{\mu}_m(s)$，Carson 域中复合材料松弛张量分量 $\hat{L}^*$ 变为[39]

$$\hat{L}^*_{11}(s) = \hat{\lambda}_m + 2\hat{\mu}_m - V_f(-a_4^2 + a_3^2)$$
$$\left(\frac{(2\hat{\mu}_m + 2\hat{\lambda}_m - C'_{33} - C'_{23})(a_4^2 - a_3^2)}{a_1} + \frac{2(a_4 - a_3)(\hat{\lambda}_m - C'_{12})^2}{a_1^2}\right)^{-1}$$

$$\hat{L}^*_{12}(s) = \hat{\lambda}_m + 2\hat{\mu}_m - V_f\left(\frac{(\hat{\lambda}_m - C'_{12})(a_4 - a_3)}{a_1}\right)$$
$$\left(\frac{(2\hat{\mu}_m + 2\hat{\lambda}_m - C'_{33} - C'_{23})(a_3^2 - a_4^2)}{a_1} + \frac{2(a_4 - a_3)(\hat{\lambda}_m - C'_{12})^2}{a_1^2}\right)^{-1}$$

$$\hat{L}^*_{22}(s) = \hat{\lambda}_m + 2\hat{\mu}_m - V_f\left(\frac{(2\hat{\mu}_m + 2\hat{\lambda}_m - C'_{33} - C'_{23})a_3}{a_1} - \frac{(\hat{\lambda}_m - C'_{12})^2}{a_1^2}\right)$$
$$\left(\frac{(2\hat{\mu}_m + 2\hat{\lambda}_m - C'_{33} - C'_{23})(a_3^2 - a_4^2)}{a_1} + \frac{2(a_4 - a_3)(\hat{\lambda}_m - C'_{12})^2}{a_1^2}\right)^{-1}$$

$$\hat{L}^*_{23}(s) = \hat{\lambda}_m + V_f\left(\frac{(2\hat{\mu}_m + 2\hat{\lambda}_m - C'_{33} - C'_{23})a_4}{a_1} - \frac{(\hat{\lambda}_m - C'_{12})^2}{a_1^2}\right)$$
$$\left(\frac{(2\hat{\mu}_m + 2\hat{\lambda}_m - C'_{33} - C'_{23})(a_3^2 - a_4^2)}{a_1} + \frac{2(a_4 - a_3)(\hat{\lambda}_m - C'_{12})^2}{a_1^2}\right)^{-1}$$

$$\hat{L}^*_{44}(s) = \hat{\mu}_m - V_f\left(\frac{2}{2\hat{\mu}_m - C'_{22} + C'_{23}} - \left(2S_3 - \frac{4S_7}{2-2\nu_m}\right)\hat{\mu}_m^{-1}\right)^{-1}$$

$$\hat{L}^*_{66}(s) = \hat{\mu}_m - V_f\left((\hat{\mu}_m - C'_{66})^{-1} - \frac{S_3}{\hat{\mu}_m}\right)^{-1} \tag{7.73}$$

其中

$$a_1 = 4\hat{\mu}_m^2 - 2\hat{\mu}_m C'_{33} + 6\hat{\lambda}_m \hat{\mu}_m - 2C'_{11}\hat{\mu}_m - 2\hat{\mu}_m C'_{23} + C'_{23}C'_{11} + 4\hat{\lambda}_m C'_{12}$$
$$- 2C'^2_{12} - \hat{\lambda}_m C'_{33} - 2C'_{11}\hat{\lambda}_m + C'_{11}C'_{33} - \hat{\lambda}_m C'_{23}$$

$$a_2 = 8\hat{\mu}_m^3 - 8\hat{\mu}_m^2 C'_{33} + 12\hat{\mu}_m^2 \hat{\lambda}_m - 4\hat{\mu}_m^2 C'_{11}$$
$$- 2\hat{\mu}_m C'^2_{23} + 4\hat{\mu}_m \hat{\lambda}_m + 4\hat{\mu}_m C'_{11}C'_{33}$$
$$- 8\hat{\mu}_m \hat{\lambda}_m C'_{33} - 4\hat{\mu}_m C'^2_{12} + 2\hat{\mu}_m C'^2_{33} - 4\hat{\mu}_m C'^2_{11}\hat{\lambda}_m + 8\hat{\mu}_m \hat{\lambda}_m C'_{12}$$
$$+ 2\hat{\lambda}_m C'_{11}C'_{33} + 4C'_{12}C'_{23}\hat{\lambda}_m - 4C'_{12}C'_{33}\hat{\lambda}_m - 2\hat{\lambda}_m C'_{12}C'_{23}$$
$$- 2C'_{23}C'^2_{12} + C'^2_{23}C'_{11} + 2C'_{33}C'^2_{12} - C'_{11}C'^2_{33} + \hat{\lambda}_m C'^2_{33} - \hat{\lambda}_m C'^2_{23}$$

$$a_3 = \frac{4\hat{\mu}_m^2 + 4\hat{\lambda}_m \hat{\mu}_m - 2C'_{11}\hat{\mu}_m - 2\hat{\mu}_m C'_{33} - C'_{11}\hat{\lambda}_m - \hat{\lambda}_m C'_{33} - C'^2_{12}}{a_2}$$
$$+ \frac{C'_{11}C'_{33} + 2\hat{\lambda}_m C'_{12}}{a_2} - \frac{S_3 - \dfrac{S_6}{2-2\nu_m}}{\hat{\mu}_m}$$

$$a_4 = -\frac{2\hat{\mu}_m C'_{23} + 2\hat{\lambda}_m \hat{\mu}_m - \hat{\lambda}_m C'_{23} - C'_{11}\hat{\lambda}_m - C'^2_{12} + 2\hat{\lambda}_m C'_{12} + C'_{11}C'_{23}}{a_2}$$
$$+ \frac{S_7}{\hat{\mu}_m(2-2\nu_m)}$$

(7.74)

系数 S_3, S_6, S_7 描述了细观结构的几何特征,包括纤维的几何形状及其几何排列[36]。对于正方形阵列排布的圆柱形纤维[37],可得

$$\begin{cases} S_3 = 0.49247 - 0.47603 V_f - 0.02748 V_f^2 \\ S_6 = 0.36844 - 0.14944 V_f - 0.27152 V_f^2 \\ S_7 = 0.12346 - 0.32035 V_f + 0.23517 V_f^2 \end{cases} \quad (7.75)$$

式(7.73)中包含了松弛张量的6个独立分量,这是因为式(7.73)代表了纤维呈正方形排布的复合材料细观结构。如果纤维排布是随机的(图1.12),那么复合材料呈现横观各向同性(1.12.4节),此时松弛张量只有5个分量独立。假设 x_1 轴是复合材料的横观各向同性轴,通过式(6.7)对横观各向同性的松弛张量进行平均化后可得

$$\begin{cases} \hat{C}_{11} = \hat{L}_{11}^* \\ \hat{C}_{12} = \hat{L}_{12}^* \\ \hat{C}_{22} = \frac{3}{4}\hat{L}_{22}^* + \frac{1}{4}\hat{L}_{23}^* + \frac{1}{2}\hat{L}_{44}^* \\ \hat{C}_{23} = \frac{1}{4}\hat{L}_{22}^* + \frac{3}{4}\hat{L}_{23}^* - \frac{1}{2}\hat{L}_{44}^* \\ \hat{C}_{66} = \hat{L}_{66}^* \end{cases} \tag{7.76}$$

由于材料的横向各向同性,其他系数使用式(1.70)可得到。例如,$\hat{C}_{44} = (\hat{C}_{22} - \hat{C}_{23})/2$。至此完成Carson域中松弛张量的推导$\hat{C}_{ij} = sC_{ij}(s)$。然后,对每个系数进行拉普拉斯逆变换,可得出时域内刚度张量系数。

$$C_{ij}(t) = L^{-1}\left[\frac{1}{s}\hat{C}_{ij}\right] \tag{7.77}$$

文献[50]中提供了拉普拉斯逆变换的MATLAB求解程序,可从网站[5]中获得,文献[8,附录D]中提供了另一种算法。

例7.4 考虑一个由横观各向同性纤维制备的体积分数为60%的复合材料,纤维的轴向性能为$E_A = 168.4\text{GPa}$,$G_A = 44.1\text{GPa}$,$\nu_A = 0.443$,横向性能为$E_T = 24.8\text{GPa}$和$\nu_T = 0.005$。环氧树脂基体由Maxwell模型式(7.12)表示,其中$E_0 = 4.08\text{GPa}$,$\tau = 39.17\text{min}$和$\nu_m = 0.311$。绘制$0 < t < 100\text{min}$区间内关于时间函数的复合材料松弛模量$E_2(t)$曲线,并与横向模量E_2进行对比。

解 此示例已使用MATLAB求解。横向模量E_2的弹性和黏弹性值见图7.4,计算过程说明如下:

(1)编写7.6.2节中的方程式程序,利用它们计算复合材料的弹性性能的弹性值,比如E_2,这些方程式的计算已编写在PMMViscoMatrix.m程序中。

(2)在Carson域中利用Maxwell模型方程式(7.15),基体的弹性模量E_0由$\hat{E}_0$替代(见PMMViscoMatrix.m),具体如下:

① 代码实现(式(7.73)~式(7.76))的部分输出为Carson域中松弛模量关于s的方程式,注意程序中需要声明变量s为符号。

② 除以s后返回到拉普拉斯域。

③ 使用文献[50]中推导的invlapFEAcomp函数将其变换至时域中。

④ 用黏弹性模型方程拟合$E_2(t)$的数值。通常,可以使用相同的模型方程进行复合材料和基体的松弛模量计算,本例选用Maxwell模型。此步的实现通过程序fitfunFEAcomp.m。

MATLAB 代码程序 PMMViscoMatrix.m、invlapFEAcomp.m 和 fitfunFEAcomp. m 可在网站[5]中获得。图 7.4 给出了计算结果。复合材料的全部 Maxwell 模型参数计算见例 7.7。

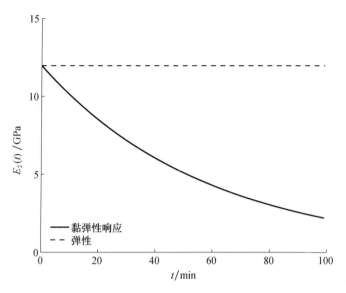

图 7.4　横向模量 E_2 的弹性和黏弹性值

7.7　黏弹性复合材料的宏观力学

7.7.1　平衡对称层合板

平衡对称层合板的面内黏弹性行为可以通过 1.15 节（表观层合材料性能）中介绍的方法获得，区别是需要在 Carson 域中进行。首先，利用式(7.76)计算每个铺层在铺层坐标下 Carson 域中的刚度。旋转每个铺层的刚度阵至层合板坐标系下，然后通过式(1.102)进行平均。通过式(1.105)在 Carson 域中得到层合板的工程参数，并除以 s 返回拉普拉斯域。最后，通过拉普拉斯逆变换得到时域的层合板刚度。然后，通过一个模型方程进行拟合，如例 7.4 所示。

7.7.2　常规层合板

得益于对应原理，经典层合理论（CLT，第 3 章）中的应力–应变关系对于 Carson 域中线性黏弹性层合复合材料是有效的。Carson 域中层合板的 A、B、D、H 矩阵可以通过一阶剪切变形理论（FSDT，3.1.1 节）计算得到。这种方法在文献[54]被采用。

7.8 黏弹性复合材料的 FEA

大多数商业软件可以进行各向同性材料的黏弹性(蠕变)分析。但对于聚合物基复合材料的黏弹性分析还是有一定的局限性。

然而,用户可以通过在商业软件中编写自定义程序,实现对本章中公式的求解。由于本章中使用的方法不依赖于应力,而是线性黏弹性方法,其实现方式并不复杂,因此通过用户定义程序求解相对简单。例 7.7 中使用了 UMAT 子程序来实现黏弹性公式求解。

例 7.5 计算 $[0/90_8]_S$ 层合板的蠕变响应。每铺层的厚度 $t_k = 1.25\mathrm{mm}$。层合板的宽度为 $2b = 20\mathrm{mm}$,长度为 $2L = 80\mathrm{mm}$。在 $x = L$ 的自由端采用耦合约束,施加 $P = 1000\mathrm{N}$ 的载荷,以实现均匀的应力 σ_x。每铺层均采用实体单元划分网格,通过施加对称边界条件建立层合板的 1/8 模型。铺层的性能为各向同性,$E_0 = 11957\mathrm{MPa}$,$\nu_{23} = 0.1886$,$g_1 = 0.999$(Maxwell),$k_1 = 0.999$(泊松常数),$\tau = 58.424\mathrm{min}$。

解 对例 5.2 模型的修改,可以得到本例的计算模型,具体步骤如下:

i. 找到例 5.2 的 cae 文件。

Menu: File,Set Work Directory,[C:\SIMULIA\User\Ex_7.5],OK
Menu: File,Open,[C:\SIMULIA\User\Ex_5.2\Ex_5.2.cae],OK
Menu: File,Save As,[C:\SIMULIA\User\Ex_7.5\Ex_7.5.cae],OK

ii. 修改部件。

Module: Part
    # change thickness of 90-deg layer editing the section solid extrude
    # on the. mdb tree,# expand: Parts(1),Part-1,Features
    # right-click: Solid extrude,Edit,Depth [11.25],OK
    # move the partition plane
    # right-click: Datum plane-1,Edit,Offset [10],OK

iii. 修改材料和截面属性。由于 Abaqus 只能进行各向同性材料的黏弹性分析,因此需删除例 5.2 中定义的材料 Material-1。

Module: Property
Menu: Material,Manager
    # select: Material-1,Delete,Yes
    # create an isotropic visco-elastic material
    Create,Name [iso-visco],Mechanical,Elasticity,Elastic

Type: Isotropic,[11975,0.1866]
Moduli time scale: Instantaneous
viscous part
Mechanical,Elasticity,Viscoelastic,Domain: Time,Time: Prony
g_1[0.999],k_1[0.999],tau_1[58.242],OK
close Material Manager pop-up window
Menu: Section,Edit,Section-1,Material: iso-visco,OK

ⅳ. 修改分析步。
Module: Step
Menu: Step,Manager
    # select: Step-1,Delete,Yes,# this also deletes the BC applied
    Create,Name[Step-1],Procedure type: General,Visco,Cont
    Tab: Basic,Time period[150]
    Tab: Incrementation,Type: Fixed,Max number of increments[200]
    Increment size[1],OK,# close Step Manager pop-up window

ⅴ. 定义约束和相互作用。
Module: Interaction
    # make surface X=40 into a rigid surface to apply a force
Menu: Tools,Reference Point,[40,10,11.25],X # RP-1
Menu: Constraint,Create
    Name[Constraint-1],Type: Coupling,Cont
    # pick: RP-1 # constant control point
    Surface,# pick the surfaces @ X=40,Done
    Constrained degrees of freedom: U1 # uncheckmark: all but U1,OK

ⅵ. 为历史输入创建一个集合。
Module: Step,Model: Model-1,Step: Step-1 # Step-1 must be selected
Menu: Tools,Set,Create
    Name[Set-1],Cont,# pick: RP-1,Done
Menu: Output,History Output Requests,Edit,H-Output-1
    Domain: Set : Set-1,Output Variables: Select from list below
    # expand: Stresses,#checkmark: S # make sure all terms are selected
    # expand: Strains,#checkmark: E # make sure all terms are selected
    # expand: Displacement/Velocity/Acceleration,# checkmark: U,OK

ⅶ. 施加荷载和边界条件。

Module: Load

  # restrain rotations UR2,UR3,of the reference point

Menu: BC,Create

  Name[BC-RP-1],Step: Initial,Type: Disp/Rota,Cont

  # pick: RP-1,Done,# checkmark: UR2,UR3,OK

  # apply load at the reference point

Menu: Load,Create

  Step: Step-1,Mechanical,Concentrated force,Cont

  # pick: RP-1,Done,CF1[1000.0],OK

ⅷ. 重新划分网格。

Module: Mesh

Menu: Mesh,Instance,Yes

ⅸ. 求解和可视化结果。

Module: Job

Menu: Job,Manager

  Create,Cont,OK,Submit,# when Completed,Results

Module: Visualization

  # plot displacement vs. time

Menu: Result,History Output,U1,Plot,# close pop-up window

Menu: Plot,Contours,On Deformed Shape

  Toolbox: Field Output,Primary,U,U1

Menu: Result,Step/Frame,Index: 0,Apply

Menu: Result,Step/Frame,Index: 1,Apply

Menu: Result,Step/Frame,Index: 150,Apply

  # analogously visualize the results of E11 and S11

结果列于表7.3。注意本例中显示的应力、应变云图均在例5.2中定义的局部坐标系下,表7.3中所列的结果提取于层合板的表层。

表7.3 例7.5的结果

时间/min	U_1/mm	ε_x/%	σ_x/MPa
0	0.0000	0.0000	0.0000
1	0.0299	0.0749	8.8889
150	0.1057	0.2643	8.8889

例 7.6　计算例 7.5 中层合板的松弛响应。在 $x=L$ 的自由端采用耦合约束,施加 $U_1=4$mm 的位移,以实现 $\varepsilon_x=0.1$ 的均匀应变。

解　对例 7.5 中的模型进行修改,可以得到大部分所需模型,具体修改步骤如下。注意,在短时间(1min)内施加初始位移通过自动增量步实现。然后,第二步通过设置固定时间增量计算 150min 内的松弛响应。

ⅰ. 找到例 7.5 的 cae 文件。
Menu: Set Work Directory,[C:\SIMULIA\User\Ex_7.6],OK
Menu: File,Open,[C:\SIMULIA\User\Ex_7.5\Ex_7.5.cae],OK
Menu: File,Save As,[C:\SIMULIA\User\Ex_7.6\Ex_7.6.cae],OK

ⅱ. 修改分析步。
Module: Step
Menu: Step,Edit,Step-1
　　Tab: Basic,Time period [1]
　　Tab: Incrementation,Type: Automatic
　　Creep/swelling/viscoelastic strain error tolerance: [1E-6],OK
Menu: Step,Create
　　Name [Step-2],Procedure type: General,Visco,Cont
　　Tab: Basic,Time period [150]
　　Tab: Incrementation,Type: Fixed,Max number of increments [200]
　　Increment Size [1],OK

ⅲ. 修改荷载和边界条件。
Module: Load
　　# delete the load at RP-1
Menu: Load,Delete,Load-1,Yes
　　# specify a displacement at RP-1
Menu: BC,Create,Step: Step-1,Disp/Rota,Cont
　　# pick: RP-1,Done,# checkmark: U1 [4.0],OK

ⅳ. 求解和可视化结果。
Module: Job
　　# create new Job. Do not overwrite the results of the creep case
Menu: Job,Manager
　　Create,Cont,OK,Submit,# when Completed,Results
Module: Visualization
　　# plot displacement vs. time

Menu: Result,History Output,U1,Plot,# close pop-up window
\# visualize the results at different times as in Example 7.5

本例的结果列于表7.4。

表7.4 例7.6的计算结果

总时间/min	步时间/min	分析步	U_1/min	ε_x	σ_x/MPa
0	0	Step-1	0.00	0.00	0.00
1	1	Step-1	4.00	0.01	1187.00
1	0	Step-2	4.00	0.01	1187.00
3	2	Step-2	4.00	0.01	1147.00
11	10	Step-2	4.00	0.01	1001.00
151	150	Step-2	4.00	0.01	91.48

例7.7 (文献[8],例7.5)计算铺层顺序为$[0/90_8]_S$层合板的松弛响应。每层铺层的厚度为$t_k=1.25$mm。层合板宽度为$2b=20$mm,长度为$2L=40$mm。在$x=L$处施加位移,以实现$\varepsilon_x=0.1$的均匀应变。每层铺层均用实体单元并应用对称条件。绘制$0<t<150$min 时间内的层合板刚度$E_x(t)$的曲线。单层板材料属性见表7.5。

表7.5 例7.7的单层板黏弹性属性

弹性模量	剪切模量	泊松比
$(E_1)_0=102417$MPa $\tau_1=16511$min	$(G_{12})_0=(G_{13})_0=5553.8$MPa $\tau_{12}=\tau_{13}=44.379$min	$\nu_{12}=\nu_{13}=0.4010$
$(E_2)_0=(E_3)_0=11975$MPa $\tau_2=\tau_3=58.424$min	$(G_{23})_0=5037.3$MPa $\tau_{23}=54.445$min	$\nu_{23}=0.1886$

解 表7.5中给出了单向板的 Maxwell 参数。在 Abaqus 中使用 UMAT 子程序,可以实现属性随时间变化的正交异性材料本构方程:

$$E_1(t)=(E_1)_0\exp(-t/\tau_1); E_2(t)=E_3(t)=(E_2)_0\exp(-t/\tau_2)$$
$$G_{12}(t)=G_{13}(t)=(G_{12})_0\exp(-t/\tau_{12})$$
$$G_{12}(t)=G_{13}(t)=(G_{12})_0\exp(-t/\tau_{12}); G_{23}(t)=(G_{23})_0\exp(-t/\tau_{23})$$

此子程序可在网站[5,umat3dvisco.for]获得。其余模型的构建步骤与例7.6类似。

i. 找到例7.6的 cae 文件,并且设置工作目录。
Menu: Set Work Directory,[C:\SIMULIA\User\Ex_7.7],OK
Menu: File,Open,[C:\SIMULIA\User\Ex_7.6\Ex_7.6.cae],OK

Menu: File,Save As,[C:\SIMULIA\User\Ex_7.7\Ex_7.7.cae],OK

ⅱ. 创建用户定义子程序中的所需的材料属性。
Module: Property
Menu: Material,Create
  Name[user],General,User Material,Type: Mechanical
  [102417 11975 0.401 0.1886 5553.8 5037.3 16551 58.424 44.379 54.445]
  # E1o,E2o,nu12o,nu23o,G12o,G23o,tau1,tau2,tau12,tau23
  # entered as a column vector
  General,Depvar,
  Number of solution-dependent state variables[6],OK

ⅲ. 修改分析步。
Module: Step
Menu: Step,Manager
  # pick: Step-1,Edit,Tab: Basic,Time period[0.001]
  Tab: Incrementation,Type: Automatic,Error tolerance[1E-6],OK
  # leave Step-2 as it is,# close Step Manager pop-up window
Step: Step-1,# selected above WS
Menu: Output,Field Output Requests,Edit,F-Output-1
  Output Variables: Edit variables[S,E,U,RF,SDV],OK
Menu: Output,History Output Requests,Edit,H-Output-1
  Output Variables: Edit variables[S11,E11,U1,RF1,SDV],OK

ⅳ. 采用二次单元重新划分网格，以避免出现沙漏刚度。
Module: Mesh
Menu: Mesh,Element Type,# select all,Done
  Family: 3D Stress,Geometric Order: Quadratic
  # checkmark: Reduced integration,OK
Menu: Mesh,Instance,Yes

ⅴ. 求解和可视化结果。
Module: Job
  # create a third Job
  # but do not to overwrite the results of previous one
Menu: Job,Manager

```
Create,Cont,Tab: General,
User subroutine file [umat3dvisco. for],OK
Submit,# when Completed,Results
```
Module: Visualization
```
# visualize the reaction force to calculate laminate stiffness
Toolbox: Field Output,Symbol,RF,RESULTANT
# to save data to Excel
# on the Left Tree
# expand: Output Databases,# expand: Job-4. odb (Job name used)
# expand: History Output,# right-click: Reaction Force
Save As,Name [XYData-RF1],OK
# expand: XYData,# right-click: XYData-RF1,Plot
# # right-click: XYData-RF1,Edit,# select: all data
# right-click: Copy,# paste in Excel
# close pop-up window
```

本例的计算结果如图 7.6 所示。

在 Abaqus 中,通常分析分为很多个分析步 $s=1,\cdots,ns$。此外,每个分析步又分为多个增量步 $i=1,\cdots,ni$。Abaqus 会记录自分析开始之后的总时间 t,包括重启动分析步的时间。此外,Abaqus 会记录每个分析步开始后的时间,称为该分析步的时间 t_s。

如图 7.5 所示,当前增量步开始时(在当前分析步)的分析步时间 t_s 传递给用户子程序中的变量 TIME(1)。在当前增量步开始时的总时间 t 传递给变量 TIME(2)。当前时间增量 Δt 传递给变量 DTIME。

这时,当前分析步时间可以如下计算:
$$t_s = \text{TIME}(1) + \text{DTIME}$$

当前(总)时间可以如下计算:
$$t = \text{TIME}(2) + \text{DTIME}$$

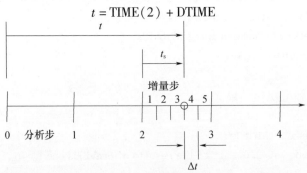

图 7.5　分析步时间 t_s,总时间 t,时间增量 Δt 的定义

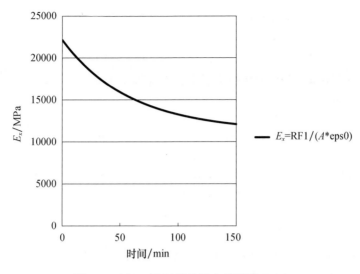

图 7.6 例 7.7 中计算的层合板刚度 $E_x(t)$

例 7.8 考虑由 40%体积分数各向同性石墨纤维和环氧树脂基体制成的复合材料,石墨纤维材料性能 $E_f = 168.4\text{GPa}$, $\nu_f = 0.443$, 环氧树脂由 Maxwell 模型表示, $E_0 = 4.082\text{GPa}$, $\tau = 39.15\text{min}$ 和 $\nu_m = 0.311$ (与时间无关)。使用六边形细观结构构建有限元细观模型(见例 6.3),在 $t = 0$ 时,施加剪切应变 $\gamma_4 = 0.02$。分别计算在 $t = 0, 20, 40, 60, 80, 100\text{min}$ 时,RVE 上的应力平均值 σ_4。

解 在 Abaqus 中使用弹性属性表示纤维。在 Abaqus 中使用黏弹性模型表示环氧树脂基体,模型定义与例 6.4 相似。求解此例的方法有两种:一种是修改 Python 脚本文件 Ex 6.4-CE.py;另一种是修改 Ex 6.4-CE.cae 文件。下面先介绍第一种方法。

首先,将 PBC_2D.py 和 srecover2D.py 文件复制到本例的工作目录。然后,将 Ex_6.4-CE.py 文件复制为 Ex_7.8.py,并打开文件进行编辑。通过修改 Ex_7.8.py 文件中的如下代码将当前目录定义为工作目录,C:\SIMULIA\User\Ex_7.8。

os.chdir(r'C:\SIMULIA\User\Ex_7.8')

然后,修改材料属性,使之包括环氧树脂基体的所有黏弹性参数。由细观结构给定的尺寸单位,其弹性模量以 TPa 为单位。

```
Ef,nuf = 168.4E-3,0.443 # TPa
Em,num,tau,g_1,k_1 = 4.082E-3,0.311,39.15,0.999,0.999 # TPa,,min,,,
```

然后,修改载荷,仅保持 $\gamma_4 = 0.02$ (模型坐标系中的 gamma_12)

```
strain = [0.000,0.0,0.01] # epsilon_11,epslion_22,gamma_12
```

接下来,需要对环氧树脂基体的粘弹性材料模型重新进行定义。纤维性能依然是弹性的。为了获取黏弹性材料的 Python 脚本,可以打开 Abaqus/CAE,然后将刚打开的模型另存为 Ex_7.8.cae。然后,按照例 7.5 的步骤定义黏弹性材料。以下是 Abaqus/CAE 中执行上述操作的详细步骤:

```
Module: Property
Menu: Material,Create
    Name [matrix],Mechanical,Elasticity,Elastic,Type: Isotropic
    Moduli time scale: Instantaneous,Data [4.082E-3,0.311]
    Mechanical,Elasticity,Viscoelastic,Domain: Time,Time: Prony
    g_1,k_1,tau_1 [0.999,0.999,39.15],OK
```

保存模型之前,将以下命令行从 Ex_7.8.rec 文件复制到 Ex_7.8.py 文件,替换以#Materials 开头的定义环氧树脂基体材料属性的两行。

```
mdb.models['Model-1'].Material(name = 'matrix')
mdb.models['Model-1'].materials['matrix'].Elastic(moduli = INSTANTANEOUS
,table = ((0.004082,0.311),))
mdb.models['Model-1'].materials['matrix'].Viscoelastic(domain = TIME,
table = ((0.999,0.999,39.15),),time = PRONY)
```

并且使用变量名替换具体数值(参数化脚本),如下:

```
mdb.models['Model-1'].Material(name = 'matrix')
mdb.models['Model-1'].materials['matrix'].Elastic(moduli = INSTANTANEOUS
    ,table = ((Em,num),))
mdb.models['Model-1'].materials['matrix'].Viscoelastic(domain = TIME,
    table = ((g_1,k_1,tau),),time = PRONY)
```

然后,保存模型以更新 .rec 文件。现在,创建一个黏弹性分析步 Step-1 以施加应变,然后创建一个分析步 Step-2,按例 7.6 中的过程一样计算相应的松弛响应。以下是 Abaqus/CAE 中执行上述操作的详细步骤:

```
Module: Step
Menu: Step,Create
    Name [Step-1],Procedure type: General,Visco,Cont
        Tab: Basic,Time period [0.001]
        Tab: Incrementation,Type: Automatic,viscoelastic tolerance [1E-6],OK
Menu: Step,Create
    Name [Step-2],Procedure type: General,Visco,Cont
        Tab: Basic,Time period [100]
```

Tab: Incrementation,Type: Fixed,Maximum number of increments: [200]
Increment Size [1],OK

保存模型之前,将以下命令行从文件 Ex_7.8.rec 中复制到 Ex_7.8.py,替换以#Step 开头的部分,以重新定义分析步 Step-1。

```
mdb.models['Model-1'].ViscoStep(cetol=1e-06,initialInc=0.001,maxInc=0.001,
    minInc=1e-08,name='Step-1',previous='Initial',timePeriod=0.001)
mdb.models['Model-1'].ViscoStep(cetol=0.0,initialInc=1.0,maxNumInc=200,
    name='Step-2',previous='Step-1',timeIncrementationMethod=FIXED,
    timePeriod=100.0)
```

然后,更新文件名以将边界条件添加到模型中。

```
execfile('C:/SIMULIA/User/Ex_7.8/PBC_2D.py')
```

然后,更新文件名

```
mdb.saveAs(pathName='C:/SIMULIA/User/Ex_7.8/Ex_7.8.cae')
```

然后,注释掉以下几行:

```
# Calculate Stresses and Strains
# execfile('srecover2D.py')
# visualize
# o3 = session.openOdb(name='Job-1.odb')
# session.viewports['Viewport: 1'].setValues(displayedObject=o3)
```

此时,更新后的脚本应与文献[5,Ex_7.8.py]一样。接下来,使用 Abaqus/CAE 新建一个空模型,然后运行脚本,如下所示:

Menu: File,New Model Database,With Standard/Explicit Model
Menu: File,Run Script [Ex_7.8.py],OK

按照下述步骤可视化结果:

Module: Job
Menu: Job,Results,Job-1
Module: Visualization
Menu: Plot,Contours,On Deformed Shape
Menu: Result,Step/Frame

注意,图 7.7 显示了通过 Menu:Result,Step/Frame 打开 Step/Frame 对话框的内容。通过此对话框,可以选择具体的分析步和增量步。在此例中,Step-1 是加载步,Step-2 是松弛响应分析步。由于在 Step-2 中使用了 Incrementation:Fixed,增量步号码与时间对应。对于依赖于时间的分析(如本例),分析帧和增量是同义的。

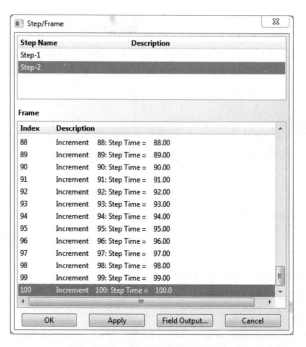

图 7.7 Step/Frame 对话框用来在结果数据中选择具体的
分析步和增量步进行输出显示

为了计算平均应力 $\bar{\sigma}_4$(模型中为 S_{12})和平均应变 $\bar{\gamma}_4$(模型中为 E_{12}),按下述步骤修改 srecover2D.py 文件。

ⅰ. 选择分析步,本例中 step-1 是施加载荷步,step-2 是松弛响应分析步。选择分析步的脚本如下:

```
# In Ex. 7.8,Step - 2 is the relaxation step
frameRepository = odb.steps['Step - 2'].frames;
```

ⅱ. 为选择增量步保存的分析帧,对脚本文件进行参数化。

```
# Get the results for frame [i],where i is the increment number
i = 0
frameS = frameRepository[i].fieldOutputs['S'].values;
```

```
frameE = frameRepository[i].fieldOutputs['E'].values;
frameIVOL = frameRepository[i].fieldOutputs['IVOL'].values;
```

其余的脚本保持不变,全部脚本文件如下(另见文献[5,srecover2D.py])

```
# srecover2D.py modified for Ex. 7.8
from visualization import *
# Open the output data base for the current Job
odb = openOdb(path = 'Job-1.odb');
myAssembly = odb.rootAssembly;


# Temporary variable to hold the frame repository speeds up the process
# In Ex. 7.8,Step-2 is the relaxation step
frameRepository = odb.steps['Step-2'].frames;

# Get the results for frame [i],where i is the increment number
i = 80
frameS = frameRepository[i].fieldOutputs['S'].values;
frameE = frameRepository[i].fieldOutputs['E'].values;
frameIVOL = frameRepository[i].fieldOutputs['IVOL'].values;

Tot_Vol = 0. ; # Total Volume
Tot_Stress = 0. ; # Stress Sum
Tot_Strain = 0. ; # Strain Sum

# Calculate Average
for II in range(0,len(frameS)):
Tot_Vol + = frameIVOL[II].data;
Tot_Stress + = frameS[II].data * frameIVOL[II].data;
Tot_Strain + = frameE[II].data * frameIVOL[II].data;

Avg_Stress = Tot_Stress/Tot_Vol;
Avg_Strain = Tot_Strain/Tot_Vol;

# from Abaqus Analysis User's Manual - 1.2.2 Conventions -
# Convention used for stress and strain components
print '2D Abaqus/Standard Stress Tensor Order: 11-22-33-12'
print 'Average stresses Global CSYS: 11-22-33-12';
print Avg_Stress;
```

```
print 'Average strain Global CSYS: 11 - 22 - 33 - 12';
print Avg_Strain;
odb.close()
```

在分析步 Step-1,增量步/分析帧 $i=0$ 时,平均应变 $\overline{\gamma}_4 = 2.0 \times 10^{-2}$,平均应力 $\overline{\sigma}_4 = 6.26 \times 10^{-5}$ TPa。实际上,平均应变在松弛响应分析中是恒定的。通过更改脚本中的分析步(Step-1 或 Step-2)和分析帧(i)的值,所有结果均可显示,具体见表7.6。

对于用 Maxwell 模型(图 7.8)表示的复合材料 23 - 平面剪切方向松弛响应,使用指数回归方法可以计算得到初始剪切模量 $G_{23}^0 = 3.13$ GPa 和松弛时间 $\tau = 39.97$ min。

表7.6 不同时间的平均应力 σ_4

帧:i	0	20	40	60	80	100
时间/min	0	20	40	60	80	100
平均应力 σ_4/MPa	62.6	38.3	23.4	14.4	8.8	5.4

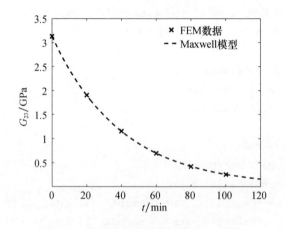

图 7.8 随时间变化的剪切模量 G_{23} 函数

习题

习题 7.1 考虑一个由 60% 体积分数各向同性石墨纤维和环氧树脂基体制成的复合材料,石墨纤维材料性能 $E_f = 168.4$ GPa,$\nu_f = 0.443$,环氧树脂性能由幂指数模型式(7.24)表示,$D_0 = 0.222$ GPa^{-1},$D_1 = 0.0135$ (GPa·min)$^{-1}$,$m = 0.17$ 和 $\nu_m = 0.311$。绘制 $0 < t < 100$ min 时间内随时间变化的复合材料松弛刚度

$C_{22}(t)$,并与复合材料弹性刚度值 C_{22} 以及基体的弹性刚度值 C_{22} 对比。

习题 7.2 考虑一个由 60% 体积分数横观各向同性石墨纤维和环氧树脂基体制成的复合材料,石墨纤维材料性能 $E_A = 168.4\text{GPa}$,$E_T = 24.82\text{GPa}$,$\nu_A = 0.443$,$\nu_T = 0.005$,$G_A = 44.13\text{GPa}$,环氧树脂性能由 Maxwell 式(7.15)模型表示,$E_0 = 4.082\text{GPa}$,$\tau = 39.15\text{min}$ 和 $\nu_m = 0.311$。绘制 $0 < t < 100\text{min}$ 时间内复合材料松弛张量的拉伸刚度分量 $C(t)$,并与复合材料弹性刚度 C 以及基体材料弹性刚度 C_m 对比。

习题 7.3 计算例 7.2 中石墨纤维/环氧树脂单向单层板(1.14 节)的 Maxwell 模型参数,绘制并对比弹性和黏弹性性能:$E_1(t)$,$E_2(t)$ 和 $G_{12}(t)$。

习题 7.4 通过用户自定义程序定义平面应力状态下横向正交各向异性单层板的 Maxwell 模型本构方程。使用例 7.3 中的黏弹性材料性能,计算铺层为 $[\pm45/90_2]_S$ 层合板的响应。每个铺层的厚度 $t_k = 1.25\text{mm}$。在层合板端部施加 $N_x = 1\text{N/mm}$ 的载荷。绘制 $0 < t < 300\text{min}$ 时间内层合板蠕变柔度 $J_x(t)$。

习题 7.5 计算铺层为 $[0/90]_S$ 层合板的 Maxwell 模型全部 9 个工程常数。每个铺层厚度为 1.25mm。层合板材料为 T300 碳纤维和 934(NR)环氧树脂,纤维体积含量 $V_f = 0.62$。环氧树脂性能由 Maxwell(7.15)模型表示 $E_0 = 4.082\text{GPa}$,$\tau = 39.15\text{min}$ 和 $\nu_m = 0.311$。T300 碳纤维为横观各向同性材料,轴向模量 $E_A = 202.8\text{GPa}$,横向模量 $E_T = 25.3\text{GPa}$,$\nu_A = 0.443$,$\nu_T = 0.005$,$G_A = 44.13\text{GPa}$,下标 A 和 T 分别表示纤维的轴向和径向(横向)。

第8章　连续介质损伤力学

复合材料中存在多种损伤模式,包括基体开裂、纤维断裂、纤维-基体脱粘等。已有大量研究试图量化每一种损伤模式,损伤随着载荷、应变、时间、循环次数等因素的演化规律以及损伤对刚度和剩余寿命的影响。连续损伤力学(CDM)通过研究损伤对材料在中尺度(单层板尺度)上的影响来描述所有这些失效模式。也就是说,CDM可以根据连续损伤变量计算每层单层板和整体层合板的退化模量。然后,使用强度或断裂力学失效准则来判断损伤的萌生。最后,根据附加参数,例如描述金属塑性的硬化指数,建立经验的硬化方程预测损伤的演化。例如,在第10章中使用一种CDM形式表现层合板的层间界面刚度退化。

硬化方程需要通过一些非标实验来调整附加参数和经验参数。由于参数针对模型进行了调整,所以模型本身的一些缺点可能会因为参数的拟合被掩盖掉。从热力学的角度来看,这些描述材料内部损伤状态的参数变量是内部状态变量,它们无法被直接测定。这与细观力学描述的损伤模型(第9章)和金属塑性模型中的状态变量相反,这些状态变量,即裂纹密度和塑性应变,都是可测的。从应用的角度来看,CDM主要缺点是需要通过附加实验来确定每个损伤模型特有的参数。此外,由于内部状态变量不可测,因此需要采用降低刚度的方法对模型中的这些附加参数进行调整,而这可能对损伤并不足够敏感[55]。

损伤的一个显著影响是刚度的降低,可利用这一点在宏观层面上定义损伤[56]。8.1节中将使用一维模型介绍这些概念。三维情况下的理论公式推导和介绍见8.2~8.4节。

8.1　一维损伤力学

一维损伤力学求解的发展涉及3个主要概念的定义:①合适可测的损伤变量;②合适的损伤激活函数;③简便的损伤演化方程或动力学方程。

8.1.1　损伤变量

考虑一根名义横截面积为$\tilde{A}$的复合材料杆件,空载,无任何损伤(图8.1(a))。当施加足够大的载荷P时,损伤开始出现(图8.1(b))。宏观层面上,损

第8章 连续介质损伤力学

伤可以通过材料刚度的降低检测到。在 CDM 中,损伤可由一个称为损伤变量的状态变量 D 表示,其代表了刚度的降低[56]。

$$D = 1 - E/\tilde{E} \tag{8.1}$$

式中: $\tilde{E}$ 为初始(无损伤状态)弹性模量; E 为损伤后模量①。早期的研究[57]将损伤概念化为由于微裂纹累积而导致的横截面积减小,即

$$D = 1 - A/\tilde{A} \tag{8.2}$$

式中: $\tilde{A}$, A 分别为初始和损伤后剩余的有效承载横截面积。

使用连续度(integrity)补充损伤定义的完整性[58],有

$$\Omega = 1 - D = A/\tilde{A} \tag{8.3}$$

其可解释为剩余的有效承载横截面积与初始横截面积的比率。注意,从原理上讲,这里的损伤是一个可测的参数,可以通过测量损伤面积、剩余面积或更实际地测量初始和剩余模量来确定。因此,从热力学的角度来说,该损伤是一个可测量的宏观状态变量,正如温度也是一个可测量的状态变量一样,它从宏观上量化了原子、分子和其他基本粒子的随机运动。虽然理论上跟踪原子和分子的搅动有可能性,但是实际操作极其困难,然而用温度计或其他装置测量温度是非常容易的。复合材料的损伤也是如此。

结构构件是根据名义横截面积 $\tilde{A}$ 进行分析的,这是设计者唯一已知的参数,而损伤状态下,剩余有效承载横截面积 $A = (1-D)\tilde{A}$ 未知。名义应力由 $\sigma = P/\tilde{A}$ 计算得到。忽略损伤结构(图 8.1(b))中代表损伤的虚拟裂纹尖端应力集中时,作用于剩余横截面积 A 上的有效应力值②为 $\tilde{\sigma} = P/A > P/\tilde{A}$。

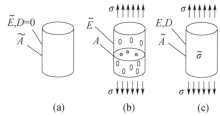

图 8.1 三种不同构型形式

(a)空载材料构型;(b)具有分布损伤的材料受力构型;(c)有效构型。

---

① 另见式(8.10)。

② 即使考虑到应力集中,代表性体积单元(RVE,见第6章)中有效应力分布的体积平均值仍为 $\tilde{\sigma} = P/A$。

因此，可以设想一种无损伤的有效构型（图8.1(c)），名义横截面积 $\tilde{A}$，在名义应力 σ 外载荷作用下，结构内部的有效应力为 $\tilde{\sigma}$，退化后的刚度为 E。这样的话，采用有效构型不仅可以允许我们使用名义几何模型进行结构分析，而且还充分考虑了有效应力的增加和损伤引起的刚度降低。

在无损伤状态下的构型(a)中，$D=0$，$\sigma = \tilde{\sigma}$，$\varepsilon = \tilde{\varepsilon}$，胡克定律表达如下：

$$\tilde{\sigma} = \tilde{E}\tilde{\varepsilon} \tag{8.4}$$

式中：$\tilde{E}$ 为常数。

在有效构型(c)中，有

$$\sigma = E\varepsilon \tag{8.5}$$

式中：E 是关于 D 的函数。

应变等效原理假设构型(b)和(c)中的应变相同，或 $\varepsilon = \tilde{\varepsilon}$。从名义应力 $\sigma = P/\tilde{A}$ 开始，令其乘以 A/A 并利用式(8.3)，最后得到有效应力 $\tilde{\sigma}$ 与名义应力 σ（应变等效情况下）之间的关系为

$$\sigma = \tilde{\sigma}[1-D] \tag{8.6}$$

利用式(8.6)、式(8.4)和式(8.5)中的 $\varepsilon = \tilde{\varepsilon}$ 条件，可以得到表观模量是损伤变量 D 的函数，即

$$E(D) = \tilde{E}[1-D] \tag{8.7}$$

能量等效原理[59]假设(b)和(c)构型中的弹性应变能相同。也就是说 $\sigma : \varepsilon = \tilde{\sigma} : \tilde{\varepsilon}$，式中

$$\begin{cases} \sigma = \tilde{\sigma}[1-D] \\ \tilde{\varepsilon} = \varepsilon[1-D] \end{cases} \tag{8.8}$$

把式(8.8)代入式(8.5)，得

$$E(D) = \tilde{E}[1-D]^2 \tag{8.9}$$

损伤变量的另一种重新定义形式为

$$D = 1 - \sqrt{E/\tilde{E}} \tag{8.10}$$

每种状态变量均有一个共轭的热力学力驱动其增长。在塑性问题中，可测的状态变量是塑性应变张量 ε^p，由其共轭热力学力–应力张量 σ 驱动。热力学损伤力 Y 被定义为与损伤状态变量 D 共轭。

动力学方程 $\dot{D}(Y)$ 控制状态变量 D 的增长，其为与状态变量 D 共轭的热力

学力 Y 的函数。理论上,任何相关变量均可以作为独立变量 Y 用来定义动力学方程 $\dot{D}(Y)$,只需要它独立于它的共轭状态变量。当损伤变量 D 是标量并用于分析一维问题时,许多作者选择应变 $\varepsilon^{[60]}$、有效应力 $\tilde{\sigma}^{[61,62]}$、剩余能量释放率(excess energy release rate) $G - 2\gamma_c^{[63]}$ 等形式作为独立变量。然而,如 8.3 节所述,基于控制问题的热力学原理形式进行选择独立变量更合理。

8.1.2 损伤阈值和激活函数

弹性域由热力学力的一个阈值定义,低于这个阈值就不会发生损伤。当受载状态处于弹性域时,损伤不会增长。当受载状态达到弹性域极限时,则会出现新的损伤。此外,弹性域还会因此改变大小或刚度。两种材料的典型一维响应如图 8.2 所示。初始时,弹性域由初始阈值定义, $\sigma \leqslant \sigma_0$ 和 $\varepsilon \leqslant \varepsilon_0$。当加载状态在此域内时,不会发生损伤。当加载状态高于阈值时,损伤发生并阈值改变。弹性域随着硬化或软化过程而演变。对于硬化的材料,应力阈值增加(图 8.2(a)),而对于软化的材料,应力阈值减小(图 8.2(b))。另一方面,对于具有硬化和软化行为的材料,其应变或有效应力阈值总是增加的,如图 8.2 所示。

弹性域可由损伤激活函数 g 定义,即

$$g = \hat{g} - \hat{\gamma} \leqslant 0 \tag{8.11}$$

式中:$\hat{g}$ 为依赖于独立变量(在一维情况下为标量 Y)的正值函数(范数);$\hat{\gamma}$ 为各向同性硬化下的更新后的损伤阈值。根据正耗散原理(the positive dissipation principle)(见 8.3 节和式(8.82)、式(8.97)),更新后的损伤阈值 $\hat{\gamma}$ 可以写为

$$\hat{\gamma} = \gamma(\delta) + \gamma_0 \tag{8.12}$$

式中:γ_0 为初始损伤阈值;γ 为正单调函数,称为硬化(或软化)函数,取决于称为损伤硬化变量的内部变量 δ。

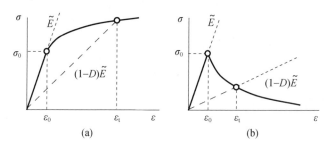

图 8.2 (a)硬化行为和(b)软化行为

在应变达到阈值 ε_0 之前不会发生损伤,卸载过程中也不会发生损伤。

8.1.3 动力学方程

损伤累积速率使用一个动力学方程表示。损伤和硬化的演化定义如下

$$\dot{D} = \dot{\lambda}\frac{\partial g}{\partial Y}; \quad \dot{\delta} = \dot{\lambda}\frac{\partial g}{\partial \gamma} \tag{8.13}$$

式中：Y 为独立变量；$\dot{\lambda} \geqslant 0$ 为损伤乘子，用来强制式(8.13)所定义的损伤与硬化演化保持一致。此外，根据 Kuhn-Tucker 条件[64]，$\dot{\lambda}$ 和 g 的值可以让我们区分两种可能的情况：无损伤增长的加载或卸载和加载时损伤增长。Kuhn-Tucker 条件如下：

$$\dot{\lambda} \geqslant 0; \quad g \leqslant 0; \quad \dot{\lambda}g = 0 \tag{8.14}$$

换句话说，Kuhn-Tucker 条件可以使我们区分如下两种不同的情况：

(1) 在弹性域范围内，无损伤状态的加载或卸载。损伤激活函数 $g<0$，由式(8.14)三式得到 $\dot{\lambda}=0$，由式(8.13)一式得到 $\dot{D}=0$。

(2) 损伤加载。在这种情况下，$\dot{\lambda}>0$，由式(8.14)三式得到 $g=0$。然后，$\dot{\lambda}$ 具体值由损伤一致性条件确定

$$g=0 \text{ 和 } \dot{g}=0 \tag{8.15}$$

例 8.1 计算拉伸载荷下一维模型的 $\dot{\lambda}$，其独立变量为有效应力 $Y=\tilde{\sigma}$，激活函数由 $\hat{g}=\tilde{\sigma}$ 定义，硬化函数由下式定义

$$\hat{\gamma}=(F_0-F_R)\delta+F_0$$

式中：F_0，F_R 分别为材料中最强细观组分单元的损伤初始阈值和强度。

解 损伤激活函数 g 定义如下：

$$g=\hat{g}-\hat{\gamma}=\tilde{\sigma}-[(F_0-F_R)\delta+F_0]\leqslant 0$$

因此，有

$$\frac{\partial g}{\partial \tilde{\sigma}}=+1; \quad \frac{\partial g}{\partial \hat{\gamma}}=-1$$

利用式(8.13)，动力学方程可以写为

$$\dot{D}=\dot{\lambda}\frac{\partial g}{\partial \tilde{\sigma}}=\dot{\lambda}; \quad \dot{\delta}=\dot{\lambda}\frac{\partial g}{\partial \hat{\gamma}}=-\dot{\lambda}$$

当新损伤出现，由一致性条件式(8.15)得

$$g=0 \quad \Rightarrow \quad \hat{\gamma}=\tilde{\sigma}$$

和

$$\dot{g}=0 \quad \Rightarrow \quad \dot{g}=\frac{\partial g}{\partial \tilde{\sigma}}\dot{\tilde{\sigma}}+\frac{\partial g}{\partial \hat{\gamma}}\dot{\hat{\gamma}}=\dot{\tilde{\sigma}}-\dot{\hat{\gamma}}=0$$

式中

$$\dot{\hat{\gamma}} = \frac{\partial \hat{\gamma}}{\partial \delta}\dot{\delta} = (F_0 - F_R)(-\dot{\lambda}) = (F_R - F_0)\dot{\lambda}$$

由式(8.15)的第二项一致性条件,得

$$\dot{\lambda} = \frac{1}{F_R - F_0}\dot{\tilde{\sigma}}$$

8.1.4 动力学方程的统计学解释

假设所有单独的损伤事件均是由材料内部的微观单元失效引起(如纤维断裂、基体裂纹、纤维-基体脱粘等)。另外,假设这些材料点的每一个均存在一个失效强度 $\tilde{\sigma}$,并且所有这些点的失效强度集合使用概率密度 $f(\tilde{\sigma})$ 表示(图8.3(b)、图8.5)。在有效应力从0增加到 $\tilde{\sigma}$ 的过程中,破坏单元的分数提供了测量损伤的一种方式,即

$$D(\tilde{\sigma}) = \int_0^{\tilde{\sigma}} f(\sigma') \mathrm{d}\sigma' = F(\tilde{\sigma}) \tag{8.16}$$

式中:$F(\tilde{\sigma})$ 为概率密度 $f(\tilde{\sigma})$ 对应的累积概率(图8.3(b));σ' 为虚拟积分变量。所以,以有效应力 $\tilde{\sigma}$ 表示的动力学方程为

$$\dot{D} = \frac{\mathrm{d}D}{\mathrm{d}\tilde{\sigma}}\dot{\tilde{\sigma}} = f(\tilde{\sigma})\dot{\tilde{\sigma}} \tag{8.17}$$

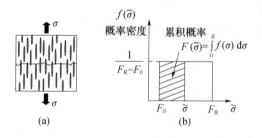

图 8.3 一维随机强度模型

8.1.5 一维随机强度模型

如8.1.3节所述,损伤累积速率由一个动力学方程表示。式(8.17)代表了一种通用的动力学方程,一旦采用特定的失效概率密度形式,该动力学方程就会具体化。

考虑一束松散的短纤维嵌入基体中,然后受到均匀的应力作用。假设所有纤维的纤维-基体界面强度相同,但纤维嵌入基体的长度是随机的。因此,纤维

拔出强度也是随机的。随机意味着纤维拔出强度在 $F_0 < \tilde{\sigma} < F_R$ 范围内任何值的概率是常数。换言之，由于纤维拔出的概率是随机的，所以不存在多纤维或少纤维拔出的应力水平。这在图 8.3 进行了说明，概率密度由式 $f(\tilde{\sigma}) = 1/(F_R - F_0)$ 表示。在式(8.17)中使用 $\tilde{\varepsilon}$ 代替 $\tilde{\sigma}$ 作为独立变量，并假设应变等效 $\varepsilon = \tilde{\varepsilon}$，得

$$f(\tilde{\varepsilon}) = \frac{\tilde{E}}{F_R - F_0}; F_0 \leq \tilde{\sigma} \leq F_R \tag{8.18}$$

文献[60]中提出了由式(8.18)表达的模型，该模型很好地表现了 Haversian 骨[65]、拉伸状态的混凝土[66]、纤维拔出损伤模式的纤维增强复合材料[67]和单向复合材料横向损伤模式的损伤累积行为。

1. 损伤激活函数

对于一维问题，选择应变作为独立变量，$\hat{g}$ 函数可以写为 $\hat{g} = \varepsilon$。因此，损伤激活函数可以写为

$$g = \varepsilon - \hat{\gamma} \leq 0 \tag{8.19}$$

式中：$\hat{\gamma}$ 为更新后的损伤阈值。假设在应变达到阈值 $\varepsilon_0 = F_0/\tilde{E}$ 之前不会发生损伤，应用一致性条件式(8.15)并利用式(8.19)，更新后的损伤阈值 $\hat{\gamma}$ 由材料中应变最大值给出，或表示为

$$\hat{\gamma} = \max(\varepsilon_0, \varepsilon) \tag{8.20}$$

2. 动力学方程

在随机强度模型的概率密度为式(8.18)条件下，动力学方程式(8.17)可以以应变 $\tilde{\varepsilon} = \varepsilon$ 的形式表达为

$$\dot{D} = \frac{dD}{d\varepsilon}\dot{\varepsilon} = \begin{cases} \tilde{E}/(F_R - F_0)\dot{\varepsilon} & (\varepsilon > \hat{\gamma}) \\ 0 & (\text{其他}) \end{cases} \tag{8.21}$$

这种情况下，独立变量为 ε，利用式(8.19)，动力学方程式(8.13)简化为

$$\dot{D} = \dot{\lambda} \tag{8.22}$$

利用 Kuhn-Tucker 条件和式(8.21)，若损伤发生，一致性条件式(8.15)简化为

$$\dot{\lambda} = \tilde{E}/(F_R - F_0)\dot{\varepsilon} \tag{8.23}$$

否则，$\dot{\lambda} = 0$。在这种特殊情况下，动力学方程已通过式(8.22)、式(8.23)明确。因此，没有必要计算硬化演变式(8.12)，因为其已由式(8.20)显式计算。注意，式(8.23)与例 8.1 的解相同，这是故意选择特殊的硬化函数产生的结果。

3. 割线本构方程

这种特例情况下,当拉伸载荷出现时,损伤变量被激活并且其值可以通过对式(8.21)积分得到,有

$$D_{\mathrm{t}} = \widetilde{E}\,\frac{\hat{\gamma} - \varepsilon_0}{F_{\mathrm{R}} - F_0}(\varepsilon > 0) \tag{8.24}$$

注意,损伤状态不取决于实际加载状态 ε;它仅取决于加载状态的历史 $\hat{\gamma}$。在本例中,假设压缩时裂纹闭合,损伤不扩展,$D_{\mathrm{c}} = 0$。数学上,单侧接触条件下的损伤可以由下列方程定义:

$$D = D_{\mathrm{t}}\,\frac{\langle\varepsilon\rangle}{|\varepsilon|} + D_{\mathrm{c}}\,\frac{\langle -\varepsilon\rangle}{|\varepsilon|} \tag{8.25}$$

式中:McCauley 运算符 $\langle x \rangle$ 定义为 $\langle x \rangle \hat{=} \frac{1}{2}(x + |x|)$。

将式(8.24)代入式(8.5),并使用应变等价关系,得

$$\sigma = E(D)\varepsilon = \begin{cases} \left(1 - \widetilde{E}\,\dfrac{\hat{\gamma} - \varepsilon_0}{F_{\mathrm{R}} - F_0}\right)\widetilde{E}\,\varepsilon\,(\varepsilon > 0) \\ \widetilde{E}\,\varepsilon\,(\varepsilon < 0) \end{cases} \tag{8.26}$$

4. 切线本构方程

在有限元计算中,有必要提供率形式的本构方程,其中应力率 $\dot{\sigma}$ 和应变率 $\dot{\varepsilon}$ 由伪时间函数表达。在上述特例情况下,可以通过对割线本构方程进行微分得到切线本构方程,即

$$\dot{\sigma} = E(D)\dot{\varepsilon} + \dot{E}(D)\varepsilon \tag{8.27}$$

当没有出现新的损伤时,即在弹性域加载或卸载时,$\dot{E}(D)$ 项为 0。当发生损伤,由式(8.20)得出 $\hat{\gamma} = \varepsilon$,并对式(8.26)中的 $E(D)$ 微分,得

$$\dot{E}(D) = -\frac{\widetilde{E}^2}{F_{\mathrm{R}} - F_0}\dot{\varepsilon} \tag{8.28}$$

如果损伤发生,把式(8.28)代入式(8.27);若未发生损伤,$\dot{E}(D) = 0$。由此,切线本构方程可以写为

$$\dot{\sigma} = \begin{cases} \left(1 - \widetilde{E}\,\dfrac{2\hat{\gamma} - \varepsilon_0}{F_{\mathrm{R}} - F_0}\right)\widetilde{E}\,\dot{\varepsilon}\,(\varepsilon > \hat{\gamma}) \\ E(D)\dot{\varepsilon}\,(\varepsilon < \hat{\gamma}) \end{cases} \tag{8.29}$$

5. 损伤模型确认

初始损伤阈值 ε_0 代表损伤初始时的最小应变,它与 F_0 成比例关系,即

$$F_0 = \widetilde{E}\,\varepsilon_0 \tag{8.30}$$

载荷作用下,拉伸试样在 $\varepsilon = \hat{\gamma} = \varepsilon_{cr}$ 时断裂,此时 $d\sigma/d\varepsilon = 0$。然后,利用式(8.29)(a)),模型中唯一的未知参数可以如下计算:

$$F_R = 2\tilde{E}\varepsilon_{cr} \quad (8.31)$$

材料参数 F_0 和 F_R 可以利用式(8.30)和式(8.31)并根据实验数据计算得到,其中 $\tilde{E}$ 为材料的未损伤状态模量。通过宏观尺度上的材料试验可以很容易得到可测值 ε_0 和 ε_{cr}。

对于 $\varepsilon_0 = 0$ 特殊情况,利用式(8.24)和 $\varepsilon = \varepsilon_{cr}$ 时的式(8.31),拉伸载荷下破坏时的临界损伤变量为

$$D_{cr} = 0.5 \quad (8.32)$$

因此,临界有效应力为

$$\tilde{\sigma}_{Tcr} = \tilde{E}\varepsilon_{cr} = 0.5F_R \quad (8.33)$$

同时利用式(8.7),临界的外加应力为

$$\sigma_{Tcr} = 0.25F_R \quad (8.34)$$

因此,在初始损伤阈值 $\varepsilon_0 = 0$ 的材料中,当 $D = 1/2$、外加应力为 $F_R/4$ 时,载荷控制下的拉伸试样发生破坏。

假设纤维-基体界面粘接强度忽略不计,可以保守计算纤维增强单层板的横向拉伸强度值。考虑极端情况,只有纤维间的基体承受横向载荷,纤维充当孔洞。在这种极端情况下,基体可以假设具有如式(8.18)所示的随机强度分布。因此,应用式(8.29)表达的随机强度模型,单层单向纤维增强板在横向拉伸载荷下的临界损伤变量可以按照式(8.32)计算得到 $D_{2t}^{cr} = 0.5$。目前,还没有合适的模型用于计算临界横向压缩损伤变量 D_{2c}^{cr}。

例 8.2 一个矩形截面梁,截面宽度 $b = 100$mm,截面高度 $2h = 200$mm,受纯弯矩载荷作用。破坏时的弯矩为 25.1MN·mm。该梁由具有随机方向的碳短纤维/环氧树脂复合材料制备,无损伤状态下的弹性模量 $\tilde{E} = 46$GPa。使用式(8.21)中的 F_R 形式得到临界弯矩表达式。假设材料在压缩状态下不发生损伤,并且在拉伸状态下具有随机强度分布,材料组分单元中最高强度未知,$F_R > 0$,$F_0 = 0$,试使用给定的数据确定 F_R 值。

解 这个问题在文献[67]中有求解过程。如图8.4所示,M 为施加的弯矩,y_c,y_t 分别为从中性轴到梁横截面相应位置的距离,在上述两个位置处产生的拉伸和压缩应力分别为 N_c,N_t。

使用 ε_t 和 ε_c 分别表示梁上下外表面的拉伸和压缩应变,y_0 表示中面到中性面的距离,并假设厚度方向上应变呈线性分布,得

第8章 连续介质损伤力学

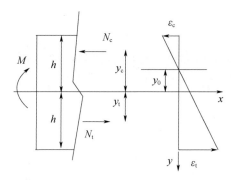

图8.4 例8.2中矩形截面梁受力图[①]

$$\varepsilon(y) = \frac{y - y_0}{h - y_0}\varepsilon_t \text{ 或 } \varepsilon(y) = \frac{-y + y_0}{h + y_0}\varepsilon_c$$

由于压缩状态下不产生损伤，压缩应力分布呈线性，由此产生的压缩应力为

$$N_c = \frac{1}{2}b(h + y_0)\widetilde{E}\varepsilon_c$$

距离 y_c 为

$$y_c = \frac{1}{3}(y_0 - 2h)$$

当梁的受拉一侧出现损伤时，中性轴会远离中面。利用式(8.26)获得拉伸应力，并在 y_0 和 h 之间对应力进行积分，有

$$N_t = \int_{y_0}^{h} dN_t = b\int_{y_0}^{h} E(D)\varepsilon(y)dy = \frac{1}{6}(h - y_0)b(3 - 2(\widetilde{E}/F_R)\varepsilon_t)\widetilde{E}\varepsilon_t$$

式中：$\widetilde{E}$ 为无损伤状态时的弹性模量。距离 y_t 为

$$y_t = \frac{1}{N_t}\int_{y_0}^{h} y dN_t = \frac{4h - 2y_0 - (\widetilde{E}/F_R)\varepsilon_t(3h + y_0)}{6 - 4(\widetilde{E}/F_R)\varepsilon_t}$$

力和力矩平衡方程为

$$N_c + N_t = 0$$
$$N_c y_c + N_t y_t = M$$

利用力平衡方程并假设应变沿厚度线性分布，就可以得到由 y_0 表示的应变 ε_t 和 ε_c，即

---

① 转载自 *Mechanics of Materials*，第8卷，第2-3期，D. Kracjcinovic，损伤力学，图2.11，第134页，版权所有(1989年)，经 Elsevier 许可。

$$\varepsilon_{\mathrm{t}} = -\frac{6hy_0}{(h-y_0)^2}\frac{F_\mathrm{R}}{\widetilde{E}}; \varepsilon_{\mathrm{c}} = \frac{6hy_0(h+y_0)}{(h-y_0)^3}\frac{F_\mathrm{R}}{\widetilde{E}}$$

利用上述关系,可以将力矩平衡方程简化为一个关于 y_0 的三次方程,即

$$M = \frac{-y_0(4h^2 + 9hy_0 + 3y_0^2)}{(h-y_0)^3}bh^2 F_\mathrm{R}$$

临界弯矩由对 y_0 微分的下式确定

$$\frac{\mathrm{d}M}{\mathrm{d}y_0} = 0$$

由上式可获得梁的破坏位置为 $y_{0\mathrm{cr}} = -0.175h$。由此,梁破坏的临界弯矩为

$$M_{\mathrm{cr}} = 0.2715 bh^2 F_\mathrm{R}$$

采用简单的测试(按照 ASTM D790 或 D6272 标准)就可以获得破坏时的弯矩;本例中 $M_{\mathrm{cr}} = 25.1 \times 10^6 \mathrm{N} \cdot \mathrm{mm}$。因此,$F_\mathrm{R}$ 可以计算得到为 $F_\mathrm{R} = 92\mathrm{MPa}$。按照结构工程中的惯例,等效弯曲强度定义为

$$\sigma_{\mathrm{Bcr}} = \frac{M_{\mathrm{cr}}}{S} = 0.407 F_\mathrm{R}$$

式中: S 为截面模量(对于矩形截面梁 $S = \frac{2}{3}bh^2$)。注意到根据式(8.34),假设采用相同动力学方程式(8.26)且材料相同的拉伸强度为 $\sigma_{\mathrm{Tcr}} = 0.25 F_\mathrm{R}$。这给出了等效弯曲强度与拉伸强度之比 $\sigma_{\mathrm{Bcr}}/\sigma_{\mathrm{Tcr}} = 1.63$,其与无钢筋混凝土的实验数据 $\sigma_{\mathrm{Bcr}}/\sigma_{\mathrm{Tcr}} = 1.6$ 吻合很好[68],其也与 ACI 规范[69]中推荐值 $\sigma_{\mathrm{Bcr}}/\sigma_{\mathrm{Tcr}} = 1.5$ 吻合较好。

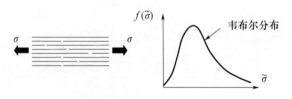

图 8.5 韦布尔分布

8.1.6 纤维方向的拉伸损伤

如果一个单层板在纤维方向上受到拉伸应力作用,则可以合理地假设基体仅分担施加荷载的一小部分,并且在加载过程中不会对基体造成损伤。那么,通过计算其中纤维束的强度,就可以准确地预测复合材料单层板的极限拉伸强度。

纤维的抗拉强度是纤维强度测试中使用的试样标距的函数。标距的长度是无效长度(也称有效载荷长度)δ,其决定了复合材料中实际的纤维强度值。无

效长度是指纤维断裂点为完全丧失承载,从断裂点开始,距离断裂点一段距离后纤维恢复承担其大部分载荷的长度(如90%)。Rosen[70]基于这一点,提出了嵌入韧性基体中的纤维纵向极限强度可以通过长度为δ的干纤维束强度来精确预测的观点。

干纤维束定义为由给定长度和直径的一定数目的平行纤维组成,如果每一根均连续,它们会承担相同的载荷。纤维束中的某一根纤维失效后,剩余的未失效的纤维平均分担载荷。干纤维束通常是指尚未与基体结合的多根纤维。当拉伸载荷缓慢施加到干纤维束中时,较弱(具有较大的缺陷尺寸)的纤维先开始失效,剩余未失效纤维上的应力相应增加。Weibull表达式[71]为

$$F(\tilde{\sigma}) = 1 - \exp\left(-\frac{\delta}{L_0}\left(\frac{\tilde{\sigma}}{\tilde{\sigma}_0}\right)^m\right) \tag{8.35}$$

累积强度概率$F(\tilde{\sigma})$表示长度为δ的纤维在等于或低于有效应力$\tilde{\sigma}$时会失效的概率。$\tilde{\sigma}_0$和m的值分别代表纤维的特征强度和纤维强度的离散程度,可通过标距为L_0的试样进行纤维强度试验确定。式(8.35)可以简化为

$$F(\tilde{\sigma}) = 1 - \exp(-\delta\alpha\tilde{\sigma}^m) \tag{8.36}$$

其中

$$\alpha = \frac{1}{L_0 \tilde{\sigma}_0^m} = \left[\frac{\Gamma(1+1/m)}{\tilde{\sigma}_{av}}\right]^m \frac{1}{L_0} \tag{8.37}$$

式中:$\Gamma(x)$为Γ(Gamma)函数[46];$\tilde{\sigma}_{av}$为标距L_0时的平均强度。

式(8.36)以未破坏纤维中真实(或表观)应力的函数形式提供了纤维束中断裂纤维的百分比,未破坏纤维百分比为$1 - F(\tilde{\sigma})$。表观应力或纤维束应力$\sigma = \sigma_b$等于施加的载荷除以纤维束总横截面积,它也等于未破坏纤维中的应力与未破坏纤维百分比的乘积,即

$$\sigma = \sigma_b = \tilde{\sigma}\exp(-\delta\alpha\tilde{\sigma}^m) \tag{8.38}$$

$\tilde{\sigma}_{max}$为式(8.38)的最大值,其很容易确定,有

$$\tilde{\sigma}_{max} = (\delta\alpha m)^{-1/m} \tag{8.39}$$

将式(8.39)代入式(8.38)得到最大(或临界)纤维束强度σ_{cr},即

$$\sigma_{cr} = (\delta\alpha m)^{-1/m}\exp(-1/m) = (\delta\alpha me)^{-1/m} \tag{8.40}$$

式中:e为自然对数的底数。复合材料的纵向拉伸强度(文献[1],式(4.82))为

$$F_{1t} = \left[V_f + \frac{E_m}{E_f}(1-V_f)\right]\sigma_{cr} \tag{8.41}$$

式中:V_f为纤维体积含量;E_f,E_m分别为纤维和基体的弹性模量。

合并式(8.36)和式(8.39),得

$$D_{1t}^{cr} = 1 - \exp(-1/m) \tag{8.42}$$

因此,纵向拉伸载荷下的临界或最大损伤变量 D_{1t}^{cr} 可由单层板破坏断裂之前断裂纤维的截面积分数计算得到[61-62],其变为仅是韦布尔模数 m 的函数。

例 8.3 由 Toraytm 碳纤维公司提供的性能数据表,其 T300 碳纤维平均拉伸强度 $\sigma_{av}=3.53\text{GPa}$,拉伸模量 $E_f=230\text{GPa}$。此外,数据表中还提供了基体环氧树脂 $E_m=4.5\text{GPa}$、纤维体积含量 $V_f=0.6$ 时单向(UD)复合材料拉伸试验结果。数据中拉伸强度为 $F_{1t}=1860\text{MPa}$。利用以上实验数据,假设韦布尔模数 $m=8.9$,请确认拉伸载荷下的损伤模型。然后,建立损伤模型本构关系并在 Abaqus 中使用一维杆单元进行计算。最后,得到单向(UD)载荷下复合材料的变-应力响应。

解

(1)损伤模型确认。根据式(8.41)并利用现有的实验数据,得

$$\sigma_{cr} = \frac{F_{1t}}{V_f + \dfrac{E_m}{E_f}(1-V_f)} = 3060\text{MPa}$$

然后,利用式(8.40),得

$$\delta\alpha = \frac{(\sigma_{cr})^{-m}}{me} = 3.92 \times 10^{-33}$$

其参数 $E_f=230\text{GPa}$,$m=8.9$ 和 $\delta\alpha=3.92\times10^{-33}$ 足以确认建立损伤模型。

(2)损伤模型建立。按照类似于 8.1.5 节所示的过程建立损伤模型,需要以下项目:

① 损伤激活函数。在本例中,选择有效应力作为独立变量。因此,损伤激活函数可以写为

$$g = \tilde{\sigma} - \hat{\gamma} \leq 0 \tag{8.43}$$

式中:$\hat{\gamma}$ 为更新后的损伤阈值。

假设初始阈值 $\sigma_0=0$,根据一致性条件式(8.15)和式(8.43),$\hat{\gamma}$ 由材料中最大有效应力值给出

$$\hat{\gamma} = \max(0, \tilde{\sigma}) \tag{8.44}$$

② 割线本构方程。在本例中,8.1.3 节中动力学方程可使用积分形式,并由式(8.36)明确给出,即

$$D = 1 - \exp(-\delta\hat{\alpha}\gamma^m) \quad (\tilde{\sigma}>0, \varepsilon>0) \tag{8.45}$$

其中损伤状态不取决于实际的载荷状态 $\tilde{\sigma}$,它只取决于载荷历史状态 $\hat{\gamma}$。

将式(8.45)代入式(8.5)和式(8.7),并利用应变等效,得到本构方程为

$$\sigma = E(D)\varepsilon = \exp(-\delta\hat{\alpha}\gamma^m)\tilde{E}\varepsilon \quad (\varepsilon > 0) \tag{8.46}$$

③ 切线本构方程。切线本构方程可以通过对割线本构方程进行微分得到,即

$$\dot{\sigma} = E(D)\dot{\varepsilon} + \dot{E}(D)\varepsilon \tag{8.47}$$

当未有新的损伤时,即在弹性域内加载或卸载,系数 $\dot{E}(D)$ 为 0。当损伤发生时,由式(8.44)得到 $\hat{\gamma} = \tilde{E}\varepsilon$,对式(8.46)中的 $E(D)$ 微分,得

$$\dot{E}(D) = -\delta\alpha\hat{\gamma}^{m-1}\exp(-\delta\alpha\hat{\gamma}^m)\tilde{E}^2\dot{\varepsilon} \tag{8.48}$$

当损伤发生时,将式(8.48)代入式(8.47)即可得到切线本构方程;当无新损伤出现时,$\dot{E}(D) = 0$。因此,切线本构方程可以写为

$$\dot{\sigma} = \begin{cases} (1 - \delta\alpha m\hat{\gamma}^m)\exp(-\delta\alpha\hat{\gamma}^m)\tilde{E}\dot{\varepsilon} & (\varepsilon > \hat{\gamma}/\tilde{E}) \\ E(D)\dot{\varepsilon} & (\varepsilon < \hat{\gamma}/\tilde{E}) \end{cases} \tag{8.49}$$

(3)数值算法。在 Abaqus 中使用 UMAT 子程序 umat1d83.for[5]实现一维损伤模型的建模。以下步骤描述了损伤本构方程显式计算的过程①。

① 在时刻 t 读出应变数据。
② 计算有效应力(假设应变等效)

$$\tilde{\sigma}_t = \tilde{E}\varepsilon_t$$

③ 更新阈值

$$\hat{\gamma}_t = \max(\hat{\gamma}_{t-1}, \tilde{\sigma}_t)$$

④ 计算损伤变量

$$D_t = 1 - \exp(-\delta\alpha(\hat{\gamma}_t)^m)$$

⑤ 计算名义应力

$$\sigma_t = (1 - D_t)\tilde{E}\varepsilon_t$$

⑥ 计算切线刚度

$$E_t^{dam} = \begin{cases} (1 - \delta\alpha m(\hat{\gamma}_t)^m)\exp(-\delta\alpha(\hat{\gamma}_t)^m)\tilde{E} & (\hat{\gamma}_t > \hat{\gamma}_{t-1}) \\ (1 - D_t)\tilde{E} & (\hat{\gamma}_t = \hat{\gamma}_{t-1}) \end{cases}$$

(4)模型响应。下面的操作过程代码和用户子程序 umat1d83.for(可在文献

---

① 对于无法对本构方程显式积分的情况,见 8.4.1 节。

[5]中获得)用于对碳(UD)纤维复合材料的一维杆件模型进行建模。名义应力-应变响应如图 8.6 中实线所示。UD 复合材料在 $\varepsilon_{cr} = 1.5\%$ 时失效,与 Toray 测试报告中的失效应变吻合较好。

ⅰ. 检索 Ex_2.4.cae 文件或按照例 2.6 中给出的操作过程重新创建模型。

```
Menu: File,Open,[C:\SIMULIA\User\Ex_2.4\Ex_2.4.cae],OK
Menu: File,Save As,[C:\SIMULIA\User\Ex_8.3\Ex_8.3.cae],OK
Menu: File,Set Work Directory,[C:\SIMULIA\User\Ex_8.3],OK
```

ⅱ. 几何修改。

```
Module: Part
    # on the left tree,expand (+) the following items:
    # expand: Model-1,# expand: Parts (1)
    # expand: Part-1,# expand: Features,# expand: Wire-1
    # right-click: Sketch,Edit
    # use dimension tool to change the length to [10]
Menu: Add,Dimension
    # pick: Line,# create the dimension below the line,[10]
    X # to close the command,Done
Menu: Feature,Regenerate
Module: Assembly
Menu: Feature,Regenerate
```

ⅲ. 删除现有的 Material-1,并根据用户材料所需的属性和参数创建新的 Material-1。修改截面属性。

```
Module: Property
Menu: Material,Manager
    # select: Material-1,Delete,Yes
    Create,Name [Material-1],General,User Material
    Mechanical Constants [203.0E3 8.9 3.92E-33]
    # or right click Read from File,File: [props.txt]
    General,Depvar,Number of state variables [2],
    # SDV1: effective stress,see umat1d83.for
    # SDV2: damage,see umat1d83.for
    OK,# close Material Manager pop-up window
Menu: Section,Edit,Section-1,Cross-sectional area: [1.0],OK
```

ⅳ. 删除载荷并添加位移载荷。
Module: Load
Menu: Load,Delete,Load-1,Yes
Menu: BC,Create
  Name [BC-2],Step: Step-1,Disp/Rota,Cont
  # pick: free-end node,Done,# checkmark: U1 [0.25],OK

ⅴ. 使用5个单元重新划分网格。
Module: Mesh
  # to make the instance "independent",on the left tree
  # expand: Assembly,# expand: Instances
  # right-click: Part-1-1,Make Independent
Menu: Seed,Instance,Approximate global size [2],Apply,OK
Menu: Mesh,Element Type
  Element Library: Standard,Geometric Order: Linear
  Family: Trus # verify T2D2 is chosen,OK
Menu: Mesh,Instance,Yes
Menu: View,Assembly Display Options
  Tab: Mesh,# checkmark: Show node labels,OK
  # create a set to track the history of some variables
Menu: Tools,Set,Create
  Name [Set-1],Type: Element,Cont
  # pick: element between nodes 5 and 6,Done

ⅵ. 更改分析步中的最多增量步数为50。伪时间域设置为50,其不是真实的时间域。另外,在输出请求中添加解相关状态变量(SDV)。
Module: Step
Step: Step-1
  # set up an incremental analysis over 50 units of pseudo time
Menu: Step,Edit,Step-1
  Tab: Basic,Time Period [50]
  Tab: Incrementation,Type: Fixed,Maximum number of incr. [50]
  Increment size [1],OK
Menu: Output,Field Output Requests,Edit,F-Output-1
  # expand: State/Field/User/Time,# checkmark: SDV,OK
Menu: Output,History Output Requests,Edit,H-Output-1
  Domain: Set: Set-1

```
# uncheckmark: all
# expand: Stresses,# expand: S,# checkmark: S11
# expand: Strains,# expand: E,# checkmark: E11
# expand: State/Field/User/Time,# checkmark: SDV,OK
```

Ⅶ. 编辑任务以调用用户材料子程序。
```
Module: Job
Menu: Job,Manager
    # select: Job-1,Edit
    Tab: General,User Subroutine,Select,[umat1d83.for],OK,OK
    Submit,# when Completed,Results
```

Ⅷ. 结果可视化。失效发生在增量步:30,此时应变 $E = 1.5 \times 10^{-2}$,Cauchy 应力 $\sigma = 3059.0$MPa,有效应力 SDV1 $= 3450$MPa,SDV2 — 损伤状态变量 $D = 0.1132$。

```
Module: Visualization
Menu: Plot,Contours,On Deformed Shape
Menu: Result,Step/Frame
    # select: Increment 30 and verify your results
    # close Step/Frame pop-up window
    # plotting the stress vs. strain results
Menu: Result,History Output
    # select: E11,Save As [e11],OK
    # select: S11,Save As [s11],OK
    # close History Output pop-up window
Menu: Tools,XYData,Create
    Source: Operate on XY data,Cont,Operators: Combine
    # select: e11,Add to Expr. ,# select: s11,Add to Expr.
    Save As [s11-vs-e11],OK,Plot Expression
```

计算结果应与图 8.6 所示的结果相似。

8.1.7 纤维方向的压缩损伤

Rosen[72]首先总结介绍了试图提高复合材料抗压强度的预测精度而提出的许多损伤模型。其文献中包括纤维屈曲模式[21,25,73-74]、折带模型[75]和屈曲引起的折带褶皱[76]。在纤维屈曲模型中,假设纤维屈曲导致整体材料破坏[72]。Rosen

模型对此进行了改进,加入了初始纤维偏折和非线性剪切刚度影响因素[73]。实验结果表明,理想排列状态下的纤维屈曲(Rosen 模型)属于缺陷敏感性问题(见4.1.1 节)。因此,即使少量的缺陷(如纤维偏折)也会导致屈曲荷载的大幅降低,从而压缩强度的实测值相对于 Rosen 的预测值偏低。复合材料中,每根纤维均存在各自不同的纤维偏折量。纤维出现偏折角为 α 的概率由高斯分布给出[25,77]。

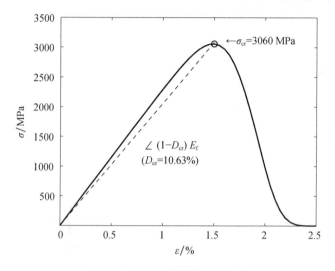

图 8.6　纤维拉伸损伤模型响应

光学测试技术[13]可用于测量横截面中每根纤维的偏折角。通过使用累积分布函数(CDF)图和概率图[25],可以证明纤维偏折分布符合是高斯分布。因此,纤维偏折的概率密度为

$$f(z) = \frac{\exp(-z^2)}{\Lambda\sqrt{2\pi}}; z = \frac{\alpha}{\Lambda\sqrt{2}} \tag{8.50}$$

式中:Λ 为标准方差;α 为连续随机变量,该例中等于偏折角。

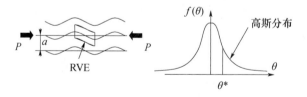

图 8.7　纤维失稳形式及纤维偏折高斯分布

CDF 给出了获得一个小于或等于给定 α 值的概率,即

$$F(z) = \operatorname{erf}(z) = \frac{2}{\sqrt{\pi}} \int_0^z \exp(-z'^2) \, dz' \tag{8.51}$$

式中:erf(z)为误差函数。

屈曲应力值与缺陷(纤维偏折)之间的关系曲线在稳定性理论中称为缺陷敏感性曲线。文献中提出的一些模型主要就是用来研究这类曲线。确定性模型在文献[21]中得到了一定发展,其与 Wang[73]提出的模型较为相似,但是确定性模型使用由式(8.52)给出的剪切响应表示。

聚合物基复合材料的剪切应力-应变响应可表示为[25,76]

$$\sigma_6 = F_6 \tanh\left(\frac{G_{12}}{F_6}\gamma_6\right) \tag{8.52}$$

式中:γ_6 为面内剪切应变;G_{12} 为初始剪切刚度,F_6 为剪切强度,两者可通过拟合应力-应变实验数据曲线得到。其完整的多项式展开[78]与实验数据吻合很好,但它们关于原点不为反对称,这导致在稳定性分析时引入了一个不对称分岔[74]。剪切试验数据可以通过多种测试方法获得,包括±45°拉伸剪切、10°偏轴拉伸、轨道剪切、Iosipescu 剪切、Arcan 剪切和扭转试验方法[79]。对于压缩状态下进行复合材料,应测量获得其非线性的剪切应力-应变曲线。

Barbero[21]推导出了关于剪切应变和偏折角的平衡应力 σ_{eq} 函数为

$$\begin{cases} \tilde{\sigma}_{eq}(\alpha,\gamma_6) = \dfrac{F_6}{2(\gamma_6+\alpha)} \dfrac{(\sqrt{2}-1)a+(\sqrt{2}+1)(b-1)}{1-a+b} \\ a = \exp(\sqrt{2}g) - \exp(2g) \\ b = \exp(2g+\sqrt{2}g) \\ g = \dfrac{\gamma_6 G_{12}}{F_6} \end{cases} \tag{8.53}$$

式(8.53)以 G_{12} 和 F_6 为未知参数。注意,如果假设剪切响应为线性关系 $\tilde{\sigma}_6 = G_{12}\gamma_6$[80],则式(8.53)关于 γ_6 没有最大值,因此嵌入线弹性基体中的具有偏折角的纤维不会屈曲。相反,若采用式(8.52)的双曲正切形式表示剪切响应,关于 γ_6 的最大值可按式(8.53)计算得到。$\tilde{\sigma}(\gamma_6)$ 曲线最大值为关于偏折角 α 的函数曲线即为缺陷敏感性曲线,其表示一根纤维(和周围的基体)的抗压强度,该抗压强度为该纤维偏折角的函数。对于偏差角为负值的情况,其满足于函数是对称的假设,$\tilde{\sigma}(-\alpha) = \tilde{\sigma}(\alpha)$。

单根纤维在达到承载极值后所承载的应力会迅速减小,因为屈曲后纤维的承载量级远低于施加的载荷。基于纤维屈曲后假设的承载大小,可以建立几种模型。假设纤维屈曲后没有后屈曲强度,不再承受载荷,则可以找到一个承载的下限。根据缺陷敏感性方程式(8.53),大偏折角的纤维屈曲应力值偏低。如果假设后屈曲强度为0,则施加的应力将重新分配到剩余的未屈曲的纤维上,这些纤维随后承受更高的有效应力 $\tilde{\sigma}(\alpha)$。在试样加载过程中的任一时刻,施加的载荷 σ(施加的应力乘以初始纤维截面积)等于有效应力乘以未屈曲的纤维截面积,即

$$\sigma = \overline{\sigma}(\alpha)[1 - D(\alpha)] \tag{8.54}$$

式中:$0 \leqslant D(\alpha) \leqslant 1$ 为每单位初始纤维截面积中的屈曲纤维截面积。对于任何有效应力值,所有超过相应偏折角的纤维都会发生屈曲。屈曲纤维的截面积 $D(\alpha)$ 与正态分布下,位于偏折角 $\pm \alpha$ 以外的纤维截面积成正比。

式(8.54)的最大值即为复合材料可承受的最大应力。因此,复合材料的抗压强度为

$$\sigma_c = \max\left[\overline{\sigma}(\alpha) \int_{-\alpha}^{\alpha} f(\alpha') \mathrm{d}\alpha'\right] \tag{8.55}$$

式中:$\overline{\sigma}(\alpha)$ 由式(8.53)给出;$f(\alpha')$ 由式(8.50)给出。式(8.55)给出了式(8.54)的最大值,其为复合材料抗压强度的唯一值,该值包含了缺陷敏感性和纤维偏折分布的影响。注意,标准方差 Λ 是一个参数,其描述的是真实的、实测的纤维偏折分布,不能任意取值以表示所有纤维的偏折。

式(8.50)中给出的分布形式不能进行闭式积分,所以式(8.55)需采用数值方法进行计算。然而,若想容易地预测抗压强度,需要推导一个显式计算公式。参考文献[21],单向纤维复合材料抗压强度的显式计算公式可由纤维偏折的标准方差 Λ、平面内剪切刚度 G_{12} 和剪切强度 F_6 共同推导出,即

$$\frac{F_{1c}}{G_{12}} = (1 + 4.76 B_a)^{-0.69} \tag{8.56}$$

式中:4.76 和 -0.69 为选择的两个用来将数值解拟合精确解的常数[21];无量纲参数 B_a 为

$$B_a = \frac{G_{12}\Lambda}{F_6} \tag{8.57}$$

在压缩破坏之前屈曲纤维的偏折角由文献[21,式(23)]给出,有

$$\begin{cases} \alpha_{cr} = a/b \\ a = 1019.011 G_{12} C_2^2 \Lambda^3 - 375.3162 C_2^3 \Lambda^4 - 845.7457 G_{12}^2 C_2 \Lambda^2 \\ \quad + g(282.1113 G_{12} C_2 \Lambda^2 - 148.1863 G_{12}^2 \Lambda - 132.6943 C_2^2 \Lambda^3) \\ b = 457.3229 C_2^3 \Lambda^3 - 660.77 G_{12} C_2^2 \Lambda^2 - 22.43143 G_{12}^2 C_2 \Lambda \\ \quad + g(161.6881 C_2^2 \Lambda^2 - 138.3753 G_{12} C_2 \Lambda - 61.38939 G_{12}^2) \\ g = \sqrt{C_2 \Lambda (8.0 C_2 \Lambda - 9.424778 G_{12})} \\ C_2 = -G_{12}^2/(4 F_6) \end{cases} \tag{8.58}$$

另外,破坏时的剪切应变为

$$\gamma_{cr} = -\alpha_{cr} + \sqrt{\alpha_{cr}^2 + \frac{3}{2}\frac{\pi F_6 \alpha_{cr}}{G_{12}}} \tag{8.59}$$

总之,当纤维增强的单层板受压时,主要的损伤模式是纤维屈曲。然而,由于纤维存在偏折缺陷,纤维的屈曲载荷比理想情况下的要低得多,以至于存在少量的纤维偏折就会导致整体屈曲载荷的大幅度降低。对于每个偏折角 α,屈曲纤维(偏折角大于 α 的纤维)的截面积分数 $D(\alpha)$ 可作为损伤状态的度量。若假设纤维没有后屈曲强度,则施加的应力将会重新分布到剩余的未屈曲纤维上,这些纤维将承受更高的有效应力。施加的应力等于有效应力与系数 $(1-D)$ 的乘积,其最大值对应于复合材料的抗压强度。因此,纵向压缩荷载下的临界损伤变量 D_{1c}^{cr} 为

$$D_{1c}^{cr} = 1 - \Omega_{1c} = 1 - \mathrm{erf}\left(\frac{\alpha_{cr}}{\Lambda\sqrt{2}}\right) \tag{8.60}$$

式中:erf 为误差函数;Λ 为实际纤维偏折高斯分布的标准方差(文献[77]中可获得试验数据);α_{cr} 为失效时的临界偏折角。

在接下来的 3 个章节中将阐述三维情况下的理论公式。

8.2 多维损伤和有效空间

建立通用多维损伤模型的第一步是定义损伤变量以及有效应力、应变空间,本节将进行介绍。第二步是定义 Helmholtz 自由能或 Gibbs 自由能的形式,并由此推导出与代表损伤和硬化的状态变量共轭的热力学力,该部分内容见 8.3 节。第三步是推导出控制损伤硬化率的动力学定律,其为关于损伤和硬化势能的函数,该部分介绍见 8.4 节。

刚度退化和随后材料响应的实验数据可用于指导选择代表损伤的变量。按照 Kachanov-Rabotnov 的方法[58,81],可以使用二阶损伤张量 D 表示正交各向异性纤维增强复合材料的损伤状态。对于高刚、高强纤维增强的复合材料,损伤状态可以使用主方向与材料方向(1、2、3)一致的二阶张量①准确表示[82-86]。这基于以下事实:主要的损伤模式是基体的微裂纹,纤维断裂和纤维-基体脱粘,所有这些损伤模式均可以概念化为平行或垂直于纤维方向的裂纹②。因此,损伤张量可以写为

$$\boldsymbol{D} = D_{ij} = D_{ij}\delta_{ij}(\text{不对 } i \text{ 求和}) \tag{8.61}$$

---

① 张量用黑体字表示,或使用下标的分量表示。
② 严格来说,损伤是横观各向同性的,因为裂纹可以在 2-3 平面中的任何方向延伸。

式中:D_i 为 $\boldsymbol{D}$ 的特征值,表示沿主材料方向 n_i 的净刚度折减;δ_{ij} 为 Kroneckerδ 函数(如果 $i=j$,则 $\delta_{ij}=1$,否则为0)。完整性张量(The integrity tensor)也为对角形式,利用能量等效原理式(8.8),得

$$\boldsymbol{\Omega}=\Omega_{ij}=\sqrt{1-D_i}\delta_{ij}(\text{不对 } i \text{ 求和}) \tag{8.62}$$

完整性张量总是对称的且为正,因为在损伤演化过程中,净减少面积必然是正值[87]。当在主体系中进行表示时,两个张量都是对角形式。下面引入一个称为损伤效应张量的对称四阶张量 $\boldsymbol{M}$,表达如下:

$$\boldsymbol{M}=M_{ijkl}=\frac{1}{2}(\Omega_{ik}\Omega_{jl}+\Omega_{il}\Omega_{jk}) \tag{8.63}$$

有效构型($\tilde{\ }$)和损伤构型(σ)之间的应力和应变转换关系如下:

$$\tilde{\boldsymbol{\sigma}}=\boldsymbol{M}^{-1}:\boldsymbol{\sigma} \quad \begin{aligned} \tilde{\boldsymbol{\varepsilon}} &= \boldsymbol{M}:\boldsymbol{\varepsilon} \\ \tilde{\boldsymbol{\varepsilon}}^e &= \boldsymbol{M}:\boldsymbol{\varepsilon}^e \\ \tilde{\boldsymbol{\varepsilon}}^p &= \boldsymbol{M}:\boldsymbol{\varepsilon}^p \end{aligned} \tag{8.64}$$

式中:上方波浪线代表在有效构型中计算的量;上标 e,p 分别表示弹性域和塑性域中的量。

根据能量等效假设[59],有效构型(图 8.1(c))中定义的本构方程形式为

$$\tilde{\boldsymbol{\sigma}}=\tilde{\boldsymbol{C}}:\tilde{\boldsymbol{\varepsilon}}^e;\ \tilde{\boldsymbol{\varepsilon}}^e=\tilde{\boldsymbol{C}}^{-1}:\tilde{\boldsymbol{\sigma}}=\tilde{\boldsymbol{S}}:\tilde{\boldsymbol{\sigma}} \tag{8.65}$$

式中:四阶张量 $\boldsymbol{C},\boldsymbol{S}$ 分别为割线刚度张量和柔度张量。

将式(8.65)代入式(8.64)得到损伤构型(图 8.1(b))中的应力 – 应变关系方程为

$$\begin{cases} \boldsymbol{\sigma}=\boldsymbol{M}:\tilde{\boldsymbol{\sigma}}=\boldsymbol{M}:\tilde{\boldsymbol{C}}:\tilde{\boldsymbol{\varepsilon}}^e,\boldsymbol{\varepsilon}^e=\boldsymbol{M}^{-1}:\tilde{\boldsymbol{\varepsilon}}^e=\boldsymbol{M}^{-1}:\tilde{\boldsymbol{S}}:\tilde{\boldsymbol{\sigma}} \\ \boldsymbol{\sigma}=\boldsymbol{M}:\tilde{\boldsymbol{C}}:\boldsymbol{M}:\boldsymbol{\varepsilon}^e,\boldsymbol{\varepsilon}^e=\boldsymbol{M}^{-1}:\tilde{\boldsymbol{S}}:\boldsymbol{M}^{-1}:\boldsymbol{\sigma} \\ \boldsymbol{\sigma}=\boldsymbol{C}:\boldsymbol{\varepsilon}^e,\boldsymbol{\varepsilon}^e=\boldsymbol{S}:\boldsymbol{\sigma} \end{cases} \tag{8.66}$$

其中

$$\boldsymbol{C}=\boldsymbol{M}:\tilde{\boldsymbol{C}}:\boldsymbol{M} \quad \boldsymbol{S}=\boldsymbol{M}^{-1}:\tilde{\boldsymbol{S}}:\boldsymbol{M}^{-1} \tag{8.67}$$

这些张量的显式表达形式见附录 B。可以看出,张量 $\boldsymbol{M}$ 是对称的,割线刚度张量和柔度张量也是对称的。

8.3 热力学公式

本节通过一系列平衡状态来描述当系统穿越由损伤和塑性的不可逆性导致

的非平衡路径时的损伤过程。通常,一个系统的当前状态(如应力、刚度、柔度)取决于当前状态信息(如应变)以及系统所经历的历程信息。第7章讨论的黏弹性材料就是这种情况。然而,对于损伤和弹塑性状态下的材料,当前状态可以用当前应变和对材料的历程效应来描述,在本章中采用损伤张量 D 和塑性应变张量 ε^p 进行表征。

8.3.1 第一定律

热力学第一定律指出,系统内能的增加等于对系统施加的热能减去系统对周围环境所做的功,即

$$\delta U = \delta Q - \delta W \tag{8.68}$$

本节中考虑的系统是一个代表性体积单元(RVE),它是最小的包含足够细观结构和不可逆过程(如损伤和塑性)特征的体积单元,并以此代表材料的整体结构。关于 RVE 的更多介绍见第 6 章。

式(8.68)的率形式表达如下:

$$\dot{U} = \dot{Q} - \dot{W} \tag{8.69}$$

式中

$$\dot{U} = \frac{\mathrm{d}}{\mathrm{d}t} \int_\Omega \rho u \mathrm{d}V \tag{8.70}$$

其中:ρ 为密度;Ω 为 RVE 的体积;u 为内能密度,它是一个内部变量且是一个势函数[①]。

对于一个可变形实体,系统做功的率等于施加在系统上的应力与应变率乘积的负值,即

$$\dot{W} = -\int_\Omega \boldsymbol{\sigma} : \dot{\boldsymbol{\varepsilon}} \mathrm{d}V \tag{8.71}$$

式中:ε 为总应变(见式(8.125))。

流入 RVE 的热量为

$$\dot{Q} = \int_\Omega \rho r \mathrm{d}V - \int_{\partial\Omega} \boldsymbol{q} \cdot \boldsymbol{n} \mathrm{d}A \tag{8.72}$$

式中:r 为单位质量的辐射热量;q 为单位面积上的热流量矢量;n 为包含体积 Ω 的表面 $\partial\Omega$ 的外法向矢量。

由于 RVE 的体积 Ω 不随时间变化,且利用散度定理[②],局部的第一定律变为

---

① 势函数的值取决于状态,而与到达该状态的路径或过程无关[88]。

② $(\int_{\partial\Omega} \boldsymbol{q} \cdot \boldsymbol{n} \mathrm{d}A = \int_\Omega \nabla \cdot \boldsymbol{q} \mathrm{d}V); \mathrm{div}(\boldsymbol{q}) = \nabla \cdot \boldsymbol{q} = \partial q_i / \partial x_i$。

$$\rho \dot{u} = \boldsymbol{\sigma} : \dot{\boldsymbol{\varepsilon}} + \rho r - \nabla \cdot \boldsymbol{q} \tag{8.73}$$

内能是储存在系统中的所有能量。例如，一个系统在发生弹性变形 $\delta\boldsymbol{\varepsilon}^e$、升温 δT 和开裂扩展 δA_c 面积损伤时，其内能 U 的变化通过下式给出①

$$\delta u = \boldsymbol{\sigma} : \delta\boldsymbol{\varepsilon}^e + C_p \delta T - (G - G_c)\delta A_c \tag{8.74}$$

式中：G 为应变能量释放率；G_c 为产生新裂纹增量的两个表面所需的表面能；$C_p = C_V$ 为实体的比热容。

一般来说，$\boldsymbol{\varepsilon} = \boldsymbol{\varepsilon}(\boldsymbol{\sigma}, u, s_\alpha)$，$s_\alpha$ 是内部变量。让我们暂时假设系统是绝热的，即 $\rho r - \nabla \cdot \boldsymbol{q} = 0$。此外，如果没有能量耗散或热传导，则 u 仅是 $\boldsymbol{\varepsilon}$ 的函数，$u = u(\boldsymbol{\varepsilon}^e)$，这里 $\boldsymbol{\varepsilon}^e$ 是弹性应变。在这种情况下，内能密度退化为应变能密度，其率形式为

$$\dot{\varphi}(\boldsymbol{\varepsilon}) = \boldsymbol{\sigma} : \dot{\boldsymbol{\varepsilon}}^e \tag{8.75}$$

并且应变余能密度为

$$\dot{\varphi}^*(\boldsymbol{\varepsilon}) = \boldsymbol{\sigma} : \dot{\boldsymbol{\varepsilon}}^e - \dot{\varphi} = \dot{\boldsymbol{\sigma}} : \boldsymbol{\varepsilon}^e \tag{8.76}$$

8.3.2 第二定律

热力学第二定律表述了热量自发从高温到低温的传递方向。从数学上讲，热流量 $\boldsymbol{q}$ 与温度 T 的梯度②方向相反，可写为

$$\boldsymbol{q} \cdot \nabla T \leqslant 0 \tag{8.77}$$

其中，等号仅在绝热过程中成立，即没有热交换，因而没有热不可逆性。

让我们想象一个从高温热源到低温热源的热传递过程，在此过程中与环境没有热交换，也没有热量损失到环境中。一旦热量开始向低温热源传递，就不可能在不增加外部功的情况下将其回流到高温热源中。也就是说，传热过程是不可逆的，即使它满足第一定律的能量守恒（式(8.73)），它没有能量损失。为方便后续使用，式(8.77)可以写为③

$$\boldsymbol{q} \cdot \nabla T^{-1} \geqslant 0 \tag{8.78}$$

第二定律证实了一个新的内部变量的引入，即熵密度 $s = s(u, \boldsymbol{\varepsilon})$，它也是一个势函数[90]。根据第二定律，熵密度率 $\dot{s} \geqslant 0$，其中等号仅在绝热过程中成立。

假设比熵（熵密度）$s = s(u, \boldsymbol{\varepsilon})$ 为一个可逆过程的势函数[90]，即

$$ds = \left(\frac{\delta q}{T}\right)_{\text{rev}} \tag{8.79}$$

---

① 根据热力学惯例和文献[89]，在这里用字母 u 表示内能，请与其他地方使用的位移矢量 $\boldsymbol{u}$ 区分开。
② 标量的梯度为一个矢量，$\nabla T = \partial T / \partial x_i$。
③ $\nabla T^{-1} = -T^{-2} \nabla T$。

$\delta Q = \int_{\Omega} \rho \delta q \mathrm{d}\Omega$,其中 $\delta q = r - \rho^{-1}\nabla \cdot \boldsymbol{q}$ 是单位质量中的输入热量,$S = \int_{\Omega} s\rho \mathrm{d}\Omega$ 是熵值。在这里用 δ 而不是 d,用来强调 δq 不是任何(势)函数的(全)微分。

定义共轭变量(见式(8.86),式(8.92),式(8.99)),注意到利用式(8.79),对于理想气体($pv = RT$)的一个可逆过程,第一定律可以改写为理想气体的 Gibbs 方程

$$\mathrm{d}u = T\mathrm{d}s - p\mathrm{d}v \tag{8.80}$$

式中:v 为比体积(单位质量体积)。

从式(8.80)中可以看出,对于计算理想气体的输入功,v 与 $-p$ 共轭;对于计算输入热能,s 与 T 共轭。

对于一个由状态变量(如 u,T,ε)表征的正回到初始状态的循环可逆过程,由式(8.79),可以得到 $\oint \mathrm{d}s = \oint \left(\dfrac{\delta q}{T}\right)_{\mathrm{rev}} = 0$。因为 s 为势函数,但 q 不是,所以对于不可逆过程,有 $\oint \mathrm{d}s = 0$ 但是 $\oint \left(\dfrac{\delta q}{T}\right)_{\mathrm{irrev}} < 0$,实验已验证这一点。在温度为 T_i 时流入 δq 热量所提供的熵输入 $\delta q/T_i$ 比在温度 $T_0 < T_i$ 时离开该循环所提供的熵输出 $\delta q/T_0$ 少(参见文献[91],例 6-2)。由于熵是一个势函数,因此可作为一个状态变量,且它总是满足 $\oint \mathrm{d}s = 0$。由此,负的净熵一定是由内部熵产来补偿。可以通过增加或减少热量(以 $\delta q/T$ 的形式)来提高或降低一个系统的熵,但是由于内部不可逆过程(如裂纹形成等),系统的熵总是增加的(正耗散原理)。

绝热系统不与周围环境热交换($\delta q = 0$),因此熵的唯一变化是由于内部不可逆性 $\dot{s} \geqslant 0$,其中等号仅在可逆过程中成立。请注意,任何系统及其周围环境均可以通过选择足够大的环境条件(如宇宙)来实现绝热条件。对于一个任意系统,总熵率总是大于(或等于)由热交换引起的输入净熵,即

$$\dot{s} \geqslant \frac{r}{T} - \frac{1}{\rho}\nabla \cdot \left(\frac{\boldsymbol{q}}{T}\right) \tag{8.81}$$

式(8.81)的左边表示系统的总熵率。式(8.81)的右边表示外部熵输入率。由此,内部熵产率为

$$\dot{\gamma}_s = \dot{s} - \frac{r}{T} + \frac{1}{\rho}\nabla \cdot \left(\frac{\boldsymbol{q}}{T}\right) \geqslant 0 \tag{8.82}$$

等式(8.82)称为局部 Clausius-Duhem 不等式。注意到 $\nabla \cdot (T^{-1}\boldsymbol{q}) = T^{-1}\nabla \cdot \boldsymbol{q} + \boldsymbol{q} \cdot \nabla T^{-1}$,有

$$\dot{\gamma}_s = \dot{s} - \frac{r}{\rho T}(\rho r - \nabla \cdot \boldsymbol{q}) + \frac{1}{\rho}\boldsymbol{q}\nabla T^{-1} \geqslant 0 \tag{8.83}$$

其中,前两项表示由局部耗散现象产生的局部熵产,最后一项表示热传导产生的

熵产①(文献[90])。假设可以识别所有局部耗散现象,它们的贡献可以写成与其共轭变量的乘积形式 $p_\alpha \dot{s}_\alpha \geq 0$,式(8.83)可以写为

$$\rho T \dot{\gamma}_s = p_\alpha \dot{s}_\alpha + T\boldsymbol{q} \cdot \nabla T^{-1} \geq 0 \tag{8.84}$$

式中:$\alpha = 1, \cdots, n$,为所考虑耗散现象的总数。注意,耗散量是标量,由热力学力 p_α 与可测状态变量 S_α 的乘积给出。该状态变量,也称为热力学通量,可以明确描述历程信息(如屈服、损伤)对材料的影响。注意,γ_s 被定义为一个熵,而不是一个耗散热量,所以它是一个势函数,而 q 不是。

对于由生长自相似圆形裂纹造成损伤的特殊情况[63],状态变量为裂纹面积 A_c,热力学力为用于裂纹扩展的有效能量 $p_c = G - G_c$,其等于能量释放率(ERR)G 和临界能量释放率 $G_c = 2\gamma_c$ 的差。后者等于表面能的两倍,因为每次出现新的裂纹均会产生两个表面(见第10章)。在这种情况下,耗散(热量)为 $\rho T \dot{\gamma} = p_c \dot{A}_c$。

根据第一定律式(8.73),考虑一个绝热过程($\rho r - \nabla \cdot \boldsymbol{q}$)并利用链式法则 $\dot{u} = \partial u / \partial \boldsymbol{\varepsilon} : \dot{\boldsymbol{\varepsilon}}$,得

$$\left[\boldsymbol{\sigma} - \rho \frac{\partial u}{\partial \boldsymbol{\varepsilon}}\right] : \dot{\boldsymbol{\varepsilon}} = 0 \tag{8.85}$$

$\dot{\boldsymbol{\varepsilon}} = 0$ 是平凡解,所以与应变共轭的应力张量可以定义为

$$\boldsymbol{\sigma} = \rho \frac{\partial u}{\partial \boldsymbol{\varepsilon}} \tag{8.86}$$

对于一个等温($\nabla T = 0$)系统,Clausius-Duhem 不等式(8.83)简化为

$$\dot{\gamma}_s = \dot{s} - \frac{1}{\rho T}(\rho r - \nabla \cdot \boldsymbol{q}) \geq 0 \tag{8.87}$$

并利用第一定律,得

$$\rho T \dot{\gamma}_s = \rho T \dot{s} - (\rho \dot{u} - \boldsymbol{\sigma} : \dot{\boldsymbol{\varepsilon}}) \geq 0 \tag{8.88}$$

Helmholtz 自由能(HFE)密度定义为

$$\psi(T, \boldsymbol{\varepsilon}, s_\alpha) = u - Ts \tag{8.89}$$

它也是一个势函数。相应的泛函是 Helmholtz 自由能② $A = \int_\Omega \rho \psi \mathrm{d}V$。Helmholtz 自由能(HFE)密度的变化率为

$$\dot{\psi} = \dot{u} - \dot{T}s - T\dot{s} \tag{8.90}$$

引入式(8.88),平衡状态下 $\dot{\gamma}_s = 0$,得

$$\rho \dot{\psi} = -\rho s \dot{T} + \boldsymbol{\sigma} : \dot{\boldsymbol{\varepsilon}} \tag{8.91}$$

---

① 即使没有局部耗散现象,$\boldsymbol{q} \cdot \nabla T^{-1} \geq 0$ 也代表了这个众所周知的事实:热流与温度梯度∇T 的方向相反。

② 使用文献[89]中的命名法。

从中可以得到,与应变共轭的应力的另一个定义形式为

$$\boldsymbol{\sigma} = \rho \frac{\partial \psi}{\partial \boldsymbol{\varepsilon}} = \boldsymbol{C} : \boldsymbol{\varepsilon} \qquad (8.92)$$

式中由于耗散现象(包括损伤)所影响的割线刚度定义为

$$\boldsymbol{C}(s_\alpha) = \rho \frac{\partial^2 \psi}{\partial \varepsilon^2} \qquad (8.93)$$

利用第一定律式(8.73)或局部 Clausius-Duhem 不等式(8.83),并展开 $\nabla \cdot (\boldsymbol{q} T^{-1}) = T^{-1} \nabla \cdot \boldsymbol{q} + \boldsymbol{q} \cdot \nabla T^{-1}$,得

$$\rho T \dot{\gamma}_s = \frac{\boldsymbol{q}}{T} \cdot \nabla T^{-1} - \rho(\dot{\psi} + s\dot{T} - \rho^{-1}\boldsymbol{\sigma}:\dot{\boldsymbol{\varepsilon}}) \geq 0 \qquad (8.94)$$

因为 $\nabla T^{-1} = -\nabla T / T^2$,Clausius-Duhem 不等式变为

$$T\rho \dot{\gamma}_s = \boldsymbol{\sigma}:\dot{\boldsymbol{\varepsilon}} - \rho(\dot{\psi} + s\dot{T}) - \frac{\boldsymbol{q}}{T} \cdot \nabla T \geq 0 \qquad (8.95)$$

因为 HFE 密度是内部变量 $\boldsymbol{\varepsilon}, T$ 和 S_α 的函数,有

$$\dot{\psi} = \frac{\partial \psi}{\partial \boldsymbol{\varepsilon}}\bigg|_{T,s_\alpha} : \dot{\boldsymbol{\varepsilon}} + \frac{\partial \psi}{\partial T}\bigg|_{\boldsymbol{\varepsilon},s_\alpha} \dot{T} + \frac{\partial \psi}{\partial s_\alpha}\bigg|_{\boldsymbol{\varepsilon},T} \dot{s}_\alpha ; \alpha = 1, \cdots, n \qquad (8.96)$$

式中:$\frac{\partial}{\partial y}\bigg|_x$ 为在常数 x 处对 y 的偏导数。

把式(8.96)代入式(8.95),并利用式(8.89)、式(8.92)和 $\nabla T^{-1} = -\nabla T / T^2$,第二定律可以重新写为

$$\dot{\gamma} = \rho T \dot{\gamma}_s = -\rho \frac{\partial \psi}{\partial s_\alpha} s_\alpha + T\boldsymbol{q} \cdot \nabla T^{-1} \geq 0 \qquad (8.97)$$

式中:$\dot{\gamma}$ 为单位体积的热耗散率。

对比式(8.97)和式(8.84),可以清楚地看出,$-\rho \frac{\partial \psi}{\partial s_\alpha} = p_\alpha$ 为与 s_α 共轭的热力学力,这也提供了热力学力的一个定义形式。

自由余能密度或 Gibbs 能量密度,定义为

$$\chi = \rho^{-1} \boldsymbol{\sigma} : \boldsymbol{\varepsilon} - \psi \qquad (8.98)$$

其也为一个势函数。相应的泛函是 Gibbs 能[①]$G = \int_\Omega \rho \chi \mathrm{d}V$。从式(8.98)中看出,它遵循应变的定义,与应力共轭,以及遵循热力学力的定义,与状态变量 s_α 共轭,有

$$\boldsymbol{\varepsilon} = \rho \frac{\partial \chi}{\partial \boldsymbol{\sigma}}; \quad p_\alpha = \rho \frac{\partial \chi}{\partial s_\alpha} = -\rho \frac{\partial \psi}{\partial s_\alpha} \qquad (8.99)$$

---

① 使用文献[89]中的命名法。

式中：s_α 包含损伤变量，因此 p_α 包含热力学损伤力（见例 8.4）。

受耗散现象（包括损伤）影响的割线弹性柔度，定义为

$$S(s_\alpha) = \rho \frac{\partial^2 \chi}{\partial \boldsymbol{\sigma}^2} \tag{8.100}$$

例 8.4 使用下面的 Gibbs 自由能来表示由横向拉伸和面内剪切载荷引起的基体裂纹的开始和累积：

$$\chi = \frac{1}{2\rho}\left[\frac{\sigma_1^2}{\tilde{E}_1} + \frac{\sigma_2^2}{(1-D_2)^2 \tilde{E}_2} + \frac{\sigma_6^2}{(1-D_6)^2 2G_{12}} - \left(\frac{\tilde{\nu}_{21}}{\tilde{E}_2} + \frac{\tilde{\nu}_{12}}{\tilde{E}_1}\right)\frac{\sigma_1 \sigma_2}{1-D_2} \right]$$

式中：$\tilde{E}_1, \tilde{E}_2, \tilde{\nu}_{12}, \tilde{\nu}_{21}, \tilde{G}_{12}$ 为单向层板未损伤状态时面内的正交各向异性弹性性能；下标 $(\ )_1$ 表示纤维方向，$(\ )_2$ 表示横向；损伤变量 D_2 和 D_6 代表基体裂纹的影响。

该方法区分了损伤激活（D_{2+}）和损伤冻结（D_{2-}），其分别对应于横向基体裂纹的张开和闭合。激活损伤变量的确定基于如下等式：

$$D_2 = D_{2+}\frac{\langle \sigma_2 \rangle}{|\sigma_2|} + D_{2-}\frac{\langle -\sigma_2 \rangle}{|\sigma_2|}$$

式中：$\langle x \rangle \hat{=} \frac{1}{2}(x + |x|)$。

对于一个平面应力状态下的单层板，仅受到面内应力作用，纤维无损伤（$D_1 = 0$），利用能量等效原理式（8.8），推导出以下表达式：ⓐ割线刚度张量；ⓑ有效应力；ⓒ与模型相关的热力学力。使用应变张量分量（$\varepsilon_1, \varepsilon_2, \varepsilon_6$）进行推导。

解 Gibbs 自由能对应力张量求导的形式定义本构模型

$$\boldsymbol{\varepsilon} = \rho \frac{\partial \chi}{\partial \boldsymbol{\sigma}} = \boldsymbol{S} : \boldsymbol{\sigma}$$

其中，$\varepsilon_6 = \gamma_6/2$，柔度张量 $\boldsymbol{S}$ 定义为

$$\boldsymbol{S} = \frac{\partial^2 \chi}{\partial \boldsymbol{\sigma}^2}$$

平面应力状态下的 Voigt 缩写形式柔度张量 $\boldsymbol{S}$ 为

$$\boldsymbol{S} = \begin{bmatrix} \dfrac{1}{\tilde{E}_1} & -\dfrac{\tilde{\nu}_{21}}{\tilde{E}_2(1-D_2)} & 0 \\ -\dfrac{\tilde{\nu}_{12}}{\tilde{E}_1(1-D_2)} & \dfrac{1}{\tilde{E}_2(1-D_2)^2} & 0 \\ 0 & 0 & \dfrac{1}{2\tilde{G}_{12}(1-D_6)^2} \end{bmatrix}$$

损伤变量出现在了 S_{12}, S_{21}, S_{22} 和 S_{66} 中。使用矩阵乘法计算张量积,请参见附录(A.14)和附录(A.20)。利用能量等效原理和式(8.67),柔度阵可以写为

$$S = M^{-1} : \tilde{S} : M^{-1}$$

式中未损伤状态的柔度阵为

$$\tilde{S} = \begin{bmatrix} \dfrac{1}{\tilde{E}_1} & -\dfrac{\tilde{\nu}_{21}}{\tilde{E}_2} & 0 \\ -\dfrac{\tilde{\nu}_{12}}{\tilde{E}_1} & \dfrac{1}{\tilde{E}_2} & 0 \\ 0 & 0 & \dfrac{1}{2\tilde{G}_{12}} \end{bmatrix}$$

缩写形式的有效损伤张量 M 与 3×3 的 Reuter 矩阵(附录(A.13))相乘后得

$$M = \begin{bmatrix} 1 & 0 & 0 \\ 0 & (1-D_2) & 0 \\ 0 & 0 & (1-D_6) \end{bmatrix}$$

刚度张量 C 可通过下式得到

$$C = M : \tilde{C} : M$$

对于该特殊例子,割线刚度张量为

$$C = \begin{bmatrix} \dfrac{\tilde{E}_1}{1-\tilde{\nu}_{21}\tilde{\nu}_{12}} & \dfrac{\tilde{\nu}_{21}\tilde{E}_2(1-D_2)}{1-\tilde{\nu}_{21}\tilde{\nu}_{12}} & 0 \\ \dfrac{\tilde{\nu}_{21}\tilde{E}_1(1-D_2)}{1-\tilde{\nu}_{21}\tilde{\nu}_{12}} & \dfrac{\tilde{E}_2(1-D_2)^2}{1-\tilde{\nu}_{21}\tilde{\nu}_{12}} & 0 \\ 0 & 0 & 2\tilde{G}_{12}(1-\tilde{D}_6)^2 \end{bmatrix}$$

有效应力 $\tilde{\sigma}$ 通过有效损伤张量 M,利用 $\tilde{\sigma} = M^{-1} : \sigma$ 关系式与名义应力 σ 联系起来,由此得到

$$\tilde{\sigma}^T = \left\{ \sigma_1, \dfrac{\sigma_2}{1-D_2}, \dfrac{\sigma_6}{1-D_6} \right\}$$

利用 $Y = \rho \partial \chi / \partial D$ 获得热力学力,对于该特殊例子,有

$$Y = \begin{Bmatrix} Y_1 \\ Y_2 \\ Y_3 \end{Bmatrix} = \begin{Bmatrix} 0 \\ \dfrac{\sigma_2^2}{(1-D_2)^3 \tilde{E}_2} - \dfrac{\sigma_1 \sigma_2 \tilde{\nu}_{12}}{(1-D_2)^2 \tilde{E}_1} \\ \dfrac{\sigma_6^2}{(1-D_6)^3 2\tilde{G}_{12}} \end{Bmatrix}$$

8.4 三维空间中的动力学定律

8.2 节中引入的损伤变量 D 是一个状态变量,其表示材料发生变化的历程信息。另外,还需要一个由热力学力表示的动力学方程来预测损伤的演化。动力学方程可以直接写成内部变量形式,如式(8.21)或用势函数的导数形式表达。对于三维空间问题,与塑性理论中使用的流动势类似,从势函数中导出动力学定律较为方便。

这里需要两个函数:一个损伤面函数 $g(Y(D), \gamma(\delta)) = 0$ 和一个凸损伤势函数 $f(Y(D), \gamma(\delta)) = 0$。损伤面在热力学力 Y 空间中界定了一个区域,热力学力 Y 在 g - 面以内时,不会发生损伤。函数 $\gamma(\delta)$ 实现了模型硬化需要的 g 和 f 函数的扩展。损伤势控制着损伤演化式(8.102)的方向。

如果损伤面和损伤势相同($g=f$),则认为该模型是关联的,并且该模型的计算会显著简化。为方便起见,假设损伤面可按照 Y 和 γ 变量分成两部分,并写成和的形式(见式(8.11)、式(8.12)),即

$$g(Y(D), \gamma(\delta)) = \hat{g}(Y(D)) - (\gamma(\delta) + \gamma_0) \tag{8.101}$$

式中:Y 为热力学力张量;$\gamma(\delta)$ 为硬化函数;γ_0 为损伤阈值;δ 为硬化变量。

由于损伤,$\hat{g}$ 可能会增长,但必须满足条件 $g<0$。可以通过逐步增加硬化函数 $\gamma(\delta)$ 来有效地实现 $\hat{g}(Y(D))$ 的增长。硬化函数 $\gamma(\delta)$ 可以按照式(8.99)、式(8.124)从耗散势中导出,前提是从硬化过程中推断出势函数的形式。或者,可以选择特定的函数的形式(如多项式、Prony 级数等),以使模型充分匹配实验数据。后一种方法在文献中更为常见。

当 $g=0$ 时,损伤发生,这时需要一个动力学方程来确定损伤 $\dot{D}$ 的大小和分量,这可以通过下式得到

$$\dot{D} = \frac{\partial D}{\partial Y} = \dot{\lambda} \frac{\partial f}{\partial Y} \tag{8.102}$$

式中:$\dot{\lambda}$ 提供了损伤增量的大小;$\partial f/\partial Y$ 为 Y - 空间中的一个方向。

为了确定损伤乘子 $\dot{\lambda}$,假设 $\dot{\lambda}$ 也与硬化变量变化率的确定关联,如下所示:

$$\dot{\delta} = \dot{\lambda}\frac{\partial g}{\partial \gamma} \tag{8.103}$$

关于 g 和 $\dot{\lambda}$ 有两种可能的情况：

(1) 如果 $g<0$，损伤不扩展且 $\dot{\lambda}=0$，所以 $\dot{D}=0$。

(2) 如果 $g=0$，损伤扩展且 $\dot{\lambda}>0$，所以 $\dot{D}>0$。

上述用 Kuhn-Tucker 条件概括如下：

$$\dot{\lambda} \geqslant 0; \quad g \leqslant 0; \quad \dot{\lambda}g = 0 \tag{8.104}$$

可根据一致性条件确定 $\dot{\lambda}$ 的值，从而得出

$$\dot{g} = \frac{\partial g}{\partial \mathbf{Y}}:\dot{\mathbf{Y}} + \frac{\partial g}{\partial \gamma}\dot{\gamma} = 0; g = 0 \tag{8.105}$$

另一方面，热力学力和硬化函数的率形式可以写为

$$\begin{cases} \dot{\mathbf{Y}} = \dfrac{\partial g}{\partial \boldsymbol{\varepsilon}}:\dot{\boldsymbol{\varepsilon}} + \dfrac{\partial \mathbf{Y}}{\partial \mathbf{D}}:\dot{\mathbf{D}} \\ \dot{\gamma} = \dfrac{\partial \gamma}{\partial \delta}\dot{\delta} \end{cases} \tag{8.106}$$

或者把式(8.103)和式(8.104)引入式(8.106)中，得到关于 $\dot{\lambda}$ 函数的表达形式：

$$\begin{cases} \dot{\mathbf{Y}} = \dfrac{\partial \mathbf{Y}}{\partial \boldsymbol{\varepsilon}}:\dot{\boldsymbol{\varepsilon}} + \dot{\lambda}\dfrac{\partial \mathbf{Y}}{\partial \mathbf{D}}:\dfrac{\partial f}{\partial \mathbf{Y}} \\ \dot{\gamma} = \dfrac{\partial \gamma}{\partial \delta}\dot{\lambda}\dfrac{\partial g}{\partial \gamma} \end{cases} \tag{8.107}$$

把式(8.107)代入式(8.105)，得

$$\dot{g} = \frac{\partial g}{\partial \mathbf{Y}}:\left[\frac{\partial \mathbf{Y}}{\partial \boldsymbol{\varepsilon}}:\dot{\boldsymbol{\varepsilon}} + \dot{\lambda}\frac{\partial \mathbf{Y}}{\partial \mathbf{D}}:\frac{\partial f}{\partial \mathbf{Y}}\right] + \frac{\partial g}{\partial \gamma}\frac{\partial \gamma}{\partial \delta}\dot{\lambda}\frac{\partial g}{\partial \gamma} = 0 \tag{8.108}$$

下一步，由 $\partial f/\partial \gamma = \partial g/\partial \gamma = -1$，式(8.108)可以进一步写为

$$\dot{g} = \frac{\partial g}{\partial \mathbf{Y}}:\frac{\partial \mathbf{Y}}{\partial \boldsymbol{\varepsilon}}:\dot{\boldsymbol{\varepsilon}} + \left[\frac{\partial g}{\partial \mathbf{Y}}:\frac{\partial \mathbf{Y}}{\partial \mathbf{D}}:\frac{\partial f}{\partial \mathbf{Y}} + \frac{\partial \gamma}{\partial \delta}\right]\dot{\lambda} = 0 \tag{8.109}$$

因此，损伤乘子 $\dot{\lambda}$ 为

$$\dot{\lambda} = \begin{cases} \mathbf{L}^{\mathrm{d}}:\dot{\boldsymbol{\varepsilon}} & (g=0) \\ 0 & (g<0) \end{cases} \tag{8.110}$$

式中

$$\mathbf{L}^{\mathrm{d}} = -\frac{\dfrac{\partial g}{\partial \mathbf{Y}}:\dfrac{\partial \mathbf{Y}}{\partial \boldsymbol{\varepsilon}}}{\dfrac{\partial g}{\partial \mathbf{Y}}:\dfrac{\partial \mathbf{Y}}{\partial \mathbf{D}}:\dfrac{\partial f}{\partial \mathbf{Y}} + \dfrac{\partial \gamma}{\partial \delta}} \tag{8.111}$$

式(8.103)、式(8.104)和式(8.110)提供了一对率形式的 D,δ，即

$$\dot{\boldsymbol{D}} = \boldsymbol{L}^{\mathrm{d}} : \frac{\partial f}{\partial \boldsymbol{Y}} : \dot{\boldsymbol{\varepsilon}}; \quad \dot{\delta} = -\dot{\lambda} \tag{8.112}$$

对本构方程 $\boldsymbol{\sigma} = \boldsymbol{C} : \boldsymbol{\varepsilon}$ 微分得到切线本构方程为

$$\dot{\boldsymbol{\sigma}} = \boldsymbol{C} : \dot{\boldsymbol{\varepsilon}} + \dot{\boldsymbol{C}} : \boldsymbol{\varepsilon} \tag{8.113}$$

其最后一项代表了刚度的降低。下一步,式(8.113)中的最后一项可以写为

$$\dot{\boldsymbol{C}} : \boldsymbol{\varepsilon} = \frac{\partial \boldsymbol{C}}{\partial \boldsymbol{D}} : \dot{\boldsymbol{D}} : \boldsymbol{\varepsilon} \tag{8.114}$$

代入式(8.112)并重新组合,有

$$\dot{\boldsymbol{C}} : \boldsymbol{\varepsilon} = \frac{\partial \boldsymbol{C}}{\partial \boldsymbol{D}} : \boldsymbol{\varepsilon} : \boldsymbol{L}^{\mathrm{d}} : \frac{\partial f}{\partial \boldsymbol{Y}} : \dot{\boldsymbol{\varepsilon}} \tag{8.115}$$

因为 $(\boldsymbol{\varepsilon}, \boldsymbol{D})$ 是状态变量,也是独立变量,所以

$$\frac{\partial \boldsymbol{\varepsilon}}{\partial \boldsymbol{D}} = 0 \tag{8.116}$$

因此,有

$$\dot{\boldsymbol{C}} : \boldsymbol{\varepsilon} = \frac{\partial \boldsymbol{\sigma}}{\partial \boldsymbol{D}} : \boldsymbol{L}^{\mathrm{d}} : \frac{\partial f}{\partial \boldsymbol{Y}} : \dot{\boldsymbol{\varepsilon}} \tag{8.117}$$

最后,将上式重新代入式(8.113),得

$$\dot{\boldsymbol{\sigma}} = \boldsymbol{C}^{\mathrm{ed}} : \dot{\boldsymbol{\varepsilon}} \tag{8.118}$$

式中:$\boldsymbol{C}^{\mathrm{ed}}$ 为损伤切线本构刚度张量,有

$$\boldsymbol{C}^{\mathrm{ed}} = \begin{cases} \boldsymbol{C} & (\dot{\boldsymbol{D}} \leqslant 0) \\ \boldsymbol{C} + \frac{\partial \boldsymbol{\sigma}}{\partial \boldsymbol{D}} : \boldsymbol{L}^{\mathrm{d}} : \frac{\partial f}{\partial \boldsymbol{Y}} & (\dot{\boldsymbol{D}} \geqslant 0) \end{cases} \tag{8.119}$$

使用数值积分方法可以得到内部变量 D、δ 和相关变量,且通常使用返回映射算法。

正如8.1.3节和8.4节所述,损伤面、损伤势和硬化函数需要许多内部的材料参数定义。这些参数不能从简单的实验中直接获得,而需要通过调整内部材料参数来确定模型,使得该模型的预测值与一些可以被实验量化的可观测的行为相吻合。综合考虑特定的模型构型、材料、实验的有效性以及开展相关实验的可行性,模型的确定会变得非常具体。因此,模型确定只能具体情况具体分析,如例8.3。

返回映射算法

返回映射算法[82,84,85]是一种求解变量 $\dot{\lambda}, \dot{\delta}, \dot{\boldsymbol{D}}, \delta$ 和 $\boldsymbol{D}$ 的数值近似算法。

内部变量通过在两个连续迭代步 $(k-1, k)$ 之间的线性求解过程中进行更新。式(8.109)的一阶线性化结果为

$$(g)_k - (g)_{k-1} = \left(\frac{\partial g}{\partial \boldsymbol{Y}} : \frac{\partial \boldsymbol{Y}}{\partial \boldsymbol{D}} : \frac{\partial f}{\partial \boldsymbol{Y}} + \frac{\partial \gamma}{\partial \delta} \right)_{k-1} \Delta \lambda_k = 0 \tag{8.120}$$

成功的迭代步得到$[g]_k = 0$
和

$$\Delta \lambda_k = \frac{-(g)_{k-1}}{\left(\frac{\partial g}{\partial \boldsymbol{Y}} : \frac{\partial \boldsymbol{Y}}{\partial \boldsymbol{D}} : \frac{\partial f}{\partial \boldsymbol{Y}} + \frac{\partial \gamma}{\partial \delta}\right)_{k-1}} \quad (8.121)$$

本构方程典型的积分完整算法如下：

(1) 通过 FEM 程序，从前一增量步中提取应变$(\boldsymbol{\varepsilon})^{n-1}$和当前增量步对应的应变增量$(\Delta\boldsymbol{\varepsilon})^n$。更新后的应变按照下式计算

$$(\boldsymbol{\varepsilon})^n = (\boldsymbol{\varepsilon})^{n-1} + (\Delta\boldsymbol{\varepsilon})^n$$

(2) 从前一增量步中提取状态变量，通过设置$k=0$开始返回映射算法的计算

$$(\boldsymbol{D})_0^n = (\boldsymbol{D})^{n-1}; (\delta)_0^n = (\delta)^{n-1}$$

(3) 更新割线刚度和 Cauchy 应力，用于计算此时的热力学力和损伤硬化

$$(\boldsymbol{C})_k^n = (\boldsymbol{M})_k^n : \tilde{\boldsymbol{C}} : (\boldsymbol{M})_k^n$$

$$(\boldsymbol{\sigma})_k^n = (\boldsymbol{C})_k^n : (\boldsymbol{\varepsilon})_k^n$$

$$(\boldsymbol{Y})_k^n; \quad (\gamma)_k^n$$

(4) 此时的损伤阈值计算

$$(g)_k = g((\boldsymbol{Y})_k^n, (\gamma(\delta))_k^n, \gamma_0)$$

这时存在两种可能的情况：
① 如果$(g)_k \leq 0$，表明没有损伤，$\Delta\lambda_k = 0$，转到步骤(8)。
② 如果$(g)_k > 0$，表明损伤扩展，$\Delta\lambda_k > 0$，转到步骤(5)。

(5) 损伤扩展。在k迭代步开始，损伤乘子由$(g)_k = 0$计算得出

$$\Delta \lambda_k = \frac{-(g)_{k-1}}{\left(\frac{\partial g}{\partial \boldsymbol{Y}}\right)_{k-1} : \left(\frac{\partial \boldsymbol{Y}}{\partial \boldsymbol{D}}\right)_{k-1} : \left(\frac{\partial f}{\partial \boldsymbol{Y}}\right)_{k-1} + \left(\frac{\partial \gamma}{\partial \delta}\right)_{k-1}}$$

(6) 利用$\Delta\lambda_k$更新状态变量

$$(D_{ij})_k^n = (D_{ij})_{k-1}^n + \Delta\lambda_k \left(\frac{\partial f}{\partial \boldsymbol{Y}}\right)_{k-1}$$

$$(\delta)_k^n = (\delta)_{k-1}^n + \Delta\lambda_k \left(\frac{\partial f}{\partial \gamma}\right)_{k-1} = (\delta)_{k-1}^n - \Delta\lambda_k$$

(7) 结束线性化的损伤过程，回到步骤(3)。

(8) 计算切线刚度张量

$$(\boldsymbol{C}^{\mathrm{ed}})^n = (\boldsymbol{C})^n + \left(\frac{\partial \boldsymbol{\sigma}}{\partial \boldsymbol{D}}\right)^n : (\boldsymbol{L}^{\mathrm{d}})^n : \left(\frac{\partial f}{\partial \boldsymbol{Y}}\right)^n$$

(9) 存储下一个载荷增量步需要使用的应力和状态变量

$$(\boldsymbol{\sigma})^n = (\boldsymbol{\sigma})^n_k; \quad (\boldsymbol{D})^n = (\boldsymbol{D})^n_k; \quad (\boldsymbol{\delta})^n = (\boldsymbol{\delta})^n_k$$

(10)积分算法结束。

例 8.5 将例 8.4 中的损伤模型应用到用户材料子程序中。如 8.4.1 节所示,使用返回映射算法实现积分计算。此外,使用以下损伤激活函数形式:

$$g = \hat{g} - \hat{\gamma} = \sqrt{\left(1 - \frac{G_{\mathrm{Ic}}}{G_{\mathrm{IIc}}}\right)\frac{Y_2 \tilde{E}_2}{F_{2t}^2} + \frac{G_{\mathrm{Ic}}}{G_{\mathrm{IIc}}}\left(\frac{Y_2 \tilde{E}_2}{F_{2t}^2}\right)^2 + \left(\frac{Y_6 \tilde{G}_{12}}{F_6^2}\right)} - \hat{\gamma} \leq 0$$

式中:G_{Ic},G_{IIc} 分别为 I 型和 II 型裂纹的临界能量释放率;F_{2t},F_6 分别为横向拉伸强度和剪切强度。

另外,使用以下损伤硬化函数形式:

$$\hat{\gamma} = \gamma + \gamma_0 = c_1 \left[\exp\left(\frac{\delta}{c_2}\right) - 1\right] + \gamma_0 \quad (\gamma_0 - c_1 \leq \hat{\gamma} \leq \gamma_0)$$

式中:γ_0 为初始损伤阈值;c_1,c_2 分别为材料参数。对于该损伤模型,所采用的 AS4/8852 碳/环氧复合材料的材料参数如表 8.1 和表 8.2 所列。

表 8.1 AS4/8852 单向层板弹性和强度参数

$\tilde{E}_1$/GPa	$\tilde{E}_2$/GPa	$\tilde{G}_{12}$/GPa	$\tilde{\nu}_{12}$	F_{2t}/MPa	F_6/MPa
171.4	9.08	5.29	0.32	62.29	92.34

表 8.2 AS4/8852 单向层板的临界能量释放率和硬化参数

G_{Ic}/(J/m^2)	G_{IIc}/(J/m^2)	γ_0	c_1	c_2
170	230	1.0	0.5	-1.8

解 该损伤模型表达了由横向拉应力和面内剪应力共同作用下的损伤。纵向拉伸/压缩对模型没有影响。因此,该模型的热力学力可以定义为 Y_2,Y_6。AS4/8852 单层板的损伤面的形状如图 8.8 所示。

为了实现 8.4.1 节中所示的返回映射算法,需要得到 $\partial f/\partial \boldsymbol{Y}$,$\partial g/\partial \boldsymbol{Y}$,$\partial f/\partial \gamma$,$\partial g/\partial \gamma$,$\partial \gamma/\partial \delta$ 和 $\partial \boldsymbol{Y}/\partial \boldsymbol{D}$ 的表达式。

假设 $f = g$,关于热力学力的势能函数和损伤面的导数由下式给出:

$$\frac{\partial g}{\partial \boldsymbol{Y}} = \frac{\partial f}{\partial \boldsymbol{Y}} = \left\{\begin{array}{c} 0 \\ \frac{1}{\hat{g}}\left(\left(1 - \frac{G_{\mathrm{Ic}}}{G_{\mathrm{IIc}}}\right)\frac{1}{4F_{2t}}\sqrt{\frac{2E_2}{Y_2}} + \frac{G_{\mathrm{Ic}}}{G_{\mathrm{IIc}}}\frac{E_2}{(F_{2t})^2}\right) \\ \frac{1}{\hat{g}} G_{12} \end{array}\right\}$$

并且损伤面对损伤硬化函数求偏导,得到

$$\frac{\partial g}{\partial \gamma} = \frac{\partial f}{\partial \gamma} = -1$$

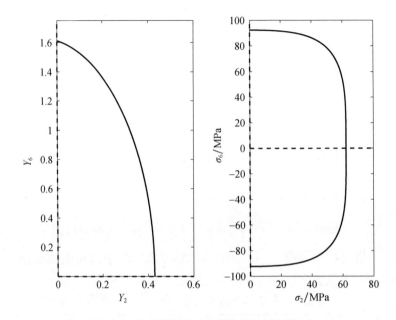

图 8.8 热力学和应力空间中的初始损伤面

此外,还需要硬化函数 γ 对其共轭变量 δ 求偏导,得

$$\frac{\partial \gamma}{\partial \delta} = \frac{c_1}{c_2}\exp\left(\frac{\delta}{c_2}\right)$$

下一步,热力学力对内部损伤变量求偏导,得

$$\frac{\partial Y}{\partial D} = \frac{\partial Y}{\partial D}\bigg|_{\sigma = \text{const}} + \frac{\partial Y}{\partial \sigma} : \frac{\partial \sigma}{\partial D}$$

另外,热力学力对应变求偏导得到

$$\frac{\partial Y}{\partial \varepsilon} = \frac{\partial Y}{\partial \sigma} : \frac{\partial \sigma}{\partial \varepsilon} = \frac{\partial Y}{\partial \sigma} : C$$

使用指标缩写形式写为

$$\frac{\partial Y}{\partial \boldsymbol{\sigma}} = \begin{bmatrix} 0 & 0 & 0 \\ \dfrac{-\sigma_2 \tilde{\nu}_{12}}{(1-D_2)^2 \tilde{E}_1} & \dfrac{2\sigma_2}{(1-D_2)^3 \tilde{E}_2} & \dfrac{\sigma_1 \tilde{\nu}_{12}}{(1-D_2)^2 \tilde{E}_1} & 0 \\ 0 & 0 & \dfrac{\sigma_6}{(1-D_6)^3 \tilde{G}_{12}} \end{bmatrix}$$

和

$$\frac{\partial \boldsymbol{\sigma}}{\partial D} = \begin{bmatrix} 0 & -\dfrac{\tilde{E}_2 \tilde{\nu}_{12}}{1-\tilde{\nu}_{12}\tilde{\nu}_{21}}\varepsilon_2 & 0 \\ 0 & -\dfrac{\tilde{E}_2 \tilde{\nu}_{21}}{1-\tilde{\nu}_{12}\tilde{\nu}_{21}}\varepsilon_1 - \dfrac{2(1-D_2)\tilde{E}_2}{1-\tilde{\nu}_{12}\tilde{\nu}_{21}}\varepsilon_2 & 0 \\ 0 & 0 & -4(1-D_6)\tilde{G}_{12}\varepsilon_6 \end{bmatrix}$$

以上方程计算在子程序 umatps85.for 中实现,其在文献[5]中可获得。子程序可以与平面应力单元和层合壳单元一起使用。横向拉伸荷载下的操作过程代码如下所示:

ⅰ. 创建工作目录。

Menu: File,Set Work Directory,[C:\SIMULIA\User\Ex_8.5],OK
Menu: File,Save As,[C:\SIMULIA\User\Ex_8.5\Ex_8.5.tra.cae],OK

ⅱ. 创建部件。

Module: Part
Menu: Part,Create
    2D Planar,Deformable,Shell,Cont
Menu:　Add,Line,Rectangle,[0,0],[10,10],X,Done

ⅲ. 定义材料、截面属性并赋予部件。

Module: Property
Menu: Material,Create
    General,User Material
    # read material properties from a file
    # right - click in Data area,Read from file,[props.txt],OK,OK
    General,Depvar
    Number of solution-dependent state variables [3],OK
Menu: Section,Create
    Solid,Homogeneous,Cont
    # checkmark: Plane stress/strain thickness [1.0],OK
Menu: Assign,Section,# pick: part,Done,OK

ⅳ. 创建装配体。

Module: Assembly
Menu: Instance,Create,Independent,OK

ⅴ. 定义分析步。
Module: Step
Menu: Step,Create
    Name[transverse],Procedure type: General,Static/General,Cont
    Tab: Basic,Time Period[50]
    Tab: Incrementation,Type: Fixed
    Maximum number of increments[50],Increment size[2.2],OK
Menu: Output,Field Output Requests,Edit,F-Output-1
    Output Variables: Preselected Defaults,# add:[SDV],OK

ⅵ. 添加载荷和 BC。
Module: Load
Menu: BC,Manager
    Create,Name[BC-1],Step: Initial,Symm/Anti/Enca,Cont
    # pick: node @ origin,Done,ENCASTRE,OK
    Create,Name[BC-2],Step: Initial,Symm/Anti/Enca,Cont
    # pick: line @ Y=0,Done,YSYMM,OK
    # transverse load imposed as displacement to create epsilon = 2%
    Create,Name[traction],Step: transverse,Disp/Rota,Cont
    # pick: line @ Y=10,Done,# checkmark: U2[0.2],OK
    # close Boundary Condition Manager pop-up window

ⅶ. 模型划分网格。因为在均匀应力/应变状态下,所以只划分一个单元。
Module: Mesh
Menu: Seed,Instance,Approximate global size[10],OK
Menu: Mesh,Element Type
    Geometric Order: Quadratic # CPS8R avoid hourglass,OK
Menu: Mesh,Instance,Yes

ⅷ. 创建一个集合用来跟踪单元中的应力和应变。
Module: Mesh
Menu: Tools,Set,Create
    Type: Element,Cont,# pick: the only element in the mesh,Done
Module: Step
Menu: Output,History Output Requests,Edit,H-Output-1
    Domain: Set: Set-1,# expand: Stresses,# checkmark: S
    # expand: Strains,# checkmark: E,OK

ix. 求解并可视化结果。
Module: Job
Menu: Job,Manager
　　Create,Cont
　　Tab: General,User subroutine file [umatps85.for],OK,OK
　　Submit,# when Completed,Results
Module: Visualization
Menu: Result,History Output
　　# select: E22 at Element 1 Int Point 1,Save As [e22],OK
　　# select: S22 at Element 1 Int Point 1,Save As [s22],OK
　　# close History Output pop-up window
Menu: Tools,XYData,Create
　　Source: Operate on XY data,Cont,Operations: Combine
　　# select: e22,Add to Expr.,# select: s22,Add to Expr.
　　Save As [s22-vs-e22],OK,Plot Expression

通过修改当前的 cae 模型,可以很容易施加剪切载荷。
ⅰ. 保存当前 cae 模型并创建一个新的 cae 文件。
Menu: File,Save
Menu: File,Save As,[C:\SIMULIA\User\Ex_8.5\Ex_8.5.she.cae],OK

ⅱ. 定义分析步。
Module: Step
Menu: Step,Manager
　　# select: transverse,Rename,[shear],OK
　　Edit,Tab: Incrementation,Type: Fixed
　　Maximum number of increments [50],Increment size [3],OK
　　# close Step Manager pop-up window

ⅲ. 施加 BC 和载荷。
Module: Load
Menu: BC,Manager
　　# shear loading imposed as displacement U1 on step: shear
　　# select: traction,Rename,[shear],OK
　　Edit,# uncheckmark: U2,# checkmark: U1 [0.4],OK # gamma_6 = 4%
　　# close BC Manager pop-up window

ⅳ. 施加约束以保证模型受纯剪切载荷。
Module: Interaction

Menu: Constraint,Create
   Type: Tie,Cont
   Surface,# pick: line @ x = 0,Done
   Surface,# pick: line @ x = 10,Done
   Discretization method: Surface to surface
   Position Tolerance: Specify distance [10]
   # uncheckmark: Adjust slave surface initial position,OK

Ⅴ. 求解并可视化结果。

Module: Job
Menu: Job,Manager
   Submit,# when Completed,Results
Module: Visualization
Menu: Result,History Output
   # select: E12 at Element 1 Int Point 1,Save As [e12],OK
   # select: S12 at Element 1 Int Point 1,Save As [s12],OK
   # close History Output pop-up window
Menu: Tools,XYData,Create
   Source: Operate on XY data,Cont,Operations: Combine
   # select: e12,Add to Expr. ,# select: s12,Add to Expr.
   Save As [s12-vs-e12],OK,Plot Expression

面内剪切应力和横向拉伸应力下的响应结果如图 8.9 所示。

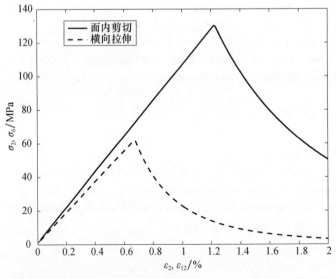

图 8.9 面内剪切应力和横向拉伸应力下的响应

8.5 损伤和塑性

对于高强、高刚纤维增强聚合物基复合材料,可以用二阶张量 $\boldsymbol{D}$ 和 $\boldsymbol{Y}$ 描述损伤状态及其共轭的热力学力。另外,在塑性和损伤过程中发生的硬化行为意味着有额外的耗散,因此

$$\rho\pi = T\rho\pi_s = \boldsymbol{\sigma}:\dot{\boldsymbol{\varepsilon}}^p + R\dot{p} + \boldsymbol{Y}:\dot{\boldsymbol{D}} + \gamma\dot{\delta} \tag{8.122}$$

式中:(R,p) 为与塑性硬化相关的热力学力 - 热力学通量对;(γ,δ) 为与损伤硬化相关的热力学力 - 热力学通量对;$\rho\pi$ 为不可逆现象产生的耗散热。

对于式(8.122)描述的特殊情况,利用式(8.92)和式(8.99),热力学力的形式定义如下:

$$\boldsymbol{\sigma} = \rho\frac{\partial\psi}{\partial\boldsymbol{\varepsilon}} = -\rho\frac{\partial\psi}{\partial\boldsymbol{\varepsilon}^p}, \boldsymbol{\varepsilon} = \rho\frac{\partial\chi}{\partial\boldsymbol{\sigma}}, \boldsymbol{Y} = -\rho\frac{\partial\psi}{\partial\boldsymbol{D}} = \rho\frac{\partial\chi}{\partial\boldsymbol{D}} \tag{8.123}$$

以及硬化方程的定义如下:

$$\gamma = \rho\frac{\partial\chi}{\partial\delta} = -\rho\frac{\partial\psi}{\partial\delta} = \rho\frac{\partial\pi}{\partial\delta}, R = \rho\frac{\partial\chi}{\partial p} = -\rho\frac{\partial\psi}{\partial p} = \rho\frac{\partial\pi}{\partial p} \tag{8.124}$$

应变可以分解如下[63]:

$$\boldsymbol{\varepsilon} = \boldsymbol{\varepsilon}^e + \boldsymbol{\varepsilon}^p \tag{8.125}$$

考虑到弹性应变部分可以由应力和柔度计算得到,上式可以改写为

$$\boldsymbol{\varepsilon} = \boldsymbol{S}:\boldsymbol{\sigma} + \boldsymbol{\varepsilon}^p \tag{8.126}$$

因此,增量中的应变 - 应力关系和率形式为

$$\begin{cases} \delta\boldsymbol{\varepsilon} = \boldsymbol{S}:\delta\boldsymbol{\sigma} + \delta\boldsymbol{S}:\boldsymbol{\sigma} + \delta\dot{\boldsymbol{\varepsilon}}^p \\ \dot{\boldsymbol{\varepsilon}} = \boldsymbol{S}:\dot{\boldsymbol{\sigma}} + \dot{\boldsymbol{S}}:\boldsymbol{\sigma} + \dot{\boldsymbol{\varepsilon}}^p \end{cases} \tag{8.127}$$

式(8.127)表明应变增量由弹性、损伤和塑性3种贡献组成。弹性应变增量是应力增量的直接结果,损伤应变增量是由于材料损伤而使柔度增加产生,塑性应变增量在恒定的柔度下产生。弹性卸载刚度不因塑性而改变,但因损伤而降低。根据这一论点,习惯上[92]假设自由能和自由余能可以分开表达如下:

$$\begin{cases} \psi(\boldsymbol{\varepsilon},\boldsymbol{\varepsilon}^p,p,\boldsymbol{D},\delta) = \psi^e(\boldsymbol{\varepsilon}^e,\boldsymbol{D},\delta) + \psi^p(\boldsymbol{\varepsilon}^p,p) \\ \chi(\boldsymbol{\sigma},\boldsymbol{\varepsilon}^p,p,\boldsymbol{D},\delta) = \chi^e(\boldsymbol{\sigma},\boldsymbol{D},\delta) + \chi^p(\boldsymbol{\varepsilon}^p,p) \end{cases} \tag{8.128}$$

习题

习题8.1 使用例8.2中的公式和材料属性,获得梁顶面上的一个点和梁底面上的另一个点的应变 - 名义应力($\varepsilon - \sigma$)曲线以及应变 - 有效应力($\varepsilon - $

$\tilde{\sigma}$)曲线。对所得曲线图进行讨论。

习题 8.2 对仅 x_1-方向运动的一维 CDM 应用一个 UMAT 程序。采用 2D 平面应力本构方程。摈弃 x_2-方向和泊松比,无损伤状态下剪切项为线弹性。重新计算例 8.2 和习题 8.1 中获得的曲线图来验证该程序。请注意,要获得相同的值,泊松比应设置为 0。

习题 8.3 Gibbs 自由能展开形式使用 Voigt 缩写形式,定义如下:

$$\chi = \frac{1}{2\rho}\left[\frac{\sigma_1^2}{(1-D_1)^2\tilde{E}_1} + \frac{\sigma_2^2}{(1-D_2)^2\tilde{E}_2} + \frac{\sigma_6^2}{(1-D_1)(1-D_2)\tilde{G}_{12}} - \left(\frac{\tilde{\nu}_{12}}{\tilde{E}_2} + \frac{\tilde{\nu}_{21}}{\tilde{E}_1}\right)\frac{\sigma_1\sigma_2}{(1-D_1)(1-D_2)}\right]$$

式中:$\tilde{E}_1$,$\tilde{E}_2$,$\tilde{\nu}_{21}$,$\tilde{\nu}_{12}$,$\tilde{G}_{12}$ 为单向层板未损伤时的面内正交各向异性弹性属性。其中,下标$()_1$ 表示纤维方向,$()_2$ 表示横向。

(1) 使用给定的 Gibbs 自由能形式,得到割线本构方程、$\boldsymbol{C}$ 和 $\boldsymbol{S}$。
(2) 获得与 D_1 和 D_2 相关的热力学力 Y_1 和 Y_2。
(3) 如果 $\boldsymbol{M}$ 用 Voigt 缩写形式表示,并乘以 Reuter 矩阵,如下:

$$\boldsymbol{M} = \begin{bmatrix} (1-D_1) & 0 & 0 \\ 0 & (1-D_2) & 0 \\ 0 & 0 & \sqrt{1-D_1}\sqrt{1-D_2} \end{bmatrix}$$

利用能量等效原理,检验 $\boldsymbol{M}$ 的定义是否可用作损伤模型中的损伤效应张量。证明并讨论你的结论。

习题 8.4 对于习题 8.3 中模型的损伤激活函数定义为

$$g = \hat{g} - \hat{\gamma} = \sqrt{Y_1^2 H_1 + Y_2^2 H_2} - (\gamma + \gamma_0)$$

式中:H_1,H_2 为依赖于弹性和强度材料属性的模型参数;Y_1,Y_2 分别为与损伤变量 D_1 和 D_2 相关的热力学力。损伤硬化取决于 δ,表达式如下:

$$\hat{\gamma} = \gamma + \gamma_0 = c_1\left[\exp\left(\frac{\delta}{c_2}\right) - 1\right] + \gamma_0$$

式中:γ_0 为初始阈值;c_1,c_2 为材料参数。所有必要的材料参数如表 8.3 和表 8.4 所列。

表 8.3 单向层板的弹性参数

$\tilde{E}_1$/GPa	$\tilde{E}_2$/GPa	$\tilde{G}_{12}$/GPa	$\tilde{\nu}_{12}$
171.4	9.08	5.29	0.32

表8.4　单向层板的损伤模型参数

H_1	H_2	γ_0	c_1	c_2
0.024	8.36	1.0	1.5	-2.8

(1)使用流程图描述算法,包括所有必要的步骤,将其实现为有限元软件中的本构子程序。

(2)计算在 UMAT 中实现计算模型所需的解析表达式。

(3)使用 UMAT 功能为平面应力本构方程编写算法。

(4)最后,使用 Abaqus 绘制仅加载 ε_2 时的 RVE 表观应力 σ_2 与表观应变 ε_2 的关系曲线。

(5)用代码描述在 Abaqus/CAE 中解决该问题的操作过程。

第9章 离散损伤力学

在第8章中,使用连续介质损伤力学方法预测复合材料初始损伤和损伤扩展。除了 CDM 方法外,还有如下的分析方法:损伤的细观力学方法,裂纹张开位移方法,细观计算力学方法和协同分析方法。连续介质损伤力学(第8章)将损伤均匀化并采用唯象理论进行处理,而其他方法则试图描述损伤的实际几何形状和特征。

在面内载荷作用下的铺层顺序为 $[0_m/90_n]_S$ 的对称层合板,已充分对复合材料层合板基体横向开裂的预测进行了研究,这种情况下基体裂纹均出现在90°方向的铺层(横向铺层)中。其他铺层顺序层合板的相关研究也已有所开展,例如 $[0/\pm\theta/0]_S$ 和 $[0/\theta_1/\theta_2]_S$ 顺序铺层中,裂纹一般出现在偏轴的 θ 方向铺层中。但是这些相关研究仍旧限于承受面内载荷条件的对称层合板。

文献[93-114]中采用损伤模型的细观力学方法(MMD)得到了含有一个或多个离散裂纹缺陷层合板的近似弹性解。该弹性解是近似的,因为采用了一些运动学假设,例如层间剪切应力在每个单层板厚度方向上呈线性[115]或双线性[116]分布,另外,假设面内位移函数[113]、应力等按特定的空间分布。含裂纹的单层板中的裂纹密度设置为状态变量,其定义为垂直于裂纹面的单位距离内的裂纹数。因此,该状态变量是可测量的。MMD 方法的优势之一是层合板弹性模量的降低是裂纹密度的函数,无需使用 CDM 方法中的额外参数即可进行计算。MMD 方法的主要缺点是绝大部分求解仅限于受面内载荷作用下的只有一个或两个铺层裂纹的对称层合板。对于存在多个铺层裂纹的一般情况,本章将介绍协同分析方法并对其进行分析。

裂纹张开位移方法(COD)[117-124]基于含空隙的弹性体理论[125]。COD 模型的独特优势在于可以表征层合板中任意单层板的基体裂纹并计算任何层合板形式的层合板刚度,甚至在任意变形(包括弯曲)条件下的非对称铺层层合板[126]。

数值求解方法,如 FEA,提供了 3D 求解方法,而无需像 MMD 和 COD 模型一样进行运动学简化[104,120,123,127-130]。但是,FEA 方法需要针对不同 LSS 和裂纹方向等重新进行网格划分和边界条件定义,因此该方法在实际应用中比较麻烦。另一种数值方法是蒙特卡罗模拟方法,该方法考虑材料中缺陷的随机分布[131-133]。然而,蒙特卡罗模拟还需要其他参数,这些参数必须通过对试验损伤

第9章 离散损伤力学

演化数据的拟合结果来进行调整得到,目前相关数据较少。

协同损伤力学(SDM)方法结合了如 CDM 和 MMD[131-132,134-138]等方法的建模策略,借鉴了上述模型中的优点。例如,在本章介绍中,层合板的刚度降低通过 MMD 方法计算,多个含裂纹铺层的一般化方法借用 CDM 方法的概念,但与 CDM 方法不同,无需其他附加参数。

9.1 概述

本节中将说明如何使用两种材料特性,即 I 和 II 两种模式的断裂韧性 G_{Ic} 和 G_{IIc},预测层合板中某一单向纤维增强单层板的损伤行为和横向拉伸破坏。考虑到相邻单层板之间的约束效应,其表观横向拉伸强度 F_{2t} 是层厚度的函数。裂纹初始应变,随应力(应变)变化直至裂纹饱和的裂纹密度及相邻铺层的应力重分布均被准确预测。

在横向拉伸和面内剪力作用下基体开裂的物理过程如下:在层合板生产过程中,材料中存在缺陷是无法避免的,这些缺陷可能是空隙、微裂纹、纤维-基体脱粘等,所有这些缺陷都可以用长度为 $2a_0$ 的典型基体裂纹表示,如图 9.1 所示。

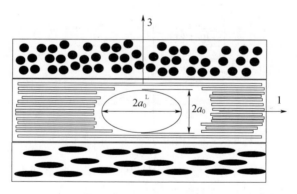

图 9.1 典型的裂纹形态

当受到载荷作用时,基体裂纹将沿着平行于纤维的方向增长,如图 9.2 所示,可以看到在 ±70°的铺层中裂纹方向与纤维方向一致。这些平行的裂纹集合降低了含裂纹单层的刚度,导致该铺层承担的载荷由其他铺层承担。在每个铺层中,由这组平行裂纹集合产生的损伤由裂纹密度表示,其定义为两个相邻裂纹之间距离的倒数 $\lambda = 1/(2l)$,如图 9.3 所示。因此,裂纹密度是表示单层板中损伤状态的唯一状态变量。真实的离散的裂纹可以通过理论方法实现模拟,该理论称为离散损伤力学(DDM)。

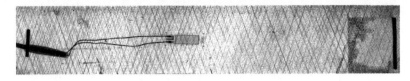

图 9.2 沿 0°方向施加拉伸应变 0.7% 后,$[0/\pm70_4/0_{1/2}]_s$
层合板中 ±70°铺层中的基体裂纹

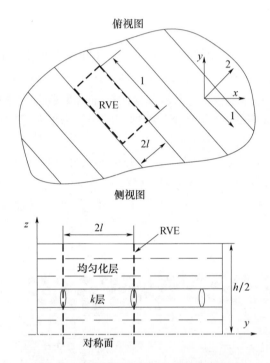

图 9.3 离散损伤力学中使用的典型胞元

对于横向拉伸和面内剪切损伤,DDM 模型的基本要素如下:

(1)对于铺层 i,其状态变量为裂纹密度 λ_i。为方便计算,定义两个损伤变量 $D_2(\lambda_i)$ 和 $D_6(\lambda_i)$,但它们不是独立的变量,相反,它们由裂纹密度计算得到。层合板中不同铺层的裂纹密度集合可以表示为 $\lambda=\lambda_i, i=1,\cdots,N$,其中 N 是层合板中的铺层数。

(2)独立变量为中面①应变 $\epsilon=\{\epsilon_1,\epsilon_2,\gamma_{12}\}^T$。

(3)损伤激活函数,用于区分损伤状态和未损伤状态,可以写为

---

① 本节中分析对象为在面内载荷下的对称铺层层合板。非对称层合板和(或)弯曲载荷作用下的层合板情况见文献[126]。

$$g = \max\left[\frac{G_{\mathrm{I}}(\lambda,\epsilon,\Delta T)}{G_{\mathrm{Ic}}}, \frac{G_{\mathrm{II}}(\lambda,\epsilon,\Delta T)}{G_{\mathrm{IIc}}}\right] - 1 \leqslant 0 \tag{9.1}$$

当 $g \leqslant 0$ 时表示材料未损伤。临界能量释放率参数很难在文献中找到,但可以通过相应方法对现有的试验数据拟合得到,见文献[55,140]中的实例。

(4) 参数 g 包含损伤阈值,其可用材料性能(不变量)G_{Ic}、G_{IIc} 表示。在损伤初始之前,$\lambda = 0$,式(9.1)为损伤初始准则。该准则与文献[141]类似,但损伤模式不耦合。$\lambda = 0$,当 $g = 0$ 时的应变即为损伤初始应变。一旦损伤发生,依据如下的自动硬化描述,式(9.1)变为损伤激励函数。

(5) 硬化函数已包含在损伤激活函数 g 中,对于给定的应变值,能量释放率 $G_{\mathrm{I}}(\lambda)$,$G_{\mathrm{II}}(\lambda)$ 的计算值是 λ 的单调递减函数。因此,随着 λ 的增大,$G_{\mathrm{I}}(\lambda)$,$G_{\mathrm{II}}(\lambda)$ 减小,从而使 $g < 0$,并且阻止进一步损伤,直到驱动的热力学力,即应变,因施加的外载而进一步增加。

(6) 无需假设损伤演化函数,其优点就是无需新的经验参数。仅使得裂纹密度 λ 自行调整到一个值,该值使得当前外载条件下层合板平衡,同时当前应变满足 $g = 0$ 的条件。返回映射算法(8.4.1节)可通过迭代计算直到满足 $g = 0$,并根据迭代增量 $\Delta\lambda = -g \big/ \dfrac{\partial g}{\partial \lambda}$ 更新裂纹密度。

(7) 裂纹密度会不断增加,直到该单层板铺层中裂纹达到饱和($\lambda \to \infty$)。此时,该铺层将失去横向刚度和剪切刚度($D_2 \approx 1, D_6 \approx 1$),因此其所负担的载荷由层合板其余铺层分担。当裂纹密度达到 $\lambda_{\lim} = 1/h_k$ 时,该含裂纹的铺层分析即停止;式中 h_k 是该铺层 k 的厚度,此时裂纹分布的间距等于铺层厚度。

介绍了模型的组成部分后,现在对各种量的计算进行说明。求解开始于对层合板退化①刚度 $Q = [A]/h$ 的计算,该层合板的第 k 铺层含有裂纹,并给定裂纹密度 λ_k;上式中 $[A]$ 是层合板面内刚度矩阵②,h 是层合板厚度。

本节中使用以下约定:

(1) (i) 表示层合板中的任一铺层。

(2) (k) 表示含裂纹的铺层。

(3) (m) 表示除含裂纹铺层以外的任一铺层($m \neq k$)。

(4) 上标/下标 (i) 表示铺层的层数;而不是幂指数或微分阶数。

(5) $x_j, j = 1,2,3$ 表示坐标 x_1, x_2, x_3,或 $j = 1,2,6$ 表示 Voigt 指标缩写形式下的量。

(6) $u(x_j), v(x_j), w(x_j), j = 1,2,3$ 表示位移的 3 个分量。

---

① 也称"损伤后的""降低的"或者"均匀化的"。

② 不要与裂纹面积 A 混淆。

(7) $\hat{p}$ 表示参数 p 的厚度平均值,厚度给定或者可从上下文中得到。

(8) $\tilde{p}$ 表示参数 p 的初始值。

(9) $\bar{p}$ 表示参数 p 的体积平均值。

9.2 近似值

实际应用中层合板大多数是对称的,而且层合板在设计时多用于承担面内载荷(文献[1],第12章)。因此,此处介绍的计算方法是针对受面内载荷作用下的对称层合板。在这种情况下,有

$$\frac{\partial w^{(i)}}{\partial x_1} = \frac{\partial w^{(i)}}{\partial x_2} = 0 \tag{9.2}$$

式中:$u(x_j)$,$v(x_j)$,$w(x_j)$,$j=1,2,3$,为第 i 铺层中一点的位移,该位移为坐标 x_j,$j=1,2,3$ 的函数。此外,铺层的厚度 h_i 假设很小,因此平面应力假设成立。

$$\sigma_3^{(i)} = 0 \tag{9.3}$$

由于所有裂纹均与纤维方向平行,且实际设计时一般避免使用较厚的铺层,所以可以认为裂纹在单层铺层中是贯穿厚度的。任何小于铺层厚度的裂纹缺陷,其在铺层厚度上和纤维方向上都是不稳定的(文献[1],7.2.1节)。

由于分析的主要目的是计算由裂纹导致的层合板刚度降低,因此只需使用变量的厚度平均值即可。厚度平均值可表示为

$$\hat{\phi} = \frac{1}{h'} \int_{h'} \phi \mathrm{d}x_3 ; \quad h' = \int \mathrm{d}x_3 \tag{9.4}$$

式中:h' 可以是某铺层或层合板的厚度,可以分别用 h_i,h 表示。

此外,

— $\hat{u}^{(i)}(x_j)$,$\hat{v}^{(i)}(x_j)$,$\hat{w}^{(i)}(x_j)$ 是层合板第 i 铺层的厚度平均位移值,且是平面内坐标 x_j,$j=1,2$ 的函数。

— $\hat{\epsilon}_1^{(i)}(x_j)$,$\hat{\epsilon}_2^{(i)}(x_j)$,$\hat{\gamma}_{12}^{(i)}(x_j)$ 是层合板第 i 铺层的厚度平均应变。

— $\hat{\sigma}_1^{(i)}(x_j)$,$\hat{\sigma}_2^{(i)}(x_j)$,$\hat{\tau}_{12}^{(i)}(x_j)$ 是层合板第 i 铺层的厚度平均应力。

裂纹导致的位移扰动,可能出现面外(层间)剪切应力分量。它们可以近似认为在第 i 铺层厚度上线性分布,即

$$\begin{cases} \tau_{13}^{(i)}(x_3) = \tau_{13}^{i-1,i} + (\tau_{13}^{i,i+1} - \tau_{13}^{i-1,i}) \dfrac{x_3 - x_3^{i-1,i}}{h_i} \\ \tau_{23}^{(i)}(x_3) = \tau_{23}^{i-1,i} + (\tau_{23}^{i,i+1} - \tau_{23}^{i-1,i}) \dfrac{x_3 - x_3^{i-1,i}}{h_i} \end{cases} \tag{9.5}$$

式中:x_3^{i-1} 为层合板第 i 铺层底面的厚度坐标,即第 $i-1$ 铺层和第 i 铺层之间界

面坐标;$\tau_{13}^{i-1,i}$为第$i-1$铺层和第i铺层之间界面的面外剪切应力。这个假设常用于其他几种分析模型,称为剪切滞后假设。该线性近似方法已证明可以得到准确的计算结果[130]。

剪力滞后方程可通过对面外剪切应变和应力的本构方程进行加权平均获得[附录A],

$$\begin{Bmatrix} \hat{u}^{(i)} - \hat{u}^{(i-1)} \\ \hat{v}^{(i)} - \hat{v}^{(i-1)} \end{Bmatrix} = \frac{h_{(i-1)}}{6}\begin{bmatrix} S_{45} & S_{55} \\ S_{44} & S_{45} \end{bmatrix}^{(i-1)}\begin{Bmatrix} \tau_{23}^{i-2,i-1} \\ \tau_{13}^{i-2,i-1} \end{Bmatrix}$$

$$+ \left[\frac{h_{(i-1)}}{3}\begin{bmatrix} S_{45} & S_{55} \\ S_{44} & S_{45} \end{bmatrix}^{(i-1)} + \frac{h_{(i)}}{3}\begin{bmatrix} S_{45} & S_{55} \\ S_{44} & S_{45} \end{bmatrix}^{(i)}\right]\begin{Bmatrix} \tau_{23}^{i-1,i} \\ \tau_{13}^{i-1,i} \end{Bmatrix} \quad (9.6)$$

$$+ \frac{h_{(i)}}{6}\begin{bmatrix} S_{45} & S_{55} \\ S_{44} & S_{45} \end{bmatrix}^{(i)}\begin{Bmatrix} \tau_{23}^{i,i+1} \\ \tau_{13}^{i,i+1} \end{Bmatrix}$$

转化式(9.6),层间应力可由界面的位移表示为

$$\begin{cases} \tau_{23}^{i,i+1} - \tau_{23}^{i-1,i} = \sum_{j=1}^{n-1}\left[[H]_{2i-1,2j-1}^{-1} - [H]_{2i-3,2j-1}^{-1}\right]\{\hat{u}^{(j+1)} - \hat{u}^{(j)}\} \\ \qquad\qquad\qquad + \left[[H]_{2i-1,2j}^{-1} - [H]_{2i-3,2j}^{-1}\right]\{\hat{v}^{(j+1)} - \hat{v}^{(j)}\} \\ \tau_{13}^{i,i+1} - \tau_{13}^{i-1,i} = \sum_{j=1}^{n-1}\left[[H]_{2i-1,2j-1}^{-1} - [H]_{2i-2,2j-1}^{-1}\right]\{\hat{u}^{(j+1)} - \hat{u}^{(j)}\} \\ \qquad\qquad\qquad + \left[[H]_{2i,2j}^{-1} - [H]_{2i-2,2j}^{-1}\right]\{\hat{v}^{(j+1)} - \hat{v}^{(j)}\} \end{cases} \quad (9.7)$$

其中,系数矩阵H为$2(N-1)\times 2(N-1)$矩阵。

9.3 单层板本构方程

对于含裂纹的第k铺层的应力-应变关系与无缺陷材料的形式相同,即

$$\hat{\sigma}_i^{(k)} = \tilde{Q}_{ij}^{(k)}(\hat{\epsilon}_j^{(k)} - \alpha_j^{(k)}\Delta T) \quad (9.8)$$

式中:$\alpha^{(k)}$为第k铺层的 CTE 值;$\sigma_i^{(k)} = \{\sigma_1^{(k)}, \sigma_2^{(k)}, \tau_{12}^{(k)}\}^T$;波浪号上标表示初始值。应变-位移方程为

$$\epsilon^{(k)} = \begin{Bmatrix} \epsilon_1^{(k)} = u_{,1}^{(k)} \\ \epsilon_2^{(k)} = v_{,2}^{(k)} \\ \gamma_{12}^{(k)} = u_{,2}^{(k)} + v_{,1}^{(k)} \end{Bmatrix} \quad (9.9)$$

对于其余的铺层($m\neq k$),它们的本构方程可以通过式(9.8)和刚度矩阵$Q_{ij}^{(m)}$获得,$Q_{ij}^{(m)}$可由式(9.32)中定义的已计算的损伤值$D_2^{(m)}, D_6^{(m)}$计算得到,并且使用常用的转换方程旋转到第k层坐标系(文献[1],5.4节)

$$Q^{(m)} = T^{-1} \begin{bmatrix} \tilde{Q}_{11}^{(m)} & (1-D_2^{(m)})\tilde{Q}_{12}^{(m)} & 0 \\ (1-D_2^{(m)})\tilde{Q}_{12}^{(m)} & (1-D_2^{(m)})\tilde{Q}_{22}^{(m)} & 0 \\ 0 & 0 & (1-D_6^{(m)})\tilde{Q}_{66}^{(m)} \end{bmatrix} T^{-T} \quad (9.10)$$

式中:$T^{-1} = [T(-\theta)]$,$T^{-T} = [T(-\theta)]^{T}$,见(文献[1],式(5.35))。

9.4 位移场

本节主要介绍对于给定的裂纹密度集合 λ 和施加的应变 ϵ,如何求解层合板中所有含裂纹铺层 i 的平均位移 $\hat{u}^{(i)}(x_j)$,$\hat{v}^{(i)}(x_j)$;$j=1,2$。假设层间剪切应力在每个铺层的厚度上线性变化,对于每个铺层的平衡方程式(1.15)可写为

$$\hat{\sigma}_{1,1}^{(i)} + \hat{\tau}_{12,2}^{(i)} + (\hat{\tau}_{13}^{i,i+1} - \hat{\tau}_{13}^{i-1,i})/h_i = 0 \quad (9.11)$$

$$\hat{\tau}_{12,1}^{(i)} + \hat{\sigma}_{2,2}^{(i)} + (\hat{\tau}_{23}^{i,i+1} - \hat{\tau}_{23}^{i-1,i})/h_i = 0 \quad (9.12)$$

利用应变-位移方程式(9.9)并将本构方程式(9.8)代入平衡方程式(9.11)、式(9.12),可得到关于 $\hat{u}^{(i)}(x_j)$,$\hat{v}^{(i)}(x_j)$ 的 $2N$ 阶偏微分方程(PDE)。该偏微分方程有如下的特解形式:

$$\begin{cases} \hat{u}^{(i)} = a_i \sinh\lambda_e x_2 + ax_1 + bx_2 \\ \hat{v}^{(i)} = b_i \sinh\lambda_e x_2 + bx_1 + a^* x_2 \end{cases} \quad (9.13)$$

式中:e 为特征值数量,通解可以写为

$$\begin{Bmatrix} \hat{u}^{(1)} \\ \hat{u}^{(2)} \\ \vdots \\ \hat{u}^{(n)} \\ \hat{v}^{(1)} \\ \hat{v}^{(2)} \\ \vdots \\ \hat{v}^{(n)} \end{Bmatrix} = \sum_{e=1}^{2N} A_e \begin{Bmatrix} a_1 \\ a_2 \\ \vdots \\ a_n \\ b_1 \\ b_2 \\ \vdots \\ b_n \end{Bmatrix} \sinh(\eta_e x_2) + \begin{Bmatrix} a \\ a \\ \vdots \\ a \\ b \\ b \\ \vdots \\ b \end{Bmatrix} x_1 + \begin{Bmatrix} b \\ b \\ \vdots \\ b \\ a^* \\ a^* \\ \vdots \\ a^* \end{Bmatrix} x_2 \quad (9.14)$$

将式(9.14)代入 PDE 方程则变为求解特征值问题

$$\begin{bmatrix} \alpha_1 & \beta_1 \\ \alpha_2 & \beta_2 \end{bmatrix} \begin{Bmatrix} a_j \\ b_j \end{Bmatrix} + \eta^2 \begin{bmatrix} \zeta_{26} & \zeta_{22} \\ \zeta_{66} & \zeta_{26} \end{bmatrix} \begin{Bmatrix} a_j \\ b_j \end{Bmatrix} = \begin{Bmatrix} 0 \\ 0 \end{Bmatrix} \quad (9.15)$$

式中:$j=1,\cdots,2N$;η 为 $2N$ 个特征值;$\{a_j,b_j\}^{\mathrm{T}}$ 表示 $2N$ 个特征向量。

事实证明,特征值中有两个始终为零(对应于式(9.14)中的线性项),其被认为是特征值集合中的最后两个,因此仅剩下 $2N-2$ 个独立求解项。那么,PDE 系统的通解由 $2N-2$ 个独立解构建,有

$$\begin{Bmatrix} \hat{u}^{(i)} \\ \hat{v}^{(i)} \end{Bmatrix} = \sum_{e=1}^{2N-2} A_e \begin{Bmatrix} a_i \\ b_i \end{Bmatrix} \sinh(\eta_e x_2) + \begin{Bmatrix} a \\ b \end{Bmatrix} x_1 + \begin{Bmatrix} b \\ a^* \end{Bmatrix} x_2 \quad (9.16)$$

式中:A_e 为线性组合中的未知系数。可以看出,通解包含 $2N+1$ 个未知系数,其包括标量 a,b,a^* 和 A_e 集合,$e=1,\cdots,2N-2$。为了确定这些系数,需要在图 9.3 中的代表性体积单元(RVE)的边界上建立 $2N+1$ 个边界条件。需要注意的是,体积单元(RVE)沿纤维方向 x_1 为一个单位长度,连续裂纹之间的距离为 $2l$(沿 x_2 方向),厚度方向为对称层合板的整个厚度 h。

通过边界条件引入了两个非常重要的参数:裂纹密度 λ 和施加在层合板上的应力 $\hat{\sigma}=\{N\}/h$,其中 $\{N\}$ 表示单位长度上面内载荷的 3 个分量①。裂纹密度通过 RVE 单元的尺寸输入,其宽度为 $2l=1/\lambda$。施加的应力(或应变)通过 RVE 单元载荷平衡输入。总之,$2N+1$ 个边界条件产生 $2N+1$ 个代数方程系统,求解得到式(9.16)中的 $2N+1$ 个系数。因此,通过给定的裂纹密度 λ 值和施加的载荷 $\hat{\sigma}=\{N\}/h$,求解式(9.16)可得到层合板中所有铺层的平均位移。

9.4.1 $\Delta T = 0$ 时的边界条件

首先考虑有机械载荷、无温度载荷的情况。为了求解 A_e,a,a^*,b 的值,需要施加以下边界条件:(a)裂纹表面应力自由;(b)外部载荷;(c)均一的位移。上述边界条件可集成到一个代数系统中,即

$$[B]\{A_e,a,a^*,b\}^{\mathrm{T}} = \{F\} \quad (9.17)$$

式中:$[B]$ 为 $(2N+1)\times(2N+1)$ 维的系数矩阵;$\{A_e,a,a^*,b\}^{\mathrm{T}}$ 代表 $2N+1$ 个未知系数;$\{F\}$ 为 RHS 或力矢量,也是 $2N+1$ 维。

(a)裂纹表面应力自由。

裂纹的表面应力自由状态下,有

$$\int_{-1/2}^{1/2} \hat{\sigma}_2^{(k)}(x_1,l)\mathrm{d}x_1 = 0 \quad (9.18)$$

---

① 不要与铺层数 N 混淆。

$$\int_{-1/2}^{1/2} \hat{\tau}_{12}^{(k)}(x_1,l)\,\mathrm{d}x_1 = 0 \qquad (9.19)$$

(b) 外部载荷。

在平行于裂纹表面的方向(纤维方向 x_1),载荷由所有层合板的铺层承担

$$\frac{1}{2l}\sum_{i=1}^{N} h_i \int_{-l}^{l} \hat{\sigma}_1^{(i)}(1/2, x_i)\,\mathrm{d}x_2 = h\hat{\sigma}_1 \qquad (9.20)$$

在垂直于裂纹表面的方向(x_2 方向),只有无裂纹缺陷铺层承载载荷

$$\sum_{m\neq k} h_m \int_{-1/2}^{1/2} \hat{\sigma}_2^{(m)}(x_1,l)\,\mathrm{d}x_1 = h\hat{\sigma}_2 \qquad (9.21)$$

$$\sum_{m\neq k} h_m \int_{-1/2}^{1/2} \hat{\tau}_{12}^{(m)}(x_1,l)\,\mathrm{d}x_1 = h\hat{\tau}_{12} \qquad (9.22)$$

(c) 位移均匀化。

对于一个均质化的对称层合板,面内载荷下层合板的厚度上产生相同的位移场,即所有无裂纹的铺层均具有相同的位移

$$\hat{u}^{(m)}(x_1,l) = \hat{u}^{(r)}(x_1,l);\ \forall\, m\neq k \qquad (9.23)$$

$$\hat{v}^{(m)}(x_1,l) = \hat{v}^{(r)}(x_1,l);\ \forall\, m\neq k \qquad (9.24)$$

其中,r 是作为参考的无裂纹缺陷铺层。具体计算时,铺层 1 作为参考层,如果铺层 1 有裂纹,那么铺层 2 将作为参考层。

9.4.2 $\Delta T \neq 0$ 时的边界条件

接下来,考虑受温度载荷的情况,此时需要给边界条件增加一个常数项。常数项不影响矩阵 $[B]$,其产生单独的力向量 $\{F\}$,即

$$\{F\}_{\Delta T\neq 0} = \begin{Bmatrix} \Delta T \sum_{j=1,2,6} \overline{Q}_{1j}^{(k)} \overline{\alpha}_j^{(k)} \\ \Delta T \sum_{j=1,2,6} \overline{Q}_{1j}^{(k)} \overline{\alpha}_j^{(k)} \\ \Delta T \sum_{i\neq k}\sum_{j=1,2,6} \overline{Q}_{1j}^{(i)} \overline{\alpha}_j^{(i)} \\ \Delta T \sum_{i\neq k}\sum_{j=1,2,6} \overline{Q}_{2j}^{(i)} \overline{\alpha}_j^{(i)} \\ \Delta T \sum_{i\neq k}\sum_{j=1,2,6} \overline{Q}_{6j}^{(i)} \overline{\alpha}_j^{(i)} \\ 0 \\ \vdots \\ 0 \end{Bmatrix} \qquad (9.25)$$

这样,对于当前的裂纹密度集合 λ,单位温度载荷($\Delta T=1$)条件下计算得到的应变即为层合板退化的 CTE 值。

9.5 退化的层合板刚度和 CTE 值

在本节中,对于一个 k 铺层含裂纹缺陷的层合板,其给定裂纹密度为 λ_k,下面介绍如何计算退化的层合板刚度 $\boldsymbol{Q} = \boldsymbol{A}/h$。其中 $\boldsymbol{A}$ 是层合板面内刚度矩阵,h 是层合板的厚度。首先,利用式(9.9)获得所有铺层的厚度平均应变场,即对式(9.16)微分。然后,依次按 a, b 和 c 三种载荷情况(三种载荷情况的 $\Delta T = 0$),通过求解式(9.9)得到应变,按列得到在第 k 层铺层坐标系下的层合板柔度 $\boldsymbol{S}$。

$$^a\hat{\sigma} = \begin{Bmatrix} 1 \\ 0 \\ 0 \end{Bmatrix}; {}^b\hat{\sigma} = \begin{Bmatrix} 0 \\ 1 \\ 0 \end{Bmatrix}; {}^c\hat{\sigma} = \begin{Bmatrix} 0 \\ 0 \\ 1 \end{Bmatrix}; \Delta T = 0 \quad (9.26)$$

由于 3 个施加的应力状态载荷是单位值,因此对于 a, b, c 三种情况,式(9.9)的应变体积平均值即代表层合板柔度矩阵中的一列:

$$\boldsymbol{S} = \begin{bmatrix} {}^a\epsilon_x & {}^b\epsilon_x & {}^c\epsilon_x \\ {}^a\epsilon_y & {}^b\epsilon_y & {}^c\epsilon_y \\ {}^a\gamma_{xy} & {}^b\gamma_{xy} & {}^c\gamma_{xy} \end{bmatrix} \quad (9.27)$$

式中:x, y 为第 k 层铺层坐标系下的坐标(图9.3)。下一步,层合板在第 k 层坐标系下的刚度为

$$\boldsymbol{Q} = \boldsymbol{S}^{-1} \quad (9.28)$$

若要得到层合板退化后的 CTE 值,只需要设置 $\hat{\sigma} = \{0, 0, 0\}^T, \Delta T = 1$ 条件。计算得到的应变即为层合板的 CTE 值,即,$\{\alpha_x, \alpha_y, \alpha_{xy}\}^T = \{\epsilon_x, \epsilon_y, \gamma_{xy}\}^T$。

9.6 退化的单层板刚度

铺层 $m(m \neq k)$ 的单层板在第 k 层铺层的坐标系(图9.3)中的刚度可通过式(9.10)得到,其中 $D_2^{(m)}$ 和 $D_6^{(m)}$ 由式(9.32)给出。含裂纹缺陷的铺层刚度 $Q^{(k)}$ 还未知。注意,上述所有量均是在铺层 k 坐标系中进行表示的。

层合板的刚度定义为含裂纹缺陷铺层 k 的刚度加上剩余的 $N-1$ 个铺层的刚度,具体如下:

$$Q = Q^{(k)} \frac{h_k}{h} + \sum_{m=1}^{n} (1 - \delta_{mk}) Q^{(m)} \frac{h_m}{h} \quad (9.29)$$

其中,当 $m = k$ 时,$\delta_{mk} = 1$,否则 $\delta_{mk} = 0$。式(9.29)左侧通过式(9.28)计算,$Q^{(m)}$ 的所有值均可以很容易计算得到,因为此时 m 铺层均没有裂纹缺陷。因此,对于含裂纹铺层 k 的退化刚度 $Q^{(k)}$ 计算如下:

$$Q^{(k)} = \frac{h_k}{h}\Big[Q - \sum_{m=1}^{n}(1-\delta_{mk})Q^{(m)}\frac{h_m}{h}\Big] \tag{9.30}$$

其中,没有上标的 Q 为层合板的刚度。

为了便于后面的计算,使用连续损伤力学概念(8.2 节),刚度 $Q^{(k)}$ 可以由无损伤铺层刚度和损伤变量 $D_2^{(k)}$, $D_6^{(k)}$ 表示,即

$$\boldsymbol{Q}^{(k)} = \begin{bmatrix} \tilde{Q}_{11}^{(k)} & (1-D_2)\tilde{Q}_{12}^{(k)} & 0 \\ (1-D_2)\tilde{Q}_{12}^{(k)} & (1-D_2)\tilde{Q}_{22}^{(k)} & 0 \\ 0 & 0 & (1-D_6)\tilde{Q}_{66}^{(k)} \end{bmatrix} \tag{9.31}$$

其中,$D_j^{(k)}$ 由给定裂纹密度 λ_k 和施加应变 ϵ^0 计算得到,有

$$D_j^{(k)}(\lambda_k,\epsilon^0) = 1 - Q_{jj}^{(k)}/\tilde{Q}_{jj}^{(k)}; j=2,6 \quad (\text{不对}j\text{求和}) \tag{9.32}$$

式中:$\tilde{Q}^{(k)}$ 为未损伤状态的初始值;$Q^{(k)}$ 为式(9.30)计算的退化值①,上述参数均在第 k 铺层坐标系中表示。

含裂纹缺陷 k 铺层的热膨胀系数通过类似的方法计算如下:

$$\alpha^{(k)} = \frac{1}{h_k}S^{(k)}\Big(hQ\alpha - \sum_{m\neq k}h_m Q^{(m)}\alpha^{(m)}\Big) \tag{9.33}$$

其中,$S = [Q^{(k)}]^{-1}$。相应的温度损伤变量计算如下:

$$D_j^{\alpha(k)} = 1 - \alpha_j^{(k)}/\tilde{\alpha}_j^{(k)}; j=2,6 \tag{9.34}$$

9.7 断裂能

位移控制下,能量释放率(ERR)定义为应变能 U 关于裂纹面积 A(见式(10.1))的偏导数。根据试验观察,脆性基体复合材料(如环氧类基体),其裂纹会以超过有限长度的形式突然扩展。那么,为了描述观察到的裂纹扩展的现象,可应用 Griffith 能量原理的离散(有限)形式,即

$$\begin{cases} G_{\mathrm{I}} = -\dfrac{\Delta U_{\mathrm{I}}}{\Delta A} \\ G_{\mathrm{II}} = -\dfrac{\Delta U_{\mathrm{II}}}{\Delta A} \end{cases} \tag{9.35}$$

式中:ΔU_{I},ΔU_{II} 分别为 I 型和 II 型有限裂纹扩展时层合板应变能的变化量;ΔA 为新生成的(有限)裂纹面积,即新生成裂纹表面的一半。裂纹面积为裂纹表面

---

① 均匀化后的。

的一半符合经典断裂力学约定,因为断裂韧度 G_c 是 Griffith 表面能 γ_c 的两倍。

为了计算能量释放率,方便起见,最好使用在含裂纹缺陷铺层坐标系下的层合板刚度 Q,因为这时,可以将 ERR 分解成张开和剪切两种裂纹模式。对于确定的应变(载荷),层合板刚度为关于裂纹密度 λ 的函数,ERR 值可计算得到,利用文献([110]和[142,3.2.10 节])相关公式,代入式(9.35)后,得

$$G_{\mathrm{I}} = -\frac{V}{2\Delta A}(\epsilon_2 - \alpha_2\Delta T)\Delta Q_{2j}(\epsilon_j - \alpha_j\Delta T)(张开型) \qquad (9.36)$$

$$G_{\mathrm{II}} = -\frac{V}{2\Delta A}(\epsilon_6 - \alpha_6\Delta T)\Delta Q_{6j}(\epsilon_j - \alpha_j\Delta T)(剪切型) \qquad (9.37)$$

式中:$V,\Delta A$ 分别为 RVE 的体积和裂纹面积增量;ΔQ_{ij} 为对应于裂纹面积变化产生的层合板刚度变化量;考虑到 ERR 的模式分解,所有量均表示为在含裂纹缺陷铺层坐标系下的层合板平均量[110]。

例 9.1 中使用的模型,其 $\Delta A = h_k$ 是两个已有裂纹之间新出现的裂纹面积。在这种情况下,裂纹密度增加了 1 倍,并且有 $\Delta Q = Q(2\lambda) - Q(\lambda) < 0$。文献[126]中给出了另一种裂纹扩展方案。

可以看出,上述提出方法提供了计算 ERR 的关键要素,即层合板退化刚度和退化的 CTE,这两个参数都是裂纹密度的函数。

现在损伤激活函数(式(9.1))可以根据任意裂纹密度 λ 和施加的应变 ϵ_x,ϵ_y,γ_{xy} 进行计算。注意,对于一个离散的裂纹(图 9.3),ERR 分量的计算可以直接从位移解式(9.16)中推导得出。当利用有限元方法(FEM)时,其计算结果对网格无依赖性且不需指定较为随意的特征长度[137],而基于弥散裂纹模型的近似计算与此相反[55]。残余热应力的影响也被考虑到该求解方程中。该计算代码在 ANSYS 可被当作用户定义材料[137],用户定义壳单元[138],在 Abaqus 中可被当作用户定义通用截面(UGENS)[140]。参见例 9.1。

9.8 求解算法

求解算法由应变施加步、层合板迭代和铺层迭代组成。层合板的状态变量是所有铺层 i 的裂纹密度阵列和面内应变 ϵ。对于每个载荷(应变)步,层合板应变逐渐增加,同时对层合板所有铺层进行损伤检查。

9.8.1 铺层迭代

当在层合板第 k 铺层中出现基体开裂时,将调用返回映射算法(RMA)(8.4.1 节)进行迭代计算,调整第 k 铺层的裂纹密度 λ_k 使得 g_k 归零,同时保持所有铺层中外力和内力的平衡。迭代过程如下:对于给定的层合板应变 ϵ 和第 k

铺层的 λ_k，计算损伤激活函数 g_k 和损伤变量的值，激活函数和损伤变量都是 λ_k 的函数。使用返回映射算法（RMA）计算裂纹密度增量（减量），如下：

$$\Delta\lambda_k = -g_k / \frac{\partial g_k}{\partial \lambda} \quad (9.38)$$

直到在给定的容差范围内满足 $g_k=0$，其中 $k=1,\cdots,N,N$ 是层合板铺层数量。开始分析时，对所有的铺层取一个可忽略不计的裂纹密度（例 9.1 中取 $\lambda=0.02/\text{mm}$）。

9.8.2 层合板迭代

为了计算层合板中含损伤的第 k 铺层的刚度降低，需假设层合板中所有其他铺层（m – 铺层）在第 k 铺层迭代分析过程中不发生损伤，但是会根据它们的当前损伤变量 $D_i^{(m)}$ 值计算损伤后的性能。给定 λ_k 试验值后，求解会得到第 k 铺层的 $g_k,D_i^{(k)}$。由于对第 k 层的求解取决于其余铺层的刚度，所以第 k 铺层已经收敛的上一个迭代步不能保证其余铺层刚度更新后下一个迭代步也能收敛。换句话说，在给定的应变加载步中，所有铺层的刚度和损伤是相互耦合的，它们必须都满足收敛条件。为满足上述条件，可以通过层合板迭代来完成，即重复对所有铺层进行迭代分析，直到所有出现裂纹损伤的铺层 k 都满足收敛 $g=0$。与经典的针对塑性建立的 RMA 不同（其硬化参数为单调增加的（8.4.1 节）），该 RMA 中必须允许裂纹密度 λ_k 可以降低。如果在层合板迭代分析过程中，其他铺层维持增加损伤导致层合板刚度降低，为保证平衡，则需要降低 λ_k。

例 9.1 考虑一个玻纤/环氧树脂层合板，铺层顺序为 $[0/90_8/0/90_8/0]$，材料性能见表 9.1。层合板受面内应变 $\varepsilon_x \neq 0, \varepsilon_y = \gamma_{xy}=0$ 作用，试绘制第 $k=2$ 铺层的裂纹密度与应变 $0<\varepsilon_x<2\%$ 的曲线。

表 9.1 例 9.1 的材料性能

性能	值
E_1/GPa	44.7
E_2/GPa	12.7
G_{12}/GPa	5.8
ν_{12}	0.297
ν_{23}	0.41
层厚/mm	0.144
$G_{\text{Ic}}/(\text{kJ/m}^2)$	0.254
$G_{\text{IIc}}/(\text{kJ/m}^2)$	1000000
$\text{CTE}_1/(10^{-6}/\text{℃})$	3.7
$\text{CTE}_2/(10^{-6}/\text{℃})$	30
$\Delta T/\text{℃}$	0

解 由于 Abaqus 没有计算裂纹密度的能力,因此必须使用一个插件。该例子中,通过使用可编程的 User General Section(UGENS,[143],26.6.6 节)来执行离散损伤力学(DDM)分析。DDM 插件可由网站[5,ugens – std.obj]中获得,具体的理论在本章中已经进行了解释。

本例说明了如何使用 Abaqus/CAE 创建模型;写入 .inp 文件;对 .in 文件进行修改以添加 UGENS 程序所需的数据;通过插件运行计算程序;最后可视化结果。以下是具体步骤的代码和 GUI 操作窗口。

1. 从命令窗口启动 Abaqus/CAE

由于分析过程中需要用到许多生成的文件,因此我们在自己选定的工作目录中启动 Abaqus/CAE。右键单击文件 c:\SIMULIA\User\Chapter_9\Ex_9.1。这将在当前的文件夹中打开 Abaqus/CAE。

2. 创建 Part

由于壳内部的应变场是相同的(x 和 y 方向),因此使用一个 S4R 单元模拟一个尺寸为 $a \times b$ 为 10mm×10mm 的单胞,其中 a,b 分别是模型沿 x,y 方向的尺寸。此例的操作开始如下:

```
    Module: Part
Menu: Part,Create,Name;[Part-1],3D,Deformable,Shell,Planar,
    Approx. size:[20]
    Add,Lines,Rectangle,[0,0],[10,10],X,Done
```

上面的代码中的第一行与图 9.4 中对应。

3. 输入截面属性

对于 UGENS,材料属性和截面参数必须以截面属性的形式输入,但是从 6-10 版本开始,Abaqus/CAE 不再具备对 UGENS 输入数据的功能。这需要人为对 Job-1.inp 文件进行编辑输入,过程与例 4.3 类似。在此之前,我们先使用 CAE 把这些参数输入 .inp 中。DDM 插件涉及到 $4+9N+3N$ 个参数,具体如下:

(1) 层合板总厚度 t。

(2) $3+9 \times N$ 个材料属性,其中 N 是对称层合板的对称层数,即 LSS 一半的层数。属性排序如下,从第一层 $k=1$(底面)开始,直到第 N 层(中面):

G_{Ic}:I 型断裂对应的临界 ERR 值;

G_{IIc}:II 型断裂对应的临界 ERR 值。如果没有试验数据,可以利用 $G_{IIc} > 4G_{Ic}$;

ΔT:测量 G_{Ic},G_{IIc} 时的温度与工作温度之间的温度变化;

E_1:纵向模量;

E_2:横向模量;

图 9.4　创建 Part

G_{12}：面内剪切模量；

ν_{12}：面内泊松比；

ν_{23}：层间泊松比，注意 $G_{23} = (E_2/2)/(1+\nu_{23})$；

α_1：纵向热膨胀系数；

α_2：横向热膨胀系数；

θ_k：层合板坐标系下的铺层方向角；

t_k：铺层厚度。

(3) $3 \times N$ 个状态变量，从第一层 $k=1$（底面）开始，到第 N 个铺层（中面）：

λ_k：铺层 k 的裂纹密度；

D_2：铺层 k 的横向损伤变量；

D_6 铺层 k 的剪切损伤变量。

在图 9.5 所示的材料定义对话框中输入 $3+9N+5N$ 个值（除层合板厚度以外的所有参数），这些值将在第 8 步中给出。

然后，创建一个截面并赋予壳的厚度（图 9.6）和 3 个层间剪切系数值（图 9.7）。由于本例中没有弯曲和横向剪切，所以层间剪切系数无关紧要；仅需输入数值即可。此外，层间剪切系数可按式(3.9)计算（另见例 3.1）。最后，将截面属性赋予给部件。Abaqus/CAE 中的 GUI 的操作过程具体如下：

第 9 章 离散损伤力学

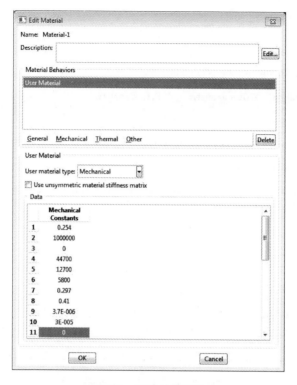

图 9.5 创建材料属性

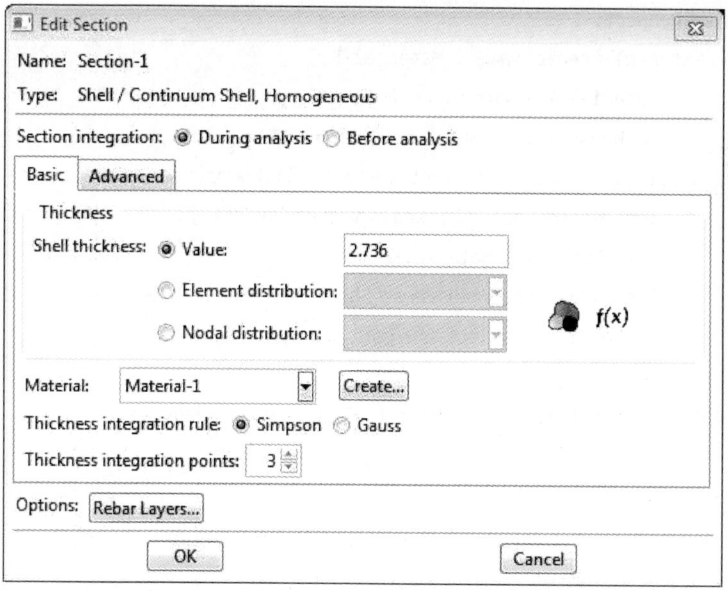

图 9.6 定义截面属性

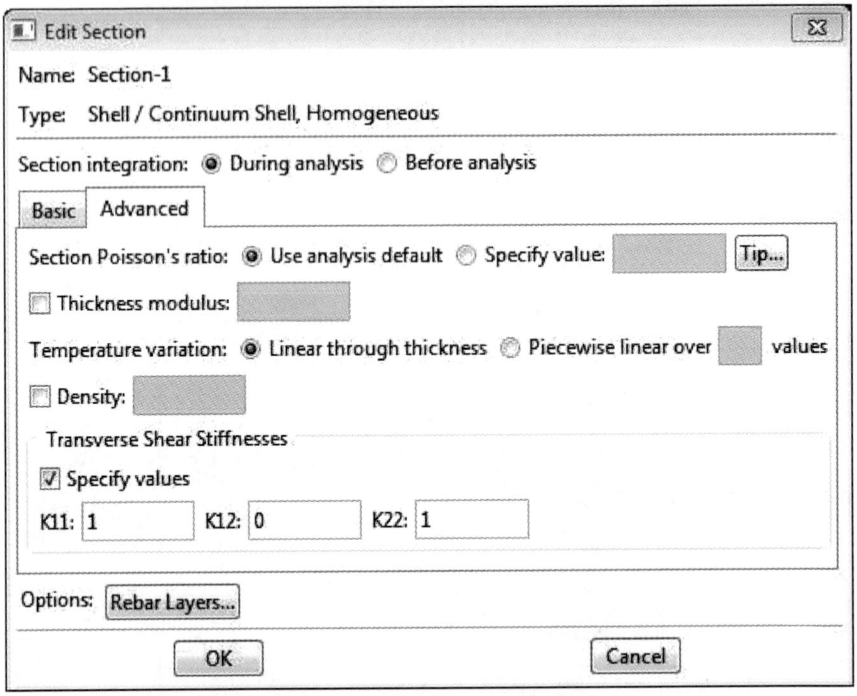

图 9.7 进一步定义截面属性

Module: Property
Menu: Material,Create,Name:[Material-1],
　　　Material Behavior: User Material,
　　　　Mechanical Constants:#3+9*N+5*N values
　　Section,Create,Name:[Section-1],Shell,Homogeneous,Cont
　　　Tab: Basic,Shell thickness Value;[2.736],
　　　　　Material: Material-1
　　　Tab: Advanced: Transverse Shear Stiffness:[1,0,1]
　　Assign,Section,# select the part,Done,OK
Module: Assembly
Menu: Instance,Create,Name:[Part-1],Type: Independent,OK

4. 创建分析步

接下来,需要创建分析步,并且在 Field Output Requests 对话框中定义输出数据(图 9.8)。由于分析涉及材料非线性(损伤),因此需特别注意求解策略(图 9.9)。分析步的具体定义步骤如下:

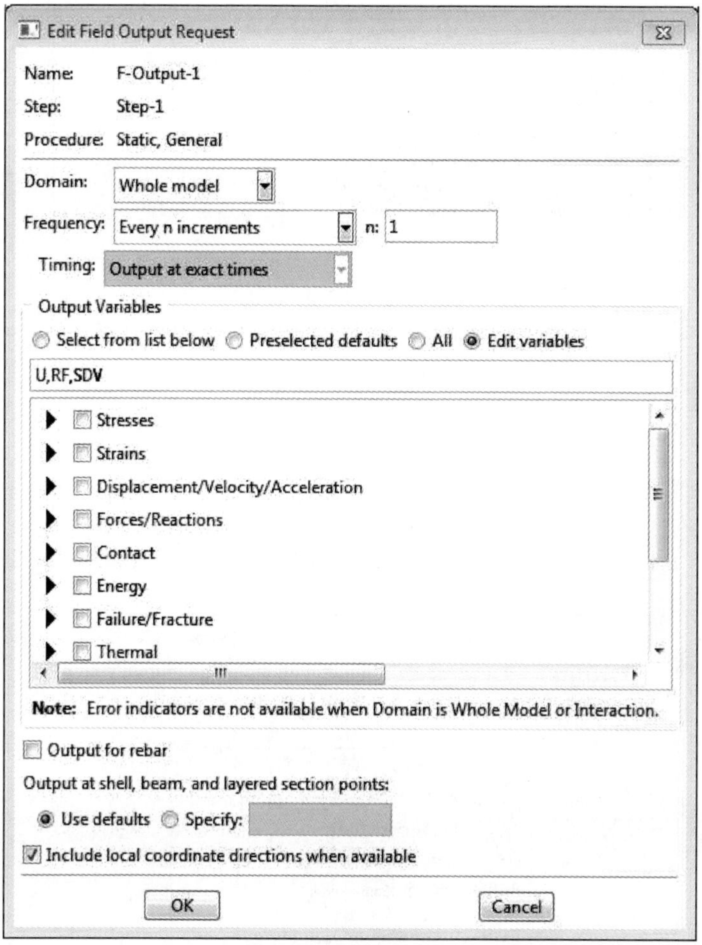

图 9.8 定义输出变量

Module: Step
Menu: Step,Create,Name:[Step - 1],After: Initial,
 Type: General,Static,General,Cont
 Tab: Basic,Time period: 1
 Tab: Advanced,max. increments:[1000],
 Incr. size: Initial:[0.01],Minimum;[1E - 005],Maximum:[0.01]

5. 定义边界条件

接下来是定义边界条件,模型关于 $x=0$ 和 $y=0$ 对称。载荷以位移形式施加,最大位移值为模型 x 向尺寸与预期最大应变之积,即 $U_1 = 0.2$(图 9.10)。具体步骤如下:

图9.9 定义分析步

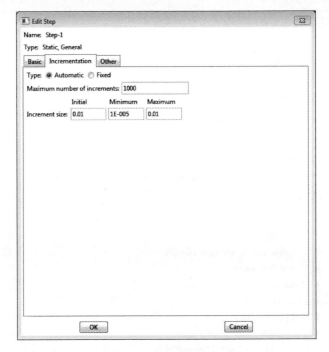

图9.10 定义边界条件

```
Module: Load
Menu: BC,Manager
    Create,name:[BC-1],Step: Initial,Type: Symmetry,Cont,
      # select vertical edge @  x=0,Done,XSYMM,OK
    Create,name:[BC-2],Step: Initial,Type: Symmetry,Cont,
      # select horizontal edge @  y=0,Done,YSYMM,OK
    Create,name:[BC-3],Step: Initial,Type: Symmetry,Cont,
      # select vertex at the origin,Done,ENCASTRE,OK
    Create,name:[BC-4],Step: Step-1,Type: Displacement,Cont,
      # select vertical edge @  x=10,Done,U1=[0.2],OK
```

6. 网格划分

然后,采用使用 S4R 单元划分网格。由于面内应变是相同的,因此只需一个单元即可。通过将网格大小定义为模型尺寸大小 $a=b$ 即可。具体步骤如下:

```
Module: Mesh
Menu: Seed: Instance,Approx. size:[10],OK
    Mesh,Element Type,Standard,Linear,Shell # S4R
    Mesh,Instance,Yes
```

7. 创建第一个计算任务

接下来,创建一个仅输出 Job-1.inp 文件的任务(图 9.11)。

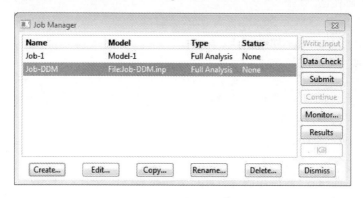

图 9.11　创建第一个计算任务(Job-1)

```
Module: Job
Menu: Job,Manager
    Create,Name:[Job-1],Source: Model,Cont,OK
    Data Check
    Write Input
```

8. 修改 .inp 文件

复制一个 Job – 1. inp 文件的副本,命名为 Job – ddm. inp,并进行如下修改。两个文件在文献[5]中均可获得。

用以下代码行:

```
* Shell General Section,
    elset = _PickedSet2,USER,PROPERTIES = 30,VARIABLES = 15
```

替换下列代码行:

```
* Shell Section,elset = _PickedSet2,material = Material - 1
```

编辑材料属性如下:

```
 0.254,    1e+06,        0.,    44700.,   12700.,    5800.,   0.297,   0.41
 3.7e-06,  3e-05,        0.,    0.144,    44700.,   12700.,   5800.,   0.297
    0.41,  3.7e-06,  3e-05,     90.,      1.152,    44700.,   12700.,   5800.
   0.297,    0.41,  3.7e-06,  3e-05,       0.,      0.072
```

并将上述材料属性代码移动至下述命令行之前:

```
* Transverse Shear
```

删除以下命令行:

```
* *  MATERIALS
* *
* Material,name = Material-1
* User Material,constants = 30
* *
```

在以下命令行之前

```
* *  STEP: Step-1
```

添加下述命令行(调用内部程序设置初始裂纹密度为 $\lambda = 0.02$)

```
* Initial Conditions,type = SOLUTION,USER
```

9. 创建第二个计算任务

接下来,创建另一个计算任务(Job-DDM)运行带插件的模型(图9.12、图9.13):

Module: Job

Menu: Job,Manager
　　Create,Name: [Job-DDM],Source: Input file,
　　Select: [C:\SIMULIA\User\Chapter_9\Ex_9.1\Job-DDM.inp],Cont,
　　Tab: General,
　　User subroutine: [C:\SIMULIA\User\Chapter_9\Ex_9.1\ugens-std.obj],
　　OK
　　Submit # wait for the run to complete
　　Results

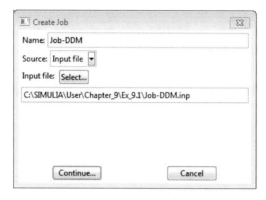

图 9.12　通过 Input File 方式创建第二个计算任务

图 9.13　编辑第二个计算任务来指定用户子程序

10. 可视化

为了实现可视化,先回顾一下状态变量的含义(见步骤 3)。本例中层合板的 2# 铺层有裂纹损伤。裂纹密度 λ_2 储存在 SDV = 4 中,初始值设置为 λ = 0.02。选择初始加载步查看初始值。在第 24 个加载步时($\varepsilon_x \approx 0.48\%$)裂纹出现。

```
Module: Visualization
Menu:   Plot,Contours,On deformed shape
        Result,Step/Frame,# select step 24
```

或者,整个可视化过程可以使用 Python 脚本文件自动实现,即 Ex_9.1.py 文件和 UgenKeyword.py 文件,可由文献[5]中获得。执行脚本可按照步骤 1 中的过程打开命令窗口,然后键入 abaqus cae nogui = Ex_9.1。然后,打开 Job.odb 可视化结果,将结果保存至 Job.csv 文件。裂纹密度与应变的曲线关系如图 9.14 所示。

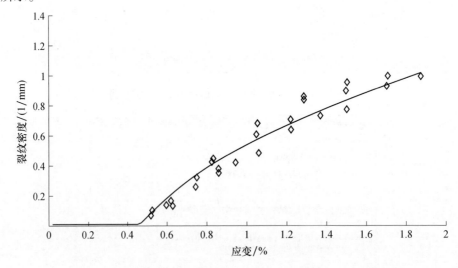

图 9.14 DDM 计算的裂纹密度-应变曲线与实验数据[144]的对比

习题

习题 9.1 计算表 9.2 中显示的层合板在 1~3 铺层中出现第一个裂纹损伤时的层合板临界应变 ε_x^c,所有层合板的材料为玻纤/环氧树脂,材料性能见表 9.1。层合板受面内应变载荷为 $\varepsilon_x \neq 0, \varepsilon_y = \gamma_{xy} = 0$。对于每种层合板,具体哪一层最先出现裂纹损伤?

第9章 离散损伤力学

表9.2 习题9.1的层合板铺层属性

层合板编号	LSS
1	$[0/90_8/0/90_8/0]$
2	$[0/70_4/-70_4/0/-70_4/70_4/0]$
3	$[0/55_4/-55_4/0/-55_4/55_4/0]$
4	$[0_2/90_8/0_2]$
5	$[15/-15/90_8/-15/15]$
6	$[30/-30/90_8/-30/30]$
7	$[40/-40/90_8/-40/40]$

习题9.2 使用习题9.1的结果,计算各单层板的横向就地强度 F_{2t}^{is}(文献[1],7.2.1节)。

习题9.3 计算表9.2中显示的层合板在4~7铺层中出现第一个裂纹损伤时的层合板临界应变 ϵ_x^c,所有层合板的材料为玻纤/环氧树脂,材料性能见表9.1。层合板受面内应变载荷为 $\epsilon_x \neq 0, \epsilon_y = \gamma_{xy} = 0$。对于每种层合板,具体哪一层最先出现裂纹损伤?

习题9.4 使用习题9.3的结果,计算各单层板的横向就地强度 F_{2t}^{is}(文献[1],7.2.1节)。

第10章 分　　层

分层是一种常见的影响复合材料层合板结构性能的失效模式。层间界面为裂纹扩展提供了一条低阻力路径,因为相邻单层之间的结合力仅取决于基体性能。分层现象可能源于初始的制造缺陷、疲劳或低速冲击产生的裂纹、几何/材料不连续(如接头和自由边)区域的应力集中或由于较高的层间应力导致。

层合板受压缩时,分层处的层板容易发生屈曲,并且,分层现象和屈曲的相互作用会进一步导致裂纹的扩展。分层现象的存在可显著降低复合材料层合板的屈曲载荷和抗压强度[145](图10.1)。在横向荷载作用下,层合板的屈曲也会驱动产生分层[146]。所以,需要联合断裂力学和几何非线性对分层屈曲破坏形式进行分析。

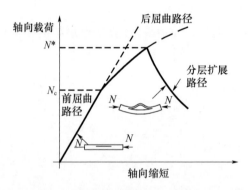

图10.1　压缩状态下层合板的分层屈曲

根据其形状,分层分为宽度贯穿形或条带形[146-153]、圆形[153-159]、椭圆形[160]、矩形[161]或任意形状[162,63]。根据在层合板厚度方向上的位置,分层又分为薄膜状态、对称裂纹状态[145,148,149]和一般状态[150,153,156,157,159]。此外,文献[164-165]针对含有多种分层状态的复合材料层合板,在面内压缩载荷作用下,开展了屈曲和分层扩展分析工作。分层屈曲的相关实验结果可参考文献[166-167]。

梁式分层试样是在一些文献中已经研究过的其他分层构型形式,文献[166-172]研究了梁式分层试样在弯曲、轴向和剪切载荷下的分层行为,利用测量复合材料、胶结接头和其他形式层合材料在Ⅰ、Ⅱ型及混合模式下的层间断裂强度,形成了一套基本的实验方法(图10.2)。

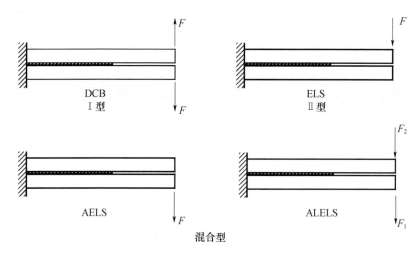

图 10.2 梁式分层试件形式

在含有压电敏感器或致动器的平板中,在机械和(或)电力加载作用下,压电层与基板之间的粘接缺陷可能会增长。因此,脱粘分层现象会导致静态或动态响应发生显著变化,这会使得智能系统的自适应特性显著降低[173-174]。另外,分层的扩展也可能由动态效应引起,如振动和冲击。例如,文献[175]中针对含有圆形分层缺陷和时变载荷条件下的复合材料层合板,研究了分层屈曲导致的层合板惯性的变化对其动力学响应的影响。

分层可以使用内聚力损伤模型(cohesive damage models)(10.1 节)和断裂力学(10.2 节)的方法进行分析。内聚力损伤模型应用了由损伤变量和损伤演化规律定义的界面本构关系。采用的内聚力损伤单元通常安插在实体单元之间[176-179]或梁/壳单元之间[178]。

在断裂力学方法中,通过比较能量释放率(ERR)与界面的断裂韧性,对已存在的分层进行分层扩展分析。当涉及混合模式时,由于界面断裂韧性的混合模式依赖性,需要将总的 ERR 分解为 I 型、II 型和 III 型三种组成部分[170,180]。一些文献中已经提出了许多基于断裂力学的模型来研究分层,包括三维模型[181-183]和简化的梁式模型[145,147,172,185]等。

断裂力学方法能够使我们预测预先存在的裂纹或缺陷的扩展规律。在一般荷载条件下,均匀的各向同性物体中,裂纹趋于向一个可以使尖端处保持纯 I 型状态的方向偏转(kinking)扩展。与此不同,层合复合材料中的分层现象被限制在其自身的平面内进行扩展,因为与相邻材料相比,界面断裂韧性相对较低。在混合模式条件下,分层裂纹随着其裂纹尖端的推进而发生扩展,其分析需要一个可以包括所有 3 种模式的断裂准则(见 10.1.2 节)。

单位体积的弹性应变能(应变能密度单位为 J/m^3)定义为 $U_0 = 1/2\sigma_{ij}\epsilon_{ij}$。应变能(单位 J)定义为体积积分形式 $U = \int_V U_0 dV$。形成或扩展一个裂纹所需的能量等于实体在裂纹形成过程中释放的弹性能。释放的能量为裂纹形成前后弹性应变能之差,即 $-\Delta U = U_{\text{after}} - U_{\text{before}}$。每单位裂纹面积 A 的能量释放率(单位 J/m^2)为

$$G = -\frac{\Delta U}{\Delta A} \tag{10.1}$$

式中:A 为新生成的裂纹表面积的一半。裂纹扩展理论可以分别由 Griffith 和 Irwin 两种方法中的一种进行发展。Griffith 能量方法一方面使用能量释放率(Energy Release Rate) G 的概念作为(可计算的)有效断裂能量,另一方面利用了材料参数 G_c,即断裂所需的临界能量。当满足如下关系时,裂纹会发生扩展:

$$G \geq G_c \tag{10.2}$$

注意到 $G_c = 2\gamma_c$,γ_c 为单位裂纹表面面积的临界断裂能,A 为新形成的裂纹面积的一半,即裂纹两个相同表面的其中一个表面面积。

Irwin(局部)方法基于应力强度因子的概念,其表示裂纹尖端附近应力场的强弱。以上两种方法是等价的,因此,能量准则可以重写成应力强度因子的形式。进一步,可以使用一些与路径无关的积分方法来计算 ERR,如 J 积分[186]。

裂纹扩展过程中释放弹性应变能 ΔU,并由此会产生新的裂纹表面,释放的弹性应变能可以通过计算闭合裂纹所需的功得到,即

$$\Delta U = W_{\text{closure}} \tag{10.3}$$

其裂纹闭合计算方法是 10.2 节所述的虚拟裂纹闭合技术(VCCT)的技术基础。

10.1 内聚力方法

内聚力方法(CZM)基于这样的假设:损伤初始时,发生分层的两个分离面之间的应力传递能力并没有完全丧失,而是通过逐渐降低分离面之间的刚度来控制应力传递能力(图 10.3)。

在 CZM 中,层合结构复合材料两个可能的分离面之间的界面采用内聚力单元(cohesive elements)进行建模[187]。Abaqus 软件中有两种类型的 CZM 单元:

(1)有限厚度的 CZM 单元,该单元基于规则的连续体($\sigma - \epsilon$)本构行为。这种类型的单元适用于粘接界面厚度较大情况下的界面分层行为的模拟。

(2)0 厚度的 CZM 单元,该单元基于牵引 - 分离间距($\sigma - \delta$)本构行为。这种类型的单元适用于粘接界面厚度可以忽略情况下的界面分层行为的模拟,如

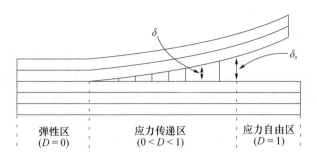

图10.3 模拟裂纹扩展的内聚力方法

复合材料层合板结构。

内聚力行为(cohesive behavior)使用牵引-分离间距方程进行描述(图10.4)。顾名思义,该方法将工程应力-应变($\sigma-\epsilon$)方程替换为牵引分离间距($\sigma-\delta$)方程。可通过将内聚力单元厚度方向对应的节点定义为重合,这样就使得内聚力单元的厚度设置为0。然而,虽然对应的节点最初是重合的,但这些节点仍然是各自独立的,并且它们在层合板复合材料变形过程中会相互分开。层合板的所有相邻的单层板之间均可以视为通过一定刚度的内聚力单元进行连接。在受力变形过程中,相邻层板之间产生的分层效应与内聚力单元的刚度正相关。

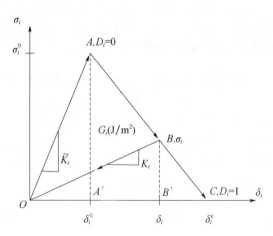

图10.4 内聚力模型的应力传递模型

0厚度COH2D4类型的CZM单元(例10.1)和通过施加接触内聚力行为(interaction cohesive behavior)模拟分层(例10.2)均是基于牵引-分离间距本构方程。该类型单元的单元刚度矩阵需要界面材料的刚度$\tilde{K}$(在Abaqus文档中称为罚(penalty)刚度),但由于内聚力单元的初始体积为零,所以该单元刚度矩阵不能像通常那样在单元体积上积分得到。CZM 单元可以想象成初始重合节点之间的一个弹簧。CZM 单元的刚度是结构刚度的一部分,并且在层合板加载过程中 CZM 单元会一起承受变形,初始重合的节点将彼此打开(I 型:打开)或滑动(II 型:剪切和III 型:撕裂)(图10.5)。CZM 单元的节点间距总是能够通过求解离散结构得到。

由于 CZM 单元的初始厚度为0,CZM 单元的变形状态不能使用传统应变的定义来描述。相反,这种情况下,变形状态变成了可由 CZM 单元连接的两个面之间的分离间距 δ 进行表达,这也使得使用牵引-分离间距方程($\sigma - \delta$)代替传统($\sigma - \epsilon$)方程成为可能[①]。

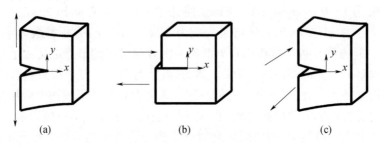

图 10.5　裂纹扩展模式
(a)I-张开型;(b)II-剪切型;(c)III-撕裂型。

10.1.1　单模式的内聚力模型

即使多种裂纹模式同时存在,CZM 均假设上述3种裂纹扩展模式之间不存在耦合,详见10.1.2节。在本节中,仅考虑层合复合材料界面处在单模式下的变形情况,即在模式 I、II 或 III(图10.5)的其中之一条件下。这3种模式的有关公式均相似。界面处的表面牵引力为 σ_i,$i =$ I,II,III 分别表示3种裂纹扩展模式。与其对应的使用 CZM 单元连接的两面之间的间距由 δ_i 表示,每一个间距均通过界面刚度 K_i 与表面牵引力 σ_i 关联起来,界面刚度也称为罚刚度(penalty stiffness)。因此,计算时,每个加载模式下的材料刚度值都是必需的,即 K_I,K_{II},K_{III}。文献[188]对于如何为 K_i 取值进行了相关讨论。

界面处的材料行为假设在损伤开始之前是线弹性的(见图10.4中的 OA 段),之后是弹性损伤(damaging-elastic)阶段(见图10.4中的 OB 段)。因此,应力-分离间距由下式描述:

---

① Abaqus 文档在描述基于牵引-分离间距方程的 CZM 单元变形时,使用应变(strain)一词与分离间距(separation)δ 同义。

第 10 章 分层

$$\sigma_i = K_i \delta_i \tag{10.4}$$

$$K_i = (1 - D_i) \tilde{K}_i \tag{10.5}$$

式中:D_i 为损伤变量;$\tilde{K}_i$ 单位为 N/mm³,其表示界面处未损伤状态材料的刚度值,该刚度值将应力 σ_i 与 δ_i 相关联,$i =$ I,II,III。$\tilde{K}_i$ 为 CZM 模型所需的附加材料属性,与描述法向变形的弹性模量 E 或剪切变形的剪切模量 G 不同。

对于每种开裂模式,当界面处开始发生损伤时(图 10.4 中的 A 点),均存在一个临界的应力 σ_i^0 值和分离间距 δ_i^0 值,此时称为损伤初始。损伤初始时,两个相邻的单层板之间并没有完全分离形成物理裂纹,而是之间的界面材料开始失去刚度。这里,σ_i^0 代表界面强度,对于每种裂纹模式,其强度值分别为:σ_I^0、σ_{II}^0、σ_{III}^0。CZM 中,这些模式称为损伤模式,因为 CZM 并不计算断裂力学问题,相反,CZM 将断裂力学问题替换成了一个连续介质损伤力学问题(见第 8 章)。界面强度也由此成为了 CZM 单元所需的附加材料参数。

因此,损伤初始准则为

$$\sigma_i = \sigma_i^0 \tag{10.6}$$

而且,损伤初始时的分离间距计算如下:

$$\delta_i^0 = \sigma_i^0 / \tilde{K}_i \tag{10.7}$$

根据式(10.5),损伤初始之后,界面材料开始逐渐丧失刚度(见图 10.4 中的 OB 段)。注意,损伤变量和损伤模式的数量一致:D_I、D_{II}、D_{III},它们是在分析过程中需要确定的状态变量(见式(8.61))。它们的物理学解释是作为刚度退化的度量,由式(10.5)给出(另见 8.2 节)。损伤变量满足以下条件:

(1)在损伤初始之前 $D = 0$(图 10.4 的 OA 段),此时界面材料未发生损伤,从而保持初始刚度值。

(2)界面材料刚度退化阶段 $0 < D < 1$(图 10.4 的 AC 段),此时界面材料逐渐丧失其刚度。

(3)在发生断裂时 $D = 1$(图 10.4 的 C 点),此时界面材料无刚度,也意味着界面没有了应力传递能力。此时 CZM 单元连接的两个面之间发生开裂。

参照图 10.4,得

$$D_i = \begin{cases} 0 & (\delta_i \leq \delta_i^0) \\ 1 & (\delta_i = \delta_i^c) \end{cases} \tag{10.8}$$

在应力 - 应变空间中,图 10.4 所示的内聚力行为显示了应力软化(stress-softening)过程。也就是说,损伤初始之后,损伤界面中的应力 σ_i 始终低于其峰值 σ_i^0。

CZM 单元采用的是连续介质损伤力学中的弹性损伤经典假设(CDM,第 8 章)。因此,从图 10.4 中 AC 线上的任何点(如 B 点)卸载时将返回原点,且不会发生永久变形。

最后,当界面的刚度降为 0 时,其内聚力连接关系会完全断裂(图 10.4 中的 C 点)。由于直到点 C 处,界面刚度才会全部丧失,从而导致内聚应力无法传递,因此 CZM 与可以和 Griffith 裂纹扩展准则式(10.2)关联起来。关联的方式为:图 10.4 中 $(\sigma-\delta)$ 曲线下的面积等于 Griffith 原理式(10.2)中的临界 ERR G_{ic}。这样,断裂时的分离间距计算如下:

$$\delta_i^c = \frac{2G_{ic}}{\sigma_i^0} \tag{10.9}$$

对于每种损伤模式均存在一个临界的 ERR 值,$G_{ic}(i=\mathrm{I},\mathrm{II},\mathrm{III})$,因此断裂时将存在 3 个分离间距,每个模式一个。除了对应 3 个模式的 3 个强度值 σ_i^0 和 3 个界面刚度值 $\tilde{K}_i$ 之外,3 个 ERR 值也是 CZM 计算所需要的材料属性。从计算分析所需实验数据量的角度来看,这是 CZM 的缺点,因为该方法需要 9 个实验值。而离散损伤力学方法仅需要 3 个 ERR 值即可预测损伤的初始和扩展(见第 9 章)。

把式(10.4)代入式(10.5),整理,得

$$D_i = 1 - \frac{\sigma_i \delta_i^0}{\sigma_i^0 \delta_i} \tag{10.10}$$

由三角形 $BB'C$ 和 $AA'C$ 的相似性,得

$$\frac{\sigma_i}{\sigma_i^0} = \frac{\delta_i^c - \delta_i}{\delta_i^c - \delta_i^0} \tag{10.11}$$

将其代入式(10.10),得

$$D_i = \frac{\delta_i^c(\delta_i - \delta_i^0)}{\delta_i(\delta_i^c - \delta_i^0)} \tag{10.12}$$

这样,损伤变量 D_i 表达成了层合板界面的分离间距 δ_i 的函数,其 δ_i 由有限元方法求解得到,且 δ_i^0 和 δ_i^c 值已事先获得。

总之,图 10.4 描述的材料行为存在 4 个不同的阶段:
(1)线弹性无损伤状态材料行为(OA 线段),相关的本构方程为式(10.4)。
(2)损伤初始点(A 点),相关的准则为式(10.6)。
(3)损伤演化(AC 线段),相关的损伤演化方程为式(10.5)和式(10.12)。
(4)断裂点(裂纹形成),相关的裂纹形成准则为式(10.2)。

如前所述,本节所述的相关公式仅适用于单模式 I、II 或 III 情况下。混合模式下的一般情况见 10.1.2 节内容。

10.1.2　混合模式的内聚力模型

当层合材料的界面处于混合模式条件下时,所有 3 个牵引力分量 σ_I、σ_II、σ_III 和所有 3 个分离间距分量 δ_I、δ_II、δ_III 均处于激活状态。换言之,混合模式意味着同时存在 2 对或 3 对 $(\sigma_i,\delta_i)\, i = \mathrm{I,II,III}$。为了减少实验的工作量,一般假设所有模式下式(10.7)中的罚刚度均相同,$\tilde{K} = \tilde{K}_i$。

混合比定义为不同模式分量之间的比率。例如,以分离间距的形式,定义如下:

$$\beta_{\delta_\mathrm{II}} = \frac{\delta_\mathrm{II}}{\delta_\mathrm{I}};\beta_{\delta_\mathrm{III}} = \frac{\delta_\mathrm{III}}{\delta_\mathrm{I}} \tag{10.13}$$

或者以 ERR 的形式定义如下:

$$\beta_{G_\mathrm{II}} = \frac{G_\mathrm{II}}{\sum_1^3 G_i};\beta_{G_\mathrm{III}} = \frac{G_\mathrm{III}}{\sum_1^3 G_i} \tag{10.14}$$

无论使用何种形式定义,混合比仅是表征混合状态的参数,这就允许我们可以通过假设界面在恒定的混合比条件下发生分层进行简化分析。而且可以进一步进行假设,即使所有模式同时发生,它们之间也不存在耦合。也就是说,每个模式的应力-分离间距关系均使用式(10.4)单独表示。

接下来,混合模式下的分离间距可以由各个模式的分离间距的范数形式 L^2 定义,即

$$\delta_\mathrm{m} = \sqrt{\sum_{i=1}^M \delta_i^2} \tag{10.15}$$

式中:M 为涉及的模式数(即两或三个模式)。下一步,替换单模式下的损伤初始准则式(10.6),例如,由二次应力准则取代

$$\sum_{i=1}^M \left(\frac{\sigma_i}{\sigma_i^0}\right)^2 = 1 \tag{10.16}$$

在 Abaqus 中还可以选择其他形式混合模式的初始损伤准则,如二次位移准则或最大应力/最大位移准则。无论选择何种损伤初始准则,其目标均为计算混合模式加载条件下损伤初始时的混合模式分离间距 δ_m^0。

在仅有 I 型和 II 型模式的情况下,$M = 2$,损伤初始时的等效混合模式分离间距 δ_m^0 按如下步骤计算得到。首先利用式(10.7)、式(10.4)和式(10.5),以分离间距的形式重写损伤初始准则式(10.16),并考虑到损伤初始前 $D_i = 0$。因此,该两种模式混合下的损伤初始由下式预测:

$$\left(\frac{\delta_\mathrm{I}}{\delta_\mathrm{I}^0}\right)^2 + \left(\frac{\delta_\mathrm{II}}{\delta_\mathrm{II}^0}\right)^2 = 1 \tag{10.17}$$

接下来,利用式(10.13)的第一项,重写式(10.15),得

$$\delta_{\mathrm{I}} = \frac{\delta_{\mathrm{m}}}{\sqrt{1+\beta^2}} \qquad (10.18)$$

再次利用式(10.13),得

$$\delta_{\mathrm{II}} = \beta\frac{\delta_{\mathrm{m}}}{\sqrt{1+\beta^2}} \qquad (10.19)$$

现在,把式(10.18)和式(10.19)代入式(10.17),考虑到式(10.17)为损伤初始的准则,应使用 δ_{m}^0 替代 δ_{m}。因此,得

$$\delta_{\mathrm{m}}^0 = \sqrt{(\delta_{\mathrm{I}}^0)^2(\delta_{\mathrm{II}}^0)^2\frac{1+\beta^2}{(\delta_{\mathrm{II}}^0)^2+\beta^2(\delta_{\mathrm{I}}^0)^2}} \qquad (10.20)$$

式(10.20)中的 δ_i^0 表示单模式加载条件下损伤初始时的分离间距,可使用式(10.7)计算得到;β 为混合比,其假定在损伤过程中为常数。

现在需要一个混合模式下的裂纹扩展准则代替单模式下的准则(式(10.2))。其中一种选择是使用ERR能量准则,如下所示:

$$\sum_{i=1}^{3}\left(\frac{G_i}{G_{ic}}\right)^{\alpha_i} = 1 \qquad (10.21)$$

该表达式试图在混合模式条件下预测断裂,类似于单模式情况下对图10.4中的 C 点的预测。为了降低实验代价,通常假设所有单模式下的指数都相同,即 $\alpha_i = \alpha$。

每个单模式下的分量 G_i 均可以通过两种方法进行计算。一些研究者[176,178]通过考虑图10.4中的 $OABB'$ 面积来计算每个单模式下的ERR分量,从而将可回收能量 OBB' 面积包含在 G_i 的定义中。这种方法与线弹性断裂力学(LEFM)有着间接的联系。其他研究者[188]使用的是损伤力学方法,其中每个单模式的ERR分量仅通过表示不可恢复能量的 OAB 面积来进行计算。这两种方法对于单模式下的分层计算结果是相同的,因为在 δ_i^c 处发生分层时,两种方法预测的 G_i 值相同(B 点达到 C 点)。然而,对于混合模式下的分层计算,由于裂纹在满足耦合准则时会扩展,因此使用这两种方法会得到不同的结果,因为耦合准则涉及 G_i/G_{ic} 比值。

理论上,基于LEFM的计算方法对混合模式下分层承载能力的预测偏保守。但是在分层过程中,每个点耗散的总能量并不是像LEFM中假设的那样瞬间释放,所以基于损伤力学的计算方法似乎是比较合适的,特别是当分层裂纹前端的非线性断裂区域的尺寸不可忽略的时候。这与层合结构复合材料的情况比较类似,其损伤区域可能会大于或等于单层的厚度,通常与尖端应力场成比例。

下面将采用损伤力学方法进行计算。也就是说,根据每个单模式下的分离

间距 δ_i,单模式下的 ERR G_i 分量均由图 10.4 中的面积 OAB 计算得到,其表示了耗散能量,即

$$A_{OAB} = A_{OAC} - A_{OBC} \tag{10.22}$$

式中: A_{OAC} 为单模式下的临界 ERR G_{ic}, A_{OBC} 可根据图 10.4 中的几何计算为

$$A_{OBC} = \frac{1}{2} BB' \times OC = \frac{1}{2} K_i \delta_i^0 \frac{\delta_i^c - \delta_i}{\delta_i^c - \delta_i^0} \delta_i^c \tag{10.23}$$

式中: $\delta_i^c = OC, K_i \delta_i^0 = \sigma_i^0, \sigma_i / \sigma_i^0$ 由式(10.11)计算得到。基于式(10.22)和式(10.23),混合模式下断裂时的单模式 ERR 分量为

$$G_i = G_{ic} - \frac{1}{2} K_i \delta_i^0 \delta_i \frac{\delta_i^c - \delta_i}{\delta_i^c - \delta_i^0} \tag{10.24}$$

模式分解式(10.24)的假设是必要的,这样对应于混合模式断裂的每个单模式 ERR 分量 G_i 可以表示为单模下分离间距 δ_i 的函数,即 $G_i = G_i(\delta_i)$。式(10.24)中的所有其他物理量均已知,如下:

(1) G_{ic} 为单模式临界 ERR(材料属性)。

(2) δ_i^0 为单模式加载条件下损伤初始时的分离间距(图 10.4 中的 A 点),由式(10.7)给出。

(3) δ_i^c 为单模式加载条件下断裂时的分离间距(图 10.4 中的 C 点),由式(10.9)给出。

式(10.24)中混合模式条件下的单模式 ERR 分量必须满足断裂时(裂纹扩展)的能量准则式(10.21)。对于两种模式联合的情况下,假设式(10.21)中 $\alpha_i = \alpha = 2$,参考文献[188,式(15)]中混合模式下断裂时(C 点)分离间距的计算如下:

$$\delta_m^c = \frac{\sqrt{1+\beta}}{\beta^2 (\delta_\mathrm{I}^{0F})^2 (\delta_\mathrm{II}^{0F})^2} \times \{\delta_\mathrm{I}^0 (\delta_\mathrm{II}^{0F})^2 + \beta \delta_\mathrm{II}^0 (\delta_\mathrm{I}^0)^2 \\ + \delta_\mathrm{I}^{0F} \delta_\mathrm{II}^{0F} \sqrt{(\delta_\mathrm{II}^{0F})^2 - (\delta_\mathrm{II}^0)^2 + 2\beta \delta_\mathrm{I}^0 \delta_\mathrm{II}^0 - \beta^2 (\delta_\mathrm{I}^0)^2 + \beta^2 (\delta_\mathrm{I}^{0F})^2}\} \tag{10.25}$$

式中: $\delta_\mathrm{I}^{0F} = \delta_\mathrm{I}^c - \delta_\mathrm{I}^0, \delta_\mathrm{II}^{0F} = \delta_\mathrm{II}^c - \delta_\mathrm{II}^0$。

Abaqus 中有 3 个能量的裂纹扩展准则:

(1) 式(10.21)中幂指数方程准则。

(2) BK 准则,其为 (G_i, G_{ic}) 另一种闭式函数。

(3) 用户输入混合模式下断裂时总的 ERR $G_c = \sum_1^3 G_{ic}$。

基于所选的准则,由式(10.25)计算出断裂时混合模式下分离间距,并且基于式(10.20),损伤初始时的混合模式下分离间距已知,则与满足式(10.8)的单模式条件下的损伤变量表达式(10.12)形式类似,混合模式条件下的损伤变量

可以表示为

$$D_m = \frac{\delta_m^c(\delta_m - \delta_m^0)}{\delta_m(\delta_m^c - \delta_m^0)} \qquad (10.26)$$

式中，可以分别基于式(10.20)和式(10.25)计算损伤初始时和断裂时的分离间距 δ_m^0 和 δ_m^c；δ_m 为混合模式条件下的当前分离间距值，由单模式条件下分离间距 δ_i 按照式(10.15)计算得到，δ_i 由有限元模型计算得到。然后根据式(10.5)计算内聚力材料的刚度退化。混合模式条件下，内聚力材料应力软化的演变与图10.4 所示相似，需要使用 δ_m^0 和 δ_m^c 代替 δ^0 和 δ^c。

除了本文提到的计算混合模式损伤演化的能量方法，其裂纹扩展准则为 ERR 的函数且 δ_m^c 基于该能量准则进行计算，在 Abaqus 软件中，还提供了一种位移方法模拟损伤演化。该位移法要求用户直接输入混合模式下断裂时的分离间距 δ_m^c，其值会直接在式(10.26)中使用。

例 10.1 一个层合材料的双悬臂梁(DCB)，长 100mm，宽 20mm，其由两层粘接而组成，粘接层厚度可忽略不计，每层厚 1.5mm。使用加载系统施加载荷使得两层之间发生分层并呈现 I 型裂纹扩展。假设处在线弹性行为范围内，创建 DCB 的 2D 模型，并使用 CZM 单元来模拟粘接层。使用 Abaqus 软件中的基于单元(element-based)的内聚力行为定义粘接层。

解 为方便求解，请在工作目录中运行 Python 脚本 ws_composites_dcb.py。该脚本文件可以从 Abaqus 帮助文档的示例中获取，如下所示。

```
# open a Windows command prompt via the Windows Taskbar and Start Menu
Start,Run,Open [cmd],OK
# Navigate to the folder C:\SIMULIA\User\Ex_10.1
# you may need to create the directory using the command: mkdir Ex_10.1
# type the following command and press the Enter - key
abaqus fetch job = ws_composites_dcb.py
# this will copy the file ws_composites_dcb.py into the Work Directory
```

该脚本文件创建了两个模型，在本例中仅使用名为 coh-els 的 2D 模型。该模型中已包含了部件、装配体、材料属性和截面属性、定义的集合和施加边界条件所需的面、约束条件和定义的场输出请求。

在模型建立过程中，采用的是有限厚度的内聚力单元模拟粘接层。这样定义只是为了便于操作和定义约束。如第 V 步骤所述，在建模过程的最后，将内聚力单元的厚度调节为 0。也就是说，在本例中使用的是厚度为 0 的内聚力单元 COH2D4。

尽管本例中施加的载荷条件仅使得双悬臂梁发生 I 型裂纹扩展，但 Abaqus

软件中要求同时定义与所有 3 种裂纹扩展模式对应的参数。
Menu: File,Set Work Directory,[C:\SIMULIA\User\Ex_10.1],OK

i. 运行 Python 脚本。
Menu: File,Run Script,[ws_composites_dcb.py],OK
 # notice that the model has been named dcb.cae
Menu: Save As,[Ex_10.1.cae],OK

ii. 定义材料和截面属性并赋予相应部件。
coh-els 模型的单位系统为[m,Pa]。名为"bulk"的材料中包含了未损伤状态下单层板材料属性:$E_1 = 135.3 \times 10^9 \mathrm{Pa}$; $E_2 = E_3 = 9 \times 10^9 \mathrm{Pa}$; $v_{12} = v_{13} = 0.24$; $v_{23} = 0.46$; $G_{12} = G_{23} = 4.5 \times 10^9 \mathrm{Pa}$; $G_{13} = 3.3 \times 10^9 \mathrm{Pa}$。使用下面的操作代码定义名为"adhesive"的新材料,并输入粘接层属性信息。式(10.4)定义的粘接层刚度值为:$K_I = E/K_{nn} = 570 \times 10^{12} \mathrm{Pa/m}$, $K_{II} = E/K_{ss} = 570 \times 10^{12} \mathrm{Pa/m}$, $K_{III} = E/K_{tt} = 570 \times 10^{12} \mathrm{Pa/m}$,其中 K_{nn}、K_{ss}、K_{tt} 为罚参数,其为 Abaqus/CAE 中标识它们的符号。式(10.6)定义的粘接层强度值为:$\sigma_I^0 = 5.7 \times 10^7 \mathrm{Pa}$, $\sigma_{II}^0 = 5.7 \times 10^7 \mathrm{Pa}$, $\sigma_{III}^0 = 5.7 \times 10^7 \mathrm{Pa}$。与 Benzeggagh-Kenane(BK)裂纹扩展准则相关的粘接参数为幂指数 2.284,断裂能 $G_{Ic} = 280 \mathrm{J/m^2}$, $G_{IIc} = 280 \mathrm{J/m^2}$, $G_{IIIc} = 280 \mathrm{J/m^2}$(见式(10.2))。

Module: Property
Model: coh-els
Part: adhesive
Menu:Material,Manager
 Create,Name [cohesive],Mechanical,Elasticity,Elastic
 Type: Traction,E/Knn [5.7E14],G1/Ktt [5.7E14],G2/Kss [5.7E14]
 # damage onset criterion
 Mechanical,Damage for Traction Separation Laws,Quads Damage
 Nominal Stress Normal-only Mode [5.7E7]
 Nominal Stress First Direction [5.7E7]
 Nominal Stress Second Direction [5.7E7]
 # crack extension criterion and required parameters
 # Benzeggagh-Kenane (BK) is used,see Abaqus documentation
 Suboptions,Damage Evolution
 Type: Energy,Mixed mode behavior: BK,# checkmark: Power [2.284]
 Normal Mode Fracture Energy [280] # GIc
 Shear Mode Fracture Energy First Direction [280] # GIIc
 Shear Mode Fracture Energy Second Direction [280] # GIIIc

```
        OK,OK,# close Material Manager pop-up window
Menu: Section,Manager
        # a section named 'bulk' defines the laminas
        # create a section for the adhesive
        Create,Name [cohesive],Category: Other,Type: Cohesive,Cont
        Material: cohesive,Response: Traction Separation
        # checkmark: Out-of-plane thickness [0.02],OK
        # close Section Manager pop-up window
Menu: Assign,Section
        # pick the part,Done,Section: cohesive,OK
```

ⅲ. 定义分析步。

```
Module: Step
Menu: Step,Create,Cont
        # geometrical non-linearity for large displacements
        Tab: Basic,Nlgeom: On
        # parameters to setup the time increments for the solution
        Tab: Incrementation,Maximum number of increments [1000]
        Increment size: Initial [0.01],Minimum [1.0E-8],OK
Menu: Output,Field Output Requests,Edit,F-Output-1
        Output Variables: # expand: State/Field/User/Time
        # checkmark: STATUS,OK
        # the STATUS variable is an on/off switch that indicates the CAE
        # postprocessor when to stop drawing the cohesive elements
        # a completely degraded (damaged) element will have STATUS 0 (off)
Menu: Output,History Output Requests,Create
        Name [H-Output-2],Step: Step-1,Cont
        Domain: Set : top
        # expand: Displ/Veloc/Accel,# expand: U,# checkmark: U2
        # expand: Forces/Reactions,# expand: RF,# checkmark: RF2,OK
        # the values of U2 (imposed displacement) and RF2 (reaction force)
        # at the specified set (node) will be stored for postprocessing
```

ⅳ. 划分网格。

使用扫描网格技术进行网格划分是因为内聚力单元的厚度是定向的。本例子中,它们需要从粘接层的底面向顶面进行定义,因为它们的延伸率代表两个层板之间的分离间距。

第 10 章 分层

黏性(viscosity)参数可以与 COH2D4 单元一起使用,以加快迭代求解的收敛速度。读者可以更改其值的大小以查看它如何影响求解速度。

Module: Mesh

Model: coh-els

Object: Part: beam

Menu: Mesh,Controls,Element Shape: Quad,Technique: Structured,OK

Menu: Mesh,Element Type

    Family: Plane Strain,# checkmark: Incompatible modes

    # Notice element CPE4I has been assigned,OK

Menu: Seed,Edges

    # pick the top horizontal line,Done

Method: By number,Number of elements [400],OK

    # pick the left vertical line,Done

Method: By number,Number of elements [2],OK

Menu: Mesh,Part,Yes

Object: Part: adhesive

Menu: Mesh,Controls,Element Shape: Quad,Technique: Sweep

    Redefine Sweep Path

    # if the highlighted path coincides with the thickness direction

    Accept Highlighted

    # otherwise change accordingly to accomplish that,OK

Menu: Mesh,Element Type

    Family: Cohesive,Viscosity: Specify [1.0E-5]

    # notice element COH2D4 has been assigned,OK

Menu: Seed,Edges

    # pick the top horizontal line,Done

    Method: By number,Number of elements [280],OK

    # pick the left vertical line,Done

    Method: By number,Number of elements [1],OK

Menu: Mesh,Part,Yes

V. 创建约束条件。

使用有限厚度部件代表粘接层方便内聚力单元属性的定义。这里施加的约束条件保证了求解开始时,内聚力单元的上下表面是重合的,这可以有效地将内聚力单元的初始厚度定义为 0。尽管没有直观进行显示,下面的操作步骤可调整从面跟随主面。

Module: Interaction

Menu: Constraint,Create,Name [top],Type: Tie,Cont
　　Choose the master type: Surface,Surfaces,top,Cont
　　Choose the slave type: Surface,coh-top,Cont
　　Position Tolerance: Specify distance [0.002],OK
Menu: Constraint,Create,Name [bot],Type: Tie,Cont
　　Choose the master type: Surface,bot,Cont
　　Choose the slave type: Surface,coh-bot,Cont
　　Position Tolerance: Specify distance [0.002],OK

ⅵ. 施加 BC 和载荷条件。

Module: Load
Menu: BC,Create,Name [top],Step: Step-1,Disp/Rota,Cont
　　Sets,top,Cont,# checkmark: U1 [0],# checkmark: U2 [0.006],OK
Menu: BC,Create,Name [bot],Step: Step-1,Disp/Rota,Cont
　　Sets,bot,Cont,# checkmark: U1 [0],# checkmark: U2 [-0.006],OK
　　# due to the symmetry of the model no additional BCs are needed

ⅶ. 求解并可视化结果。

Module: Job
Menu: Job,Manager
　　Create,Name [Ex-10-1],Cont,OK,Submit,# when completed,Results
Module: Visualization
Menu: Plot,Deformed Shape # notice the beams are now separated
Menu: Plot,Contours,On Deformed Shape
Menu: Animate,Time History # see the progression of the delamination

可视化模型中的力-分离间距行为：

Module: Visualization
Menu: Result,History Output
　　Output Variables: Reaction force: RF2 PI,Save As [coh-els-RF2],OK
　　Output Variables: Spatial displ: U2 PI,Save As [coh-els-U2],OK
　　# close the History Output pop-up window
Menu: Tools,XY Data,Create,Source: Operate on XY data,Cont
　　Operators: Combine(X,X),# pick: coh-els-U2,Add to Expression
　　# pick: coh-els-RF2,Add to Expression,
　　Save As [cohesive elements],OK
　　# close Operate on XY Data pop-up window

Menu: Tools,XY Data,Plot,cohesive elements

计算结果的曲线图应与图 10.6 中的虚线一致。当内聚力单元刚度开始折减（损伤）时,反作用力就会下降。随着内聚力单元刚度的折减,反作用力会不断降低。在可视化模块中,可以放大裂纹尖端附近并查看代表粘接层的内聚力单元形态。

例 10.2 使用基于表面(surface-based)的内聚力方法替代基于单元的内聚力方法求解例 10.1。

解 可以通过修改例 10.1 中的.cae 文件求解本例。删除例 10.1 中代表粘接层的部件,然后将粘接层定义为两个单层板重合表面之间的相互作用。这个过程即为将基于单元的内聚力方法替换为基于表面的内聚力方法。

Menu: File,Set Work Directory,[C:\SIMULIA\User\Ex_10.2],OK
Menu: File,Open,[C:\SIMULIA\User\Ex_10.1\Ex_10.1.cae],OK
    # on the Model-tree
    # right-click: coh-els,Rename,[coh-surf],OK
Menu: File,Save As,[C:\SIMULIA\User\Ex_10.2\Ex_10.2.cae],OK

ⅰ. 删除不必要的特征。

Module: Assembly
Model: coh-surf
Menu: Tools,Surface,Manager
    # pick: coh-top,Delete,Yes
    # pick: coh-bot,Delete,Yes,# close Surface Manager
Module: Interaction
Menu: Constraint,Manager
    # pick: bot,Delete,Yes,# pick: top,Delete,Yes
    # close Constraint Manager
    # on the model-tree
    # expand: coh-surf,# expand: Assembly,# expand: Instances
    # right-click: adhesive-1,Delete,Yes
Menu: Feature,Regenerate
Module: Mesh
Model: coh-surf
Object: Part: beam
Menu: Mesh,Delete Part Mesh,Yes

ⅱ. 创建分割特征。

此时,模型仅由两个层合梁组成。现在,有必要划分出两个梁之间的界面,以代表粘接的区域。下面创建这个分割特征。
 Menu: Tools,Partition
  Type: Edge,Method: Enter parameter
  # pick bottom edge of beam,Done,Normalized edge parameter [0.7]
  Create Partition,Done,# close the Create Partition pop-up window

ⅲ．划分网格。
Menu: Mesh,Part,Yes

ⅳ．创建集合和表面。
例 10.1 中名为 top 的表面将被重新定义为两个梁之间的整个交界面。由于两个梁堆叠在一起，因此选择重合表面其中的一个较为麻烦。为了便于选择所需的表面，需要引入一个有用的可视化工具——Display Group Toolbar。
 Module: Assembly
  # locate the Toolbar named Display Group,If it is not available
  # enable it using Menu: View,Toolbars,Display Group
  # click the first icon of Display Group Toolbar (Replace Selected)
  # pick the top beam,Done,# only the picked part will be visible
 Menu: Tools,Surface,Edit,top
  # pick the two lines forming the bottom edge of the beam,Done
 Menu: Tools,Set,Create
  Name [bond],Type: Geometry,Cont
  # pick the longest line at the bottom of the beam,Done
  # click on third icon of Display Group Toolbar (Replace All)
  # the full assembly will be visualized

与示例 10.1 中名为 adhesive-1 部件代表粘接层一样，该例子中，使用名为 bond 的集合代表粘接层。
ⅴ．定义粘接层的两表面之间的接触关系。
首先定义了接触的属性和特征，然后选择相互作用的两个表面。
 Module: Interaction
 Menu: Interaction,Property,Create
  Name [coh],Type: Contact,Cont
  Mechanical,Tangential Behavior,Friction formulation: Frictionless
  Mechanical,Normal Behavior,Pressure-Overclosure: Hard Contact
  Constraint enforcement method: Default
  Mechanical,Cohesive Behavior

Eligible Slave Nodes: Specify the bounding node set in surf-to-surf
Traction-separation Behavior: Specify stiffness coeffs, Uncoupled
Knn [5.7e14], Kss [5.7e14], Ktt [5.7e14] # KI, KII, and KIII
Mechanical, Geometric Properties,
 Out-of-plane surface thickness [0.02]
Mechanical, Damage, Criterion: Quadratic traction
Normal Only [5.7e7], Shear-1 Only [5.7e7], Shear-2 [5.7e7]
these define the onset of damage in the adhesive layer
checkmark: Specify damage evolution
Type: Energy,
 # checkmark: Specify mixed mode behavior: Benzeggagh-Kenane
checkmark: Specify power-law/BK exponent [2.284]
Normal Fracture Energy [280], # Critical crack propagation GIc
1st Shear Fracture Energy [280] # GIIc
2nd Shear Fracture Energy [280] # GIIIc
checkmark: Specify damage stabilization, Viscosity coeff. [1e-5]
it plays the same role as the one used for cohesive elements
OK # the properties of the adhesive layer have been defined
Menu: Interaction, Create
 Name [coh], Step: Initial, Type: Surface-to-surface (Standard), Cont
Surfaces, # pick: bot, Cont # this is the master surface
Surfaces, # pick: top, Cont # this is the slave surface
Discretization method: Node to surface
Tab: Bonding
checkmark: Limit bonding to slave nodes in subset: bond
This represents the adhesive layer bonding the beams, OK

ⅵ. 求解并可视化结果。
Module: Job
Menu: Job, Manager
 Create, Name [Ex-10-2], Cont, OK, Submit, # when completed, Results
Module: Visualization
Menu: Plot, Deformed Shape # notice the beams have been separated
Menu: Plot, Contours, On Deformed Shape
Menu: Animate, Time History # progression of the delamination

在模型中可视化力-分离间距行为:
Module: Visualization

```
Menu: Result,History Output
    Output Variables: Reaction force: RF2 PI,
        Save As [coh-surf-RF2],OK
    Output Variables: Spatial displ: U2 PI,
        Save As [coh-surf-U2],OK
    # close the History Output pop-up window
Menu: Tools,XY Data,Create,Source: Operate on XY data,Cont
    Operators: Combine(X,X),# pick: coh-surf-U2,Add to Expression
    # pick : coh-surf-RF2,Add to Expression,
        Save As [cohesive surfaces],OK
    # close Operate on XY Data pop-up window
Menu: Tools,XY Data,Plot,cohesive surfaces
```

计算结果的曲线图应与图 10.6 中的实线一致。当两表面之间的分离间距达到临界值 δ_i^0(对应临界应力值 σ_i^0)时,反力开始下降。与使用内聚力单元的例 10.1 不同,在本例中,界面的渐进损伤在 Abaqus 中计算没有使用特殊的单元类型。在可视化模块中,可以放大裂纹尖端的区域,应注意到该处没有代表黏接层的单元。

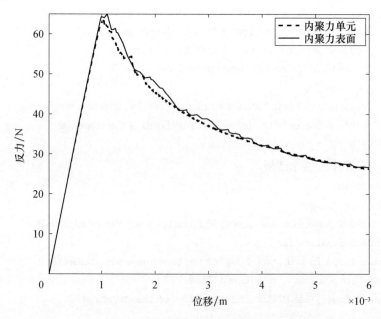

图 10.6　使用内聚力单元(例 10.1)和内聚力表面(例 10.2)
模型计算得到的力 – 分离间距曲线比较

10.2 虚拟裂纹闭合技术

虚拟裂纹闭合技术(VCCT)借助断裂力学方法用来分析层合板材料中的分层问题。该方法应用线弹性断裂力学(LEFM)理论,仅模拟脆性裂纹的扩展,不考虑裂纹尖端塑性区形成所消耗的能量。

裂纹扩展的条件基于 Griffith 原理,式(10.2)。对于 I 型裂纹条件下的单模式变形情况,裂纹扩展条件为

$$\frac{G_{\mathrm{I}}}{G_{\mathrm{Ic}}} \geqslant 1 \tag{10.27}$$

式中:G_{I} 为 I 型裂纹的 ERR;G_{Ic} 为 I 型裂纹的临界 ERR,为材料属性。

ERR 的定义见式(10.1)。在 VCCT 中,应变能的变化 ΔU 由 Irwin 原理式(10.3)计算,其被认为等于裂纹闭合所需的功 W_{closure}。

把式(10.3)和式(10.1)代入式(10.27),I 型裂纹扩展条件变为

$$\frac{W_{\mathrm{closure}}/\Delta A}{G_{\mathrm{Ic}}} \geqslant 1 \tag{10.28}$$

如图 10.7 所示,Abaqus[189] 软件根据 FE 节点位移和力计算得到 W_{closure}。最初,裂纹表面之间为刚性连接。重合节点 2-5 处的节点力由 FE 求解得到。VCCT 计算采用自相似裂纹扩展(self-similar crack propagation)假设,即在裂纹扩展过程中,认为节点 2-3-4-5 之间的裂纹形态与节点 1-2-5-6 之间的裂纹形态相似。这意味着裂纹扩展后,节点 2-5 之间的分离间距将等于裂纹扩展前节点 1-6 之间的间距:$w_{2,5} = w_{1,6}$。如果节点 2-5 打开(裂纹扩展),则使其裂纹闭合所需的弹性功为

$$W_{\mathrm{closure}} = \frac{1}{2} F_{2,5} w_{2,5} = \frac{1}{2} F_{2,5} w_{1,6} \tag{10.29}$$

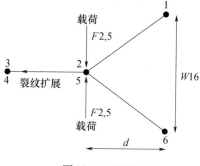

图 10.7 VCCT

把式(10.29)代入式(10.28),裂纹扩展条件变为

$$\frac{F_{2,5} w_{1,6}}{2\Delta A} \frac{1}{G_{Ic}} \geq 1 \tag{10.30}$$

新形成的裂纹面积 $\Delta A = d \times b$,其中 d 为发生裂纹扩展的单元长度,b 为裂纹宽度。

通过考虑其他对应的分离间距和节点力,VCCT 方法同样适用于计算 II 型或 III 型裂纹扩展。在文献[190]中还提供了一种称为雅可比导数法的精确算法。

裂纹扩展准则式(10.30)仅适用于单模式加载条件下,因为它是从单模式准则式(10.2)中推导出的。对于混合模式加载情况,必须使用混合模式准则替换单模式裂纹扩展准则式(10.2)。例如,可以使用幂指数方程式(10.21),其中 3 个临界 ERR G_{ic} (i = I, II, III) 为材料属性,3 个 ERR G_i 利用 VCCT 进行计算得到,与式(10.30)相似。或者,也可以使用 BK 方程[191]或 Reeder 方程[180,192]进行计算。

例 10.3 使用虚拟裂纹闭合技术(VCCT)求解例 10.1。

解 可以通过修改例 10.2 中建立的 cae 文件,求解本例题。首先,必须更改接触定义,删除基于表面的内聚力相关属性并重新定义接触属性。然后,修改输出请求,使其包含模型中的应变能释放率(StrainEnergy Release Rate) G。最后,在将模型提交求解之前,使用关键字编辑器(Keyword Editor)添加 VCCT 相关特征。

Menu: File,Set Work Directory,[C:\SIMULIA\User\Ex_10.3],OK
Menu: File,Open,[C:\SIMULIA\User\Ex_10.2\Ex_10.2.cae],OK
on the Model-tree
right-click: coh-surf,Rename,[vcct],OK
Menu: File,Save As,[C:\SIMULIA\User\Ex_10.3\Ex_10.3.cae],OK

i. 删除不需要的特征。
Module: Interaction
Menu: Interaction,Property,Edit,coh
  # pick: Cohesive Behavior,Delete
  # pick: Damage,Delete,OK

ii. 重新定义接触属性。
Menu: Interaction,Edit,coh
  Sliding formulation: Small sliding
  Tab: Clearance
  Initial clearance: Uniform value across slave surface,
   [1.0e-7],OK

初始间隙的使用可预防由于网格划分导致的初始过盈问题,方便模型求解。详细信息请参见 Abaqus 帮助文档。

ⅲ. 修改输出请求。

Module: Step

Step: Step-1

Menu: Output,Field Output Requests,Edit,F-Output-1

  Output Variables: # expand: Failure/Fracture

  # checkmark: ENRRT,# checkmark: BDSTAT,OK

  # ENRRT and BDSTAT allow tracking damage evolution

ⅳ. 添加有关 VCCT 的关键字。

Menu: Model,Edit Keywords,vcct

  # click on the line that starts with the text: * Contact Pair

  Add After,# Type in the following text:

  [* Initial Conditions,type = CONTACT

  top,bot,bond]

  # click on the line that starts with the text: * Static

  Add After,# Type in the following text:

  [* Debond,slave = top,master = bot

  * Fracture criterion,type = VCCT,mixed mode behavior = BK,

    tolerance = 0.1

  280.0,280.0,280.0,2.284]

  OK

ⅴ. 求解并可视化结果。

Module: Job

Menu: Job,Manager

  Create,Name[Ex-10-3],Cont,OK,Submit,# when completed,Results

Module: Visualization

Menu: Plot,Deformed Shape # notice the beams have been separated

Menu: Plot,Contours,On Deformed Shape

Menu: Animate,Time History # see progression of the delamination

要在模型中可视化力-分离间距曲线,请执行以下操作:

Module: Visualization

Menu: Result,History Output

  Output Variables: Reaction force: RF2 PI,Save As[vcct-RF2],OK

```
        Output Variables: Spatial displ: U2 PI,Save As[vcct-U2],OK
        # close the History Output pop-up window
Menu: Tools,XY Data,Create,Source: Operate on XY data,Cont
        Operators: Combine(X,X),# pick: vcct-U2,Add to Expression
        # pick: vcct-RF2,Add to Expression,Save As[vcct],OK
        # close Operate on XY Data pop-up window
Menu: Tools,XY Data,Plot,vcct
```

计算结果曲线应与图 10.8 中的实线一致。变量 BDSTAT 中存储了模型中两个重合节点之间关于粘接属性的损伤演化变量,其值范围从 0(粘接完全破坏)到 1(粘接未损伤)。BDSTAT 可用于在分层的任何阶段,对裂纹尖端的定位。在 Visualization Module 中,通过可视化 BDSTAT 并放大以定位裂纹尖端。

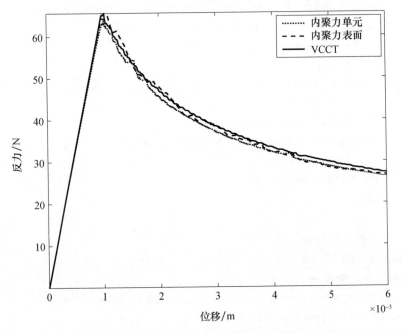

图 10.8 使用虚拟裂纹闭合技术(VCCT)、内聚力单元和内聚力表面方法得到的力—分离间距曲线对比

习题

习题 10.1 在例 10.1 和例 10.2 中,通过 Abaqus 提取反作用力与位移的值。绘制的反作用力-位移曲线应该与图 10.6 相似。要将数据提取出来作为

文本文件,请执行以下操作:

  Module: Visualization

  Menu: Report,XY

    # pick: cohesive surfaces,# cohesive elements for Example 10.1

    Tab: Setup,Name [coh_surf.rpt],# coh_ele.rpt for Example 10.1

    # un-checkmark: Append to file,OK

    # the report file will be saved in your Work Directory

习题 10.2 在例 10.3 中提取反作用力与位移的值,依照例 10.1 和例 10.2 的形式绘制它们,讨论有何不同？其对比图应与图 10.8 相似。

附录 A　张量代数

本书中一些方程的推导需要张量运算。而方程推导所涉及的大部分张量运算不易在教科书中找到,因此将其放在附录 A 中供读者参考[193]。

A.1　应力和应变的主方向

由于应力和应变张量是对称的二阶张量,因此它们有 3 个主值和 3 个正交的主方向。应力张量 σ_{ij} 的主值 λ^q 和主方向 n_i^q 满足以下条件

$$[\sigma_{ij} - \lambda^q \delta_{ij}]n_i^q = 0 \tag{A.1}$$

$$n_i^q n_j^q = 1 \tag{A.2}$$

式中:δ_{ij} 为 Kronecker delta 函数($\delta_{ij}=1, i=j$;否则为 0)。每个主方向均由其相对于初始坐标系的方向余弦进行描述。

主方向按行排列形成矩阵 $[A]$。那么,主值的对角矩阵 $[A^*]$ 为

$$[A^*] = [a][A][a]^T \tag{A.3}$$

可以证明 $[a]^{-1} = [a]^T$,其中 $[a]$ 为由式(1.21)给出的变换矩阵。

A.2　张量对称性

张量的次对称性(Minor symmetry)为使用指标符号缩写提供了依据(见 1.5 节)。次对称性是指相邻的下标交换时张量分量值相等。例如,刚度张量 C 的具有次对称性,意味着

$$A_{ijkl} = A_{jilk} = A_{\alpha\beta} \tag{A.4}$$

主对称性是指当相邻的一对下标被交换时或当缩写后的下标交换时张量分量值相等。例如

$$A_{ijkl} = A_{klij}$$
$$A_{\alpha\beta} = A_{\beta\alpha} \tag{A.5}$$

A.3　张量的矩阵表示

具有次对称性的四阶张量 A_{ijkl} 只有 36 个独立常数。因此,利用指标符号缩

写法,它可以用6×6矩阵表示。令矩阵$[a]$表示张量A的缩写后的矩阵形式,根据下列转换关系,$[a]$的每个元素分别对应于张量A中的每个分量。

$$a_{\alpha\beta} = A_{ijkl} \tag{A.6}$$

式中

$$\begin{cases} \alpha = i & (i=j) \\ \alpha = 9 - (i+j) & (i \neq j) \end{cases} \tag{A.7}$$

对β与k和l之间应用相同的变换关系,$[a]$矩阵形式表示为

$$[a] = \begin{bmatrix} A_{1111} & A_{1122} & A_{1133} & A_{1123} & A_{1113} & A_{1112} \\ A_{2211} & A_{2222} & A_{2233} & A_{2223} & A_{2213} & A_{2212} \\ A_{3311} & A_{3322} & A_{3333} & A_{3323} & A_{3313} & A_{3312} \\ A_{2311} & A_{2322} & A_{2333} & A_{2323} & A_{2313} & A_{2312} \\ A_{1311} & A_{1322} & A_{1333} & A_{1323} & A_{1313} & A_{1312} \\ A_{1211} & A_{1222} & A_{1233} & A_{1223} & A_{1213} & A_{1212} \end{bmatrix} \tag{A.8}$$

使用缩写形式进行张量运算较为方便,特别是当计算结果也可以用缩写形式表示时。这样软件计算时,会节省大量内存和计算时间,因为对 36 个元素运算比对 81 个元素运算更快。这些运算的例子包括两个 4 阶张量的内积运算和 4 阶张量的求逆运算。然而,指标符号形式下的张量进行张量运算并不能直接转换成缩写形式下的矩阵运算。例如,两个 4 阶张量的双点积表示为

$$\begin{aligned} C &= A : B \\ C_{ijkl} &= A_{ijmn} B_{mnkl} \end{aligned} \tag{A.9}$$

令6×6的矩阵$[a]$、$[b]$和$[c]$分别表示上述 3 个张量。那么可以通过证明得到

$$[a][b] \neq [c]$$

或者

$$a_{\alpha\beta} b_{\beta\gamma} \neq c_{\alpha\gamma} (\text{矩阵相乘}) \tag{A.10}$$

本附录的剩余部分章节给出了指标缩写形式下具有代表性的张量运算的公式。

A.4 双点积

式(A.9)中,以C_{1211}元素为例,其可以展开如下:

$$\begin{aligned} C_{1211} = &A_{1211}B_{1111} + A_{1222}B_{2211} + A_{1233}B_{3311} \\ &+ 2A_{1212}B_{1211} + 2A_{1213}B_{1311} + 2A_{1223}B_{2311} \end{aligned} \tag{A.11}$$

为了通过矩阵相乘也得到相同的计算结果,可以将矩阵$[a]$的最后3列乘以2,然后再执行乘法运算

$$[c]=\begin{bmatrix} A_{1111} & A_{1122} & A_{1133} & 2A_{1123} & 2A_{1113} & 2A_{1112} \\ A_{2211} & A_{2222} & A_{2233} & 2A_{2223} & 2A_{2213} & 2A_{2212} \\ A_{3311} & A_{3322} & A_{3333} & 2A_{3323} & 2A_{3313} & 2A_{3312} \\ A_{2311} & A_{2322} & A_{2333} & 2A_{2323} & 2A_{2313} & 2A_{2312} \\ A_{1311} & A_{1322} & A_{1333} & 2A_{1323} & 2A_{1313} & 2A_{1312} \\ A_{1211} & A_{1222} & A_{1233} & 2A_{1223} & 2A_{1213} & 2A_{1212} \end{bmatrix}$$
$$\begin{bmatrix} B_{1111} & B_{1122} & B_{1133} & B_{1123} & B_{1113} & B_{1112} \\ B_{2211} & B_{2222} & B_{2233} & B_{2223} & B_{2213} & B_{2212} \\ B_{3311} & B_{3322} & B_{3333} & B_{3323} & B_{3313} & B_{3312} \\ B_{2311} & B_{2322} & B_{2333} & B_{2323} & B_{2313} & B_{2312} \\ B_{1311} & B_{1322} & B_{1333} & B_{1323} & B_{1313} & B_{1312} \\ B_{1211} & B_{1222} & B_{1233} & B_{1223} & B_{1213} & B_{1212} \end{bmatrix} \quad (A.12)$$

这个变换可以利用 Reuter 矩阵计算得到,Reuter 矩阵为

$$[R]=\begin{bmatrix} 1 & 0 & 0 & 0 & 0 & 0 \\ 0 & 1 & 0 & 0 & 0 & 0 \\ 0 & 0 & 1 & 0 & 0 & 0 \\ 0 & 0 & 0 & 2 & 0 & 0 \\ 0 & 0 & 0 & 0 & 2 & 0 \\ 0 & 0 & 0 & 0 & 0 & 2 \end{bmatrix} \quad (A.13)$$

将其代入式(A.12)中得到

$$[c]=[a][R][b] \quad (A.14)$$

A.5 张量求逆

首先,为方便起见,定义为一个四阶等同张量(identity tensor)I_{ijkl},其满足

$$I_{ijmn}A_{mnkl}=A_{ijkl} \quad (A.15)$$

若A_{ijkl}具有次对称性,如下的张量可使式(A.15)成立:

$$I_{ijkl}=\frac{1}{2}(\delta_{ik}\delta_{jl}+\delta_{il}\delta_{jk}) \quad (A.16)$$

式中:δ_{ij}为 Kronecker delta 函数,定义如下

$$\begin{cases} \delta_{ij} = 1 & (i=j) \\ \delta_{ij} = 0 & (i \neq j) \end{cases} \quad (A.17)$$

Voigt 缩写形式下,四阶等同张量采用 $[i]$ 表示,它等于 Reuter 矩阵的逆矩阵,有

$$[i] = \begin{bmatrix} 1 & 0 & 0 & 0 & 0 & 0 \\ 0 & 1 & 0 & 0 & 0 & 0 \\ 0 & 0 & 1 & 0 & 0 & 0 \\ 0 & 0 & 0 & 1/2 & 0 & 0 \\ 0 & 0 & 0 & 0 & 1/2 & 0 \\ 0 & 0 & 0 & 0 & 0 & 1/2 \end{bmatrix} = [R]^{-1} \quad (A.18)$$

现在,张量的逆定义为一个与原张量相乘得到等同张量的张量,如下:

$$A_{ijmn} A_{mnkl}^{-1} = I_{ijkl} \quad (A.19)$$

下面介绍一下下面的符号:

$[a]^{-1}$——缩写形式下张量 A_{ijkl} 的逆;

$[a^{-1}]$——张量 A_{ijkl} 的逆的缩写形式。

若 A_{ijkl} 具有次对称性,其 $a_{\alpha\beta}^{-1}$ 的分量由如下步骤得到:

(1)利用矩阵 $[R]$ 将矩阵 $[a]$ 的最后 3 列乘以 2。
(2)对得到的矩阵求逆。
(3)得到的矩阵再乘以 $[i]$。

换句话说,矩阵 $[a^{-1}]$ 由下式计算

$$[a^{-1}] = [[a][R]]^{-1}[i] = [i][a]^{-1}[i] \quad (A.20)$$

A.6 张量导数

A.6.1 张量对其自身求导

任何对称二阶张量 Φ_{ij} 均满足

$$\mathrm{d}\Phi_{ij} = \mathrm{d}\Phi_{ji} \quad (A.21)$$

因此,一个二阶对称张量对其自身求偏导,得

$$\frac{\partial \Phi_{ij}}{\partial \Phi_{kl}} = J_{ijkl} \quad (A.22)$$

式中:J_{ijkl} 为一个四阶张量,定义如下

$$\begin{cases} J_{ijkl} = 1 & (i=k, j=l) \\ J_{ijkl} = 1 & (i=l, j=k) \\ J_{ijkl} = 0 & (其他) \end{cases} \quad (A.23)$$

指标缩写形式下,张量 J_{ijkl} 的矩阵形式为

$$[j] = \begin{bmatrix} 1 & 0 & 0 & 0 & 0 & 0 \\ 0 & 1 & 0 & 0 & 0 & 0 \\ 0 & 0 & 1 & 0 & 0 & 0 \\ 0 & 0 & 0 & 1 & 0 & 0 \\ 0 & 0 & 0 & 0 & 1 & 0 \\ 0 & 0 & 0 & 0 & 0 & 1 \end{bmatrix} \qquad (A.24)$$

A.6.2 张量的逆对该张量求导

二阶张量与其逆并乘得到二阶等同张量或 Kronecker delta

$$A_{ij}A_{jk}^{-1} = \delta_{ik} \qquad (A.25)$$

把式(A.25)对于 A_{mn} 求偏导并重新排列后得到

$$A_{ij}\frac{\partial A_{jk}^{-1}}{\partial A_{mn}} = -\frac{\partial A_{ij}}{\partial A_{mn}}A_{jk}^{-1} \qquad (A.26)$$

两边并乘 A_{lk}^{-1} 并重新排列,得

$$\frac{\partial A_{jk}^{-1}}{\partial A_{mn}} = -A_{ik}^{-1}\frac{\partial A_{kl}}{\partial A_{mn}}A_{lj}^{-1} \qquad (A.27)$$

最后利用式(A.22),得

$$\frac{\partial A_{ij}^{-1}}{\partial A_{mn}} = -A_{ik}^{-1}J_{klmn}A_{lj}^{-1} \qquad (A.28)$$

附录 B 二阶对角损伤模型

为了完整性,本附录给出了与二阶对角损伤模型相关的显式表达式。

B.1 有效和损伤空间

在与 D 的主方向一致的坐标系中,二阶损伤张量可以由一个对角张量(见式(8.61))表示,即

$$D_{ij} = d_i \delta_{ij} \text{(不对 } i \text{ 求和)} \tag{B.1}$$

D 的主方向可以与纤维方向,横向和厚度方向重合;d_i 为损伤张量的特征值,其代表沿着这些方向上的损伤率。损伤张量的对偶变量为完整性张量(integrity tensor),$\boldsymbol{\Omega} = \sqrt{\boldsymbol{I} - \boldsymbol{D}}$,其表示未损伤率。

二阶损伤张量 $\boldsymbol{D}$ 和完整性张量 $\boldsymbol{\Omega}$ 均为对角张量,矩阵形式为

$$\boldsymbol{D}_{ij} = \begin{bmatrix} d_1 & 0 & 0 \\ 0 & d_2 & 0 \\ 0 & 0 & d_3 \end{bmatrix} \tag{B.2}$$

$$\boldsymbol{\Omega}_{ij} = \begin{bmatrix} \sqrt{1-d_1} & 0 & 0 \\ 0 & \sqrt{1-d_2} & 0 \\ 0 & 0 & \sqrt{1-d_3} \end{bmatrix} = \begin{bmatrix} \Omega_1 & 0 & 0 \\ 0 & \Omega_2 & 0 \\ 0 & 0 & \Omega_3 \end{bmatrix} \tag{B.3}$$

现在,定义一个称为损伤效应张量(damage effect tensor)的对称四阶张量 $\boldsymbol{M}$(见式(8.63)),即

$$M_{ijkl} = \frac{1}{2}(\Omega_{ik}\Omega_{jl} + \Omega_{il}\Omega_{jk}) \tag{B.4}$$

缩写形式下损伤效应张量乘以 Reuter 矩阵后以 6×6 矩阵形式表示为

$$\boldsymbol{M} = M_{\alpha\beta} = \begin{bmatrix} \Omega_1^2 & 0 & 0 & 0 & 0 & 0 \\ 0 & \Omega_2^2 & 0 & 0 & 0 & 0 \\ 0 & 0 & \Omega_3^2 & 0 & 0 & 0 \\ 0 & 0 & 0 & \Omega_2\Omega_3 & 0 & 0 \\ 0 & 0 & 0 & 0 & \Omega_1\Omega_3 & 0 \\ 0 & 0 & 0 & 0 & 0 & \Omega_1\Omega_2 \end{bmatrix} \tag{B.5}$$

对于正交各向异性材料,与 Reuter 矩阵相乘后,损伤状态刚度张量 $\boldsymbol{C}$ 在指标缩写形式下可以用 6×6 矩阵形式表达,其为未损伤状态刚度张量 $\overline{\boldsymbol{C}}$ 的函数,有

$$C_{\alpha\beta} = \begin{bmatrix} \overline{C}_{11}\Omega_1^4 & \overline{C}_{12}\Omega_1^2\Omega_2^2 & \overline{C}_{13}\Omega_1^2\Omega_3^2 & 0 & 0 & 0 \\ \overline{C}_{12}\Omega_1^2\Omega_2^2 & \overline{C}_{22}\Omega_2^4 & \overline{C}_{23}\Omega_2^2\Omega_3^2 & 0 & 0 & 0 \\ \overline{C}_{13}\Omega_1^2\Omega_3^2 & \overline{C}_{23}\Omega_2^2\Omega_3^2 & \overline{C}_{33}\Omega_3^4 & 0 & 0 & 0 \\ 0 & 0 & 0 & 2\overline{C}_{44}\Omega_2^2\Omega_3^2 & 0 & 0 \\ 0 & 0 & 0 & 0 & 2\overline{C}_{55}\Omega_1^2\Omega_3^2 & 0 \\ 0 & 0 & 0 & 0 & 0 & 2\overline{C}_{66}\Omega_1^2\Omega_2^2 \end{bmatrix}$$

(B.6)

式中:$\overline{C}_{44} = \overline{G}_{23}, \overline{C}_{55} = \overline{G}_{13}, \overline{C}_{66} = \overline{G}_{12}$。这里对四阶弹性张量使用了 Voigt 指标符号缩写:$C_{\alpha\beta}$ 替换了 C_{ijkl},其中 α, β 取值为 1、2、3、4、5、6,对应的下标对分别为 11、22、33、23、13 和 12。

有效应力分量与实际应力分量之间的关系如下:

$$\begin{cases} \overline{\sigma}_1 = \sigma_1 \Omega_1^{-2}; & \overline{\sigma}_4 = \sigma_4 \Omega_2^{-1}\Omega_3^{-1} \\ \overline{\sigma}_2 = \sigma_2 \Omega_2^{-2}; & \overline{\sigma}_5 = \sigma_5 \Omega_1^{-1}\Omega_3^{-1} \\ \overline{\sigma}_3 = \sigma_3 \Omega_3^{-2}; & \overline{\sigma}_6 = \sigma_6 \Omega_1^{-1}\Omega_2^{-1} \end{cases}$$

(B.7)

应变分量之间关系如下:

$$\begin{cases} \overline{\varepsilon}_1 = \varepsilon_1 \Omega_1^2; & \overline{\varepsilon}_4 = \varepsilon_4 \Omega_2 \Omega_3 \\ \overline{\varepsilon}_2 = \varepsilon_2 \Omega_2^2; & \overline{\varepsilon}_5 = \varepsilon_5 \Omega_1 \Omega_3 \\ \overline{\varepsilon}_3 = \varepsilon_3 \Omega_3^2; & \overline{\varepsilon}_6 = \varepsilon_6 \Omega_1 \Omega_2 \end{cases}$$

(B.8)

式(B.8)中的上划线表示有效参数。

B.2 热力学力 Y

通过满足 Clausius–Duhem 不等式确保非负耗散,热力学力(见式(8.128))定义为

$$Y_{ij} = -\frac{\partial \psi}{\partial D_{ij}} = -\frac{1}{2}(\varepsilon_k - \varepsilon_{kl}^{\mathrm{p}})\frac{\partial C_{klpq}}{\partial D_{ij}}(\varepsilon_{pq} - \varepsilon_{pq}^{\mathrm{p}}) = -\frac{1}{2}\varepsilon_{kl}^{\mathrm{e}}\frac{\partial C_{klpq}}{\partial D_{ij}}\varepsilon_{pq}^{\mathrm{e}}$$ (B.9)

与损伤变量共轭的热力学力二阶张量矩阵形式为

$$\boldsymbol{Y} = Y_{ij} = \begin{bmatrix} Y_{11} & 0 & 0 \\ 0 & Y_{22} & 0 \\ 0 & 0 & Y_{33} \end{bmatrix} \tag{B.10}$$

或者写成 Voigt 缩写形式,即

$$\boldsymbol{Y} = Y_\alpha = \{Y_{11}, Y_{22}, Y_{33}, 0, 0, 0\}^{\mathrm{T}} \tag{B.11}$$

利用式(B.9),得到由有效应变表示的热力学力的显式表达式,即

$$\begin{cases} Y_{11} = \dfrac{1}{\Omega_1^2}(\overline{C}_{11}\overline{\varepsilon}_1^{e\,2} + \overline{C}_{12}\overline{\varepsilon}_2^e\overline{\varepsilon}_1^e + \overline{C}_{13}\overline{\varepsilon}_3^e\overline{\varepsilon}_1^e + 2\overline{C}_{55}\overline{\varepsilon}_5^{e\,2} + 2\overline{C}_{66}\overline{\varepsilon}_6^{e\,2}) \\ Y_{22} = \dfrac{1}{\Omega_2^2}(\overline{C}_{22}\overline{\varepsilon}_2^{e\,2} + \overline{C}_{12}\overline{\varepsilon}_2^e\overline{\varepsilon}_1^e + \overline{C}_{23}\overline{\varepsilon}_3^e\overline{\varepsilon}_2^e + 2\overline{C}_{44}\overline{\varepsilon}_4^{e\,2} + 2\overline{C}_{66}\overline{\varepsilon}_6^{e\,2}) \\ Y_{33} = \dfrac{1}{\Omega_3^2}(\overline{C}_{33}\overline{\varepsilon}_3^{e\,2} + \overline{C}_{13}\overline{\varepsilon}_3^e\overline{\varepsilon}_1^e + \overline{C}_{23}\overline{\varepsilon}_3^e\overline{\varepsilon}_2^e + 2\overline{C}_{44}\overline{\varepsilon}_4^{e\,2} + 2\overline{C}_{55}\overline{\varepsilon}_5^{e\,2}) \end{cases} \tag{B.12}$$

热力学力写成实际应力的形式,如下

$$\begin{cases} Y_{11} = \dfrac{1}{\Omega_1^2}\left(\dfrac{\overline{S}_{11}}{\Omega_1^4}\sigma_1^2 + \dfrac{\overline{S}_{12}}{\Omega_1^2\Omega_2^2}\sigma_2\sigma_1 + \dfrac{\overline{S}_{13}}{\Omega_1^2\Omega_3^2}\sigma_3\sigma_1 + \dfrac{2\overline{S}_{55}}{\Omega_1^2\Omega_3^2}\sigma_5^2 + \dfrac{2\overline{S}_{66}}{\Omega_1^2\Omega_2^2}\sigma_6^2\right) \\ Y_{22} = \dfrac{1}{\Omega_2^2}\left(\dfrac{\overline{S}_{22}}{\Omega_2^4}\sigma_2^2 + \dfrac{\overline{S}_{12}}{\Omega_2^2\Omega_1^2}\sigma_2\sigma_1 + \dfrac{\overline{S}_{23}}{\Omega_2^2\Omega_3^2}\sigma_3\sigma_2 + \dfrac{2\overline{S}_{44}}{\Omega_2^2\Omega_3^2}\sigma_4^2 + \dfrac{2\overline{S}_{66}}{\Omega_2^2\Omega_1^2}\sigma_6^2\right) \\ Y_{33} = \dfrac{1}{\Omega_3^2}\left(\dfrac{\overline{S}_{33}}{\Omega_3^4}\sigma_3^2 + \dfrac{\overline{S}_{13}}{\Omega_3^2\Omega_1^2}\sigma_3\sigma_1 + \dfrac{\overline{S}_{23}}{\Omega_3^2\Omega_2^2}\sigma_3\sigma_2 + \dfrac{2\overline{S}_{44}}{\Omega_3^2\Omega_2^2}\sigma_4^2 + \dfrac{2\overline{S}_{55}}{\Omega_3^2\Omega_1^2}\sigma_5^2\right) \end{cases} \tag{B.13}$$

热力学力对于损伤变量的偏导($\partial \boldsymbol{Y}/\partial \boldsymbol{D}$)由下式给出:

$$\dfrac{\partial \boldsymbol{Y}}{\partial \boldsymbol{D}} = \begin{bmatrix} \dfrac{Y_{11}}{\Omega_1^4} & 0 & 0 & 0 & 0 & 0 \\ 0 & \dfrac{Y_{22}}{\Omega_2^4} & 0 & 0 & 0 & 0 \\ 0 & 0 & \dfrac{Y_{33}}{\Omega_3^4} & 0 & 0 & 0 \\ 0 & 0 & 0 & 0 & 0 & 0 \\ 0 & 0 & 0 & 0 & 0 & 0 \\ 0 & 0 & 0 & 0 & 0 & 0 \end{bmatrix} \tag{B.14}$$

热力学力对于实际应变的偏导为

$$\frac{\partial \boldsymbol{Y}}{\partial \boldsymbol{\varepsilon}^e} = \begin{bmatrix} -\dfrac{P_{11}}{\Omega_1^2} & -\dfrac{\overline{C}_{12}\overline{\varepsilon}_1^e}{\Omega_1^2} & -\dfrac{\overline{C}_{13}\overline{\varepsilon}_1^e}{\Omega_1^2} & 0 & -2\dfrac{\overline{C}_{55}\overline{\varepsilon}_5^e}{\Omega_1^2} & -2\dfrac{\overline{C}_{66}\overline{\varepsilon}_6^e}{\Omega_1^2} \\ -\dfrac{\overline{C}_{12}\overline{\varepsilon}_2^e}{\Omega_2^2} & -\dfrac{P_{22}}{\Omega_2^2} & -\dfrac{\overline{C}_{23}\overline{\varepsilon}_2^e}{\Omega_2^2} & -2\dfrac{\overline{C}_{44}\overline{\varepsilon}_4^e}{\Omega_2^2} & 0 & -2\dfrac{\overline{C}_{66}\overline{\varepsilon}_6^e}{\Omega_2^2} \\ -\dfrac{\overline{C}_{13}\overline{\varepsilon}_3^e}{\Omega_3^2} & -\dfrac{\overline{C}_{23}\overline{\varepsilon}_3^e}{\Omega_3^2} & -\dfrac{P_{33}}{\Omega_3^2} & -2\dfrac{\overline{C}_{44}\overline{\varepsilon}_4^e}{\Omega_3^2} & -2\dfrac{\overline{C}_{55}\overline{\varepsilon}_5^e}{\Omega_3^2} & 0 \\ 0 & 0 & 0 & 0 & 0 & 0 \\ 0 & 0 & 0 & 0 & 0 & 0 \\ 0 & 0 & 0 & 0 & 0 & 0 \end{bmatrix}$$

(B.15)

式中

$$\begin{cases} P_{11} = 2\overline{C}_{11}\overline{\varepsilon}_1^e + \overline{C}_{12}\overline{\varepsilon}_2^e + \overline{C}_{13}\overline{\varepsilon}_3^e \\ P_{22} = \overline{C}_{12}\overline{\varepsilon}_1^e + 2\overline{C}_{22}\overline{\varepsilon}_2^e + \overline{C}_{23}\overline{\varepsilon}_3^e \\ P_{33} = \overline{C}_{13}\overline{\varepsilon}_1^e + \overline{C}_{23}\overline{\varepsilon}_2^e + 2\overline{C}_{33}\overline{\varepsilon}_3^e \end{cases} \quad (\text{B.16})$$

热力学力对于实际不可恢复应变的偏导由下式给出

$$\frac{\partial \boldsymbol{Y}}{\partial \boldsymbol{\varepsilon}^p} = -\frac{\partial \boldsymbol{Y}}{\partial \boldsymbol{\varepsilon}^e} \quad (\text{B.17})$$

实际应力对于损伤张量的偏导为

$$\frac{\partial \boldsymbol{\sigma}}{\partial \boldsymbol{D}} = \begin{bmatrix} P'_{11} & 0 & 0 & 0 & 0 & 0 \\ 0 & P'_{22} & 0 & 0 & 0 & 0 \\ 0 & 0 & P'_{33} & 0 & 0 & 0 \\ 0 & -\dfrac{1}{2}\dfrac{\Omega_3 \overline{C}_{44}\overline{\varepsilon}_4^e}{\Omega_2} & -\dfrac{1}{2}\dfrac{\Omega_2 \overline{C}_{44}\overline{\varepsilon}_4^e}{\Omega_3} & 0 & 0 & 0 \\ -\dfrac{1}{2}\dfrac{\Omega_3 \overline{C}_{55}\overline{\varepsilon}_5^e}{\Omega_1} & 0 & -\dfrac{1}{2}\dfrac{\Omega_1 \overline{C}_{55}\overline{\varepsilon}_5^e}{\Omega_3} & 0 & 0 & 0 \\ -\dfrac{1}{2}\dfrac{\Omega_2 \overline{C}_{66}\overline{\varepsilon}_6^e}{\Omega_1} & -\dfrac{1}{2}\dfrac{\Omega_1 \overline{C}_{66}\overline{\varepsilon}_6^e}{\Omega_2} & 0 & 0 & 0 & 0 \end{bmatrix}$$

(B.18)

式中

$$\begin{cases} P'_{11} = -\overline{C}_{11}\overline{\varepsilon}_1^e - \overline{C}_{12}\overline{\varepsilon}_2^e - \overline{C}_{13}\overline{\varepsilon}_3^e \\ P'_{22} = -\overline{C}_{12}\overline{\varepsilon}_1^e - \overline{C}_{22}\overline{\varepsilon}_2^e - \overline{C}_{23}\overline{\varepsilon}_3^e \\ P'_{33} = -\overline{C}_{13}\overline{\varepsilon}_1^e - \overline{C}_{23}\overline{\varepsilon}_2^e - \overline{C}_{33}\overline{\varepsilon}_3^e \end{cases} \quad (\text{B.19})$$

B.3 损伤面

引入两个四阶张量 $\boldsymbol{B}$ 和 $\boldsymbol{J}$ 的条件下,以张量形式表示的各向异性损伤准则,在热力学力空间 $\boldsymbol{Y}$ 中定义了一个多轴极限表面,其限定了损伤范围。损伤演化由与损伤面相关的损伤势和各向同性硬化函数定义。建议的损伤面 g^{d} 由下式给出:

$$g^{\mathrm{d}} = (\hat{Y}^N_{ij} J_{ijhk} \hat{Y}^N_{hk})^{1/2} + (Y^S_{ij} B_{ijhk} Y^S_{hk})^{1/2} - (\gamma(\delta) + \gamma_0) \qquad (\text{B.20})$$

式中:γ_0 为初始损伤阈值;$\gamma(\delta)$ 为硬化函数。

损伤面对于热力学力的偏导为

$$\frac{\partial g^{\mathrm{d}}}{\partial \boldsymbol{Y}} = \begin{bmatrix} \dfrac{J_{11} Y^N_{11}}{\Phi^N} + \dfrac{B_{11} Y^S_{11}}{\Phi^S} \\ \dfrac{J_{22} Y^N_{22}}{\Phi^N} + \dfrac{B_{22} Y^S_{22}}{\Phi^S} \\ \dfrac{J_{33} Y^N_{33}}{\Phi^N} + \dfrac{B_{33} Y^S_{33}}{\Phi^S} \\ 0 \\ 0 \\ 0 \end{bmatrix} \qquad (\text{B.21})$$

式中

$$\begin{cases} \Phi^N = \sqrt{J_{11}(Y^N_{11})^2 + J_{22}(Y^N_{22})^2 + J_{33}(Y^N_{33})^2} \\ \Phi^S = \sqrt{B_{11}(Y^S_{11})^2 + B_{22}(Y^S_{22})^2 + B_{33}(Y^S_{33})^2} \end{cases} \qquad (\text{B.22})$$

损伤面对于损伤硬化函数的偏导为

$$\frac{\partial g^{\mathrm{d}}}{\partial \gamma} = -1 \qquad (\text{B.23})$$

B.4 不可恢复应变面

在有效构型 $(\bar{\sigma}, R)$ 中,不可恢复应变(屈服)面 g^{p} 是热力学力的函数。因此,不可恢复应变面表达式为

$$g^{\mathrm{p}} = \sqrt{f_{ij} \bar{\sigma}_i \bar{\sigma}_j + f_i \bar{\sigma}_i} - (R(p) + R_0) \qquad (\text{B.24})$$

式中:$i = 1, 2, \cdots, 6$;R_0 为初始不可恢复应变阈值;R 为硬化函数。

不可恢复应变面对于有效应力的偏导为

$$\frac{\partial \boldsymbol{g}^p}{\partial \overline{\sigma}} = \begin{bmatrix} \dfrac{\frac{1}{2}f_1 + 2f_{11}\overline{\sigma}_1 + 2f_{12}\overline{\sigma}_2 + 2f_{13}\overline{\sigma}_3}{\Phi^p} \\ \dfrac{\frac{1}{2}f_2 + 2f_{22}\overline{\sigma}_2 + 2f_{12}\overline{\sigma}_1 + 2f_{23}\overline{\sigma}_3}{\Phi^p} \\ \dfrac{\frac{1}{2}f_3 + 2f_{33}\overline{\sigma}_3 + 2f_{13}\overline{\sigma}_1 + 2f_{23}\overline{\sigma}_2}{\Phi^p} \\ \dfrac{f_4\overline{\sigma}_4}{\Phi^p} \\ \dfrac{f_5\overline{\sigma}_5}{\Phi^p} \\ \dfrac{f_6\overline{\sigma}_6}{\Phi^p} \end{bmatrix} \quad (\text{B.25})$$

式中

$$\Phi^p = (f_1\overline{\sigma}_1 + f_2\overline{\sigma}_2 + f_3\overline{\sigma}_3 + f_{11}\overline{\sigma}_1^2 + f_{22}\overline{\sigma}_2^2 + f_{33}\overline{\sigma}_3^2 + 2f_{12}\overline{\sigma}_1\overline{\sigma}_2 + 2f_{13}\overline{\sigma}_1\overline{\sigma}_3 + 2f_{23}\overline{\sigma}_2\overline{\sigma}_3 + f_6\overline{\sigma}_6^2 + f_5\overline{\sigma}_5^2 + f_4\overline{\sigma}_4^2)^{1/2} \quad (\text{B.26})$$

屈服面对于不可恢复应变硬化函数的偏导为

$$\frac{\partial g^p}{\partial R} = -1 \quad (\text{B.27})$$

附录 C 使用的软件

本书仅涉及 4 个应用软件。Abaqus 软件的使用频次最多。BMI3 仅在第 4 章中使用。MATLAB 用于符号和数值计算。最后一个,必须利用 Intel Fortran 软件编写用户材料子程序,并链接到 Abaqus 软件中,其用法比较简单,因为它通过批处理文件进行调用,不需要用户介入。当然,编写新的用户材料子程序需要掌握一些 Fortran 编程知识,但本书示例中提供和使用的几个子程序例子可使读者的编程变得更容易。

本附录的主要目的是介绍本书中使用的软件,即 Abaqus 和 BMI3,以及如何使用 Intel Fortran 编写用户子程序并与 Abaqus 链接。读者可以在示例中附带的 MATLAB 代码提供的帮助下使用和学习 MATLAB,这些代码有些在本书中,有些可以从文献[5]中下载。

本书软件均在 Windows7 系统中进行操作,在 Linux 系统中的操作方法与其类似。这些应用程序的供应商自身提供大量的资料、培训课程、用户群等,读者可以使用这些学习资料和途径熟悉软件界面。读者也可以通过本书的网址:http://barbero.cadec-online.com/feacm-abaqus(文献[5])或本书的论坛(文献[194])获取相关资料。

C.1 Abaqus

Abaqus 是一个商业有限元分析(FEA)应用软件,其拥有称为 CAE 的友好图形用户界面(GUI)和内容丰富的帮助文档。一旦启动,用户就可以轻松浏览软件的菜单等。

在 GUI 中的所有操作均会生成 CAE 脚本,这些脚本保存在几个文件中,但如果没有学习过 CAE 脚本方面知识的话会较难理解这些脚本[195]。

本书中涉及的相关示例均采用 Abaqus 6.10 版本进行操作。这些示例在 Windows 7 和 Linux 系统中的操作方式基本相同。在 Windows 7 中,可通过 Start, All Programs, Abaqus 6.10, Abaqus CAE 的操作过程从"Start"菜单中打开 Abaqus 软件。

本书中,假定用户已经创建了一个 c:\Simulia\user 文件夹,所有模型文件均放在该文件夹中。

使用快捷方式 Start，All Programs，Abaqus 6.10 启动 GUI 后，所有程序的临时文件都将位于默认工作目录中，该目录是在安装 Abaqus 软件时设置的，通常为 c:\temp。可以通过编辑快捷方式中的 Start in:字段更改默认工作目录，如图 C.1 所示。另外，也可以通过执行以下操作从 Abaqus/CAE 中更改当前会话的工作目录：

```
Menu: File,Set Work Directory
```

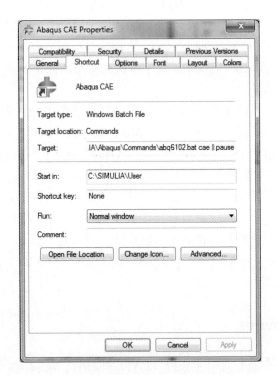

图 C.1　编辑快捷方式更改默认的工作目录

Abaqus 将默认的工作目录用作存储临时文件的位置，这些临时文件包括：

.dat 数据文件，其包含请求打印输出的结果。

.fil 结果文件，其包含如失稳模态等结果（见例 4.3）。该文件为非标准格式，只能通过 Abaqus 软件读取。

.inp 输入文件，其包含 Abaqus 求解器所需的输入信息。

.jnl 日志文件，其包含自前一个 Save 操作后对.mdb 的所有更改信息。

.log 日志文件，其包含 Abaqus 求解器最后执行的简要日志信息。

.odb 数据输出文件，其包含 Abaqus 求解器所有计算结果。

.rec 恢复文件,其包含自前一个 Save 操作以来用户在 GUI 上执行操作的记录。

.rpy 回放文件,其包含在 GUI 中用于生成模型的所有操作过程的命令。

本书中,我们倾向于在默认工作目录中存储和改写临时文件,这样不用占用用户文件夹的空间,但有时也会有意将临时文件存储在用户文件夹中。例如,如果想从.dat 文件中读取一些输出结果,我们会将其存储在用户文件夹中。

本节主要对 Abaqus 软件的安装和"系统"进行描述。而 Abaqus/CAE GUI 的相关功能在本书第 3 章之后的示例中逐步介绍。因此,强烈建议读者安装 Abaqus,然后再继续学习第 3 章,并实操第 3 章中的所有例题。第 3 章结束后,读者应该对 Abaqus/CAE 的复合材料建模方面非常熟练。通过学习和实践本书中所有例子,对此将获得更高的熟练程度。根据作者的经验,学生们通过使用本书中的示例以及 Abaqus 附带的帮助文档,可以成功自学 Abaqus/CAE。另一个帮助学习的来源是使用 Abaqus/CAE GUI 演示示例的教学视频[5]。

C.1.1 Abaqus 编程特点

在 Abaqus 中运行 UMAT 子程序和使用其他 Abaqus 编程功能之前,必须按此顺序安装以下软件:

(1) Microsoft Visual Studio© (VS) 2008。
(2) Intel Fortran© 11.1。

首先,安装 Microsoft Visual Studio 2008(VS2008),并使用网页上提供的更新将其更新到 service pack 1(SP1)。

接下来,安装 Intel Fortran 11.1。Fortran 必须在 VS 之后安装,这样 VS 就可以将 Fortran 集成进去。安装过程中,请选中"X64 Compilersand Tools"复选框,其默认情况下未勾选,如图 C.2 所示。

接下来,找到该批处理文件:

c:\SIMULIA\Abaqus\Commands\abq6102.bat

复制一份后,修改如下(也可在文献[5]中获得)。

rem Windows 64 version of.bat file to execute Abaqus
rem with User Programmable Features and Intel MKL Math libraries

SET PATH = C:\Program Files (x86)\Microsoft Visual Studio 9.0\VC\Bin;% PATH% ;
SET PATH = C:\Program Files (x86)\Intel\Compiler\11.1\048\bin\Intel64;% PATH% ;
SET PATH = C:\Program Files (x86)\Common Files\Microsoft Shared\VSA\9.0\VsaEnv;
% PATH% ;

图 C.2　Visual Studio 2008 安装页面

SET LIB = C:\Program Files (x86)\Microsoft Visual Studio 9.0\VC\Lib;%LIB%
SET LIB = C:\Program Files (x86)\Intel\Compiler\11.1\048\mkl\em64t\lib;%LIB%

call "C:\Program Files (x86)\Intel\Compiler\11.1\048\bin\intel64\ifortvars_intel64"

rem keep next lines exactly as in your original installation
@ echo off
call "C:\SIMULIA\Abaqus\6.10-1\exec\abq6101.exe" %*

请注意,安装中,文件名和位置可能略有不同,特别是在 32 位系统中使用时。修改 abq6102.bat 文件时,必须使用系统中所需文件的确切位置,否则 Abaqus 在编译/链接 UMAT 子例程时会提示错误消息。

接下来,找到环境变量文件:

C:\SIMULIA\Abaqus\6.10-2\site\abaqus_v6.env

复制一份后,修改 compile_fortran 指令以包含 3 个附加开关。前两个,/DABQWIN86_64 和/Qmkl:sequential,允许使用 Intel 数学核心数据库 MKL。第三个,free,使得 Abaqus 理解用户子程序中自由格式的 Fortran 代码,即使 Abaqus 要求用户提供 Fortran 代码为.for 格式。对于 Windows 64 位系统,环境变量文件的相关部分内容应如下所示:

compile_fortran=['ifort','/c','/DABQ_WIN86_64','/Qmkl:sequential','/free',
　　'/recursive','/Qauto-scalar','/QxW','/nologo','/Od','/include:% I']

需要使用 Visual Studio 2008 将 Fortran 代码链接到 Abaqus 之前对其进行调试。可以采用一个简单的程序,比如文献[5]中提供的 umat_tester.f90,在链接到 Abaqus 之前测试该用户材料子程序。读者可以对 umat_tester.f90 进行适应性修改以满足计算需要。

只要链接 Abaqus 的 Fortran 文件具有.for 扩展名或者利用 abaqus make 将文件编译为目标代码.obj 格式,就可以将该 Fortran 代码以自由格式进行编写。默认情况下,VS2008 读取 Fortran 文件需要.for 扩展名,其包含 72 列固定格式的 Fortran 77 代码。由于 F77 非常受限制,推荐使用 Fortran 90 的自由格式。即使.for 文件包含自由格式的代码 VS2008 也可以读取,只要将编译器指令! DEC $ FREEFORM 添加到代码的第一行即可。另外,也可以在 VS2008 内部更改其默认行为。除此之外,在环境变量文件中,"free"开关应添加到 compile_fortran 指令中,以便 Abaqus 可以处理自由格式的代码。

Abaqus 帮助文档中的 UMAT、VUMAT、UGENS 和 UEL 示例均采用 F77 编写。文献[196]中提供了一个用 F90 编写的自由格式转换器。

C.2　BMI3

只需要使用 inp2bmi3.exe 对输入文件进行过滤后,本机 BMI3 代码就可读取 Abaqus 中的输入文件.inp。因此,读者可以在 Abaqus/CAE 中创建模型,从作业管理器中写入.inp 文件,再用 inp2bmi3.exe 进行过滤,最后用 bmi3.exe 程序执行计算。本书中的可执行文件可在网站[5]中找到。输入/输出文件如表 C.1 所列。执行顺序如下:

(1) 在 Abaqus/CAE 中创建模型并输出 Job-1.inp 文件。

(2) 执行 inp2bmi3.exe 得到过滤后的文件。如需要,可修改 BMI3.dat 文件中的 mode、node 和 component。

(3) 执行 bmi3.exe。记录显示器上或 BMI3.out 中的结果。

注意,过滤器 inp2bmi3.exe 不能理解所有的.inp 中的命令。大多数的限制可以通过修改过滤器来消除,在文献[5]中提供了 inp2bmi3.f90 源代码。当前存在的限制如下:

(1) 网格必须使用 S8R 或 S8R5 单元进行划分(见例 4.2)。
(2) 所有载荷必须是集中载荷类型。
(3) 必须使用 Shell General Stiffness 指定材料属性和截面。
(4) 整体模型只能使用一种材料。
(5) 不能使用约束边界条件。
(6) 不能使用材料方向(*orientation)。

表 C.1 执行 BMI3 的输入/输出文件

程序	读取	写入	注释
inp2bmi3.exe	Job-1.inp	BMI3.inp	过滤后的模型
		BMI3.dat	材料属性
		ABQ.inp	过滤后的 Abaqus 输入文件
BMI3.exe	BMI3.inp	BMI3.out	结果
	BMI3.dat	MODES.out	失稳模态

每次提交作业或从作业管理器窗口执行 Write Input 时,Abaqus/CAE 均会写入一个 .inp 文件。然后,运行 inp2bmi3.exe 生成 BMI3.inp、ABQ.inp 和 BMI3.dat 文件。如果在 Abaqus 中使用命令模式执行计算 ABQ.inp(见例 2.1),或者将其导入 Abaqus/CAE 中再执行计算,将得到分叉载荷 $\Lambda^{(cr)}$ 和失稳模态。

材料属性和扰动参数通过 BMI3.dat 提供,其由过滤器自动创建。BMI3.dat 中最后一行包含 modenum,nodenum,component 信息,其作为扰动参数 δ。如果这 3 个值均为 0(默认值),则 BMI3 会选择第一阶模态以及产生最大模态幅值的节点和位移组合。BMI3 的结果打印输出在 BMI3.out 中,失稳模态保存在 MODES.out 文件中。

请注意,结果(分叉载荷、斜率和曲率)显示为负号。这在稳定性分析中很常见。BMI3 软件的另一个特点是横向挠度 ω(垂直于板方向)与 Abaqus 结果符号相反。由于横向挠度 ω 通常用作扰动参数,因此在进行结果的解释时必须考虑其符号的变化(见例 4.2)。

参考文献

[1] E. J. Barbero. *Introduction to Composite Materials Design-Second Edition.* http://barbero.cadec-online.com/icmd. CRC Press, Boca Raton, FL, 2010.

[2] D. Frederick and T. -S. Chang. *Continuum Mechanics.* Scientific Publishers, Cambridge, MA, 1972.

[3] F. P. Beer, E. R. Johnston Jr., and J. T. DeWolf. *Mechanics of Materials*, 3rd Edition. McGraw-Hill, Boston, MA, 2001.

[4] J. N. Reddy. *Energy and Variational Methods in Applied Mechanics.* Wiley, 1984.

[5] E. J. Barbero. Web resource: http://barbero.cadec-online.com/feacm-abaqus.

[6] S. S. Sonti, E. J. Barbero, and T. Winegardner. Mechanical properties of pultruded E-glass-vinyl ester composites. In *50th Annual Conference, Composites Institute, Society of the Plastics Industry (February) pp. 10-C/1-7*, 1995.

[7] Simulia. Abaqus documentation. The default location for the HTML documentation is http://HOME:2080/v6.10/ where HOME is the url of the server where the documentation is stored during installation. The default location for the PDF documentation is C:\SIMULIA\Documentation\docs\v6.10\pdf_books\index.pdf.

[8] E. J. Barbero. *Finite Element Analysis of Composite Materials-First Edition.* CRCPress, Boca Raton, FL, 2007.

[9] ANSYS Inc. Ansys mechanical apdl structural analysis guide, release 140.0, http://www1.ansys.com/customer/content/documentation/140/ans_str.pdf, 2011.

[10] Simulia. Abaqus keywords reference manual. The default location is http://HOME:2080/v6.10/books/key/default.htm where HOME is the url of the server where the documentation is stored during installation.

[11] SolidWorks. http://www.solidworks.com/sw/engineering-education-software.htm.

[12] E. J. Barbero. Computer aided design environment for composites. http://www.cadec-online.com, 2011.

[13] R. J. Roark and W. C. Young. *Roark's Formulas for Stress and Strain*, 6th Edition. McGraw-Hill, New York, NY, 1989.

[14] J. N. Reddy. *Mechanics of Laminated Composite Plates and Shells*, 2nd Edition. CRCPress, Boca Raton, FL, 2003.

[15] E. Hinton and D. R. J. Owen. *An Introduction to Finite Element Computations.* Pineridge Press, Swansea, UK, 1979.

[16] NAFEMS. Test R0031/3. Technical report, National Agency for Finite Element Methods and

Standards (UK),1995.

[17] E. J. Barbero. 3-d finite element for laminated composites with 2-d kinematic constraints. *Computers and Structures*,45(2):263-271,1992.

[18] E. J. Barbero. Cadec application program interface. http://www.cadec-online.com/Help/API.aspx.

[19] Simulia. Abaqus benchmarks manual. The default location is http://HOME:2080/v6.10/books/bmk/default.htm where HOME is the url of the server where the documentation is stored during installation.

[20] E. J. Barbero and J. Trovillion. Prediction and measurement of post-critical behavior of fiber-reinforced composite columns. *Composites Science and Technology*,58:1335-1341,1998.

[21] E. J. Barbero. Prediction of compression strength of unidirectional polymer matrix composites. *Journal of Composite Materials*,32(5)(5):483-502,1998.

[22] A. Puck and H. Schurmann. Failure analysis of frp laminates by means of physically based phenomenological models. *Composites Science and Technology*,62:1633-1662,2002.

[23] MIL17.org. The composite materials handbook, web resource, http://www.mil17.org.

[24] L. A. Godoy. *Theory of Stability-Analysis and Sensitivity*. Taylor and Francis, Philadelphia, PA, 2000.

[25] E. J. Barbero, L. A. Godoy, and I. G. Raftoyiannis. Finite elements for three-mode interaction in buckling analysis. *International Journal for Numerical Methods in Engineering*,39(3):469-488,1996.

[26] L. A. Godoy, E. J. Barbero, and I. G. Raftoyiannis. Finite elements for post-bucklinganalysis. i-the w-formulation. *Computers and Structures*,56(6):1009-1017,1995.

[27] E. J. Barbero, I. G. Raftoyiannis, and L. A. Godoy. Finite elements for post-buckling analysis. ii-application to composite plate assemblies. *Computers and Structures*,56(6):1019-1028,1995.

[28] I. G. Raftoyiannis, L. A. Godoy, and E. J. Barbero. Buckling mode interaction in composite plate assemblies. *Applied Mechanics Reviews*,48(11/2):52-60,1995.

[29] E. J. Barbero. Prediction of buckling-mode interaction in composite columns. *Mechanics of Composite Materials and Structures*,7(3):269-284,2000.

[30] S. Yamada and J. G. A. Croll. Buckling behavior pressure loaded cylindrical panels. *ASCE Journal of Engineering Mechanics*,115(2):327-344,1989.

[31] C. T. Herakovich. *Mechanics of Fibrous Composites*. Wiley, New York,1998.

[32] J. D. Eshelby. The determination of the elastic field of an ellipsoidal inclusion and related problems. *Proceedings of the Royal Society*,A241:376-396,1957.

[33] J. D. Eshelby. The elastic field outside an ellipsoidal inclusion. *Acta Metall.*, A252:561-569,1959.

[34] T. Mori and K. Tanaka. The elastic field outside an ellipsoidal inclusion. *Acta Metall.*,21:571-574,1973.

[35] R. Hill. A self-consistent mechanics of composite materials. *J. Mech. Phys. Solids*,13:213-222,

1965.

[36] S. Nemat-Nasser and M. Hori. Micromechanics: *Overall Properties of Heterogeneous Materials*. North-Holland, Amsterdam, 1993.

[37] R. Luciano and E. J. Barbero. Formulas for the stiffness of composites with periodic microstructure. *I. J. Solids and Structures*, 31(21):2933-2944, 1995.

[38] R. Luciano and E. J. Barbero. Analytical expressions for the relaxation moduli of linear viscoelastic composites with periodic microstructure. *ASME J. Applied Mechanics*, 62(3):786-793, 1995.

[39] E. J. Barbero and R. Luciano. Micromechanical formulas for the relaxation tensor of linear viscoelastic composites with transversely isotropic fibers. *I. J. Solids and Structures*, 32(13):1859-1872, 1995.

[40] J. Aboudi. *Mechanics of Composite Materials: A Unified Micromechanical Approach*. Vol. 29 of Studies in Applied Mechanics. Elsevier, New York, NY, 1991.

[41] Z. Hashin and S. Shtrikman. A variational approach to the elastic behavior of multiphase materials. *J. Mechanics and Physics of Solids*, 11:127-140, 1963.

[42] V. Tvergaard. Model studies of fibre breakage and debonding in a metal reinforced by short fibres. *J. Mechanics and Physics of Solids*, 41(8):1309-1326, 1993.

[43] J. L. Teply and G. J. Dvorak. Bound on overall instantaneous properties of elasticplastic composites. *J. Mechanics and Physics of Solids*, 36(1):29-58, 1988.

[44] R. Luciano and E. Sacco. Variational methods for the homogenization of periodic media. *European J. Mech. A/Solids*, 17:599-617, 1998.

[45] G. J. Creus. *Viscoelasticity: Basic Theory and Applications to Concrete Structures*. Springer-Verlag, Berlin, 1986.

[46] Eric W. Weisstein. Gamma function. MathWorld-A Wolfram Web Resource. http://mathworld.wolfram.com/GammaFunction.html.

[47] G. D. Dean, B. E. Read, and P. E. Tomlins. A model for long-term creep and the effectsof physical ageing in poly (butylene terephthalate). *Plastics and Rubber Processing and Applications*, 13(1):37-46, 1990.

[48] B. F. Oliveira and G. J. Creus. An analytical-numerical framework for the study of ageing in fibre reinforced polymer composites. *Composites B*, 65(3-4):443-457, 2004.

[49] K. Ogatha. Discrete-Time Control System. Prentice Hall, Englewood Cliffs, NJ, 1987.

[50] K. J. Hollenbeck. Invlap. m: A matlab function for numerical inversion of Laplace transforms by the de Hoog algorithm (1998), http://www.mathworks.com/.

[51] R. S. Lakes. *Viscoelastic Solids*. Boca Raton, FL, 1998.

[52] J. D. Ferry. *Viscoelastic Properties of Polymers*. 3rd. Edition. Wiley, New York, 1980.

[53] R. M. Christensen. *Theory of Viscoelasticity*. New York, 1972.

[54] P. Qiao, E. J. Barbero, and J. F. Davalos. On the linear viscoelasticity of thin-walled laminated composite beams. *Journal of Composite Materials*, 34(1):39-68, 2000.

[55] E. J. Barbero, F. A. Cosso, R. Roman, and T. L. Weadon. Determination of material parameters for Abaqus progressive damage analysis of E-Glass Epoxy laminates. *Composites Part B: Engineering*, 46(3):211-220, 2012.

[56] J. Lemaitre and A. Plumtree. Application of damage concepts to predict creep-fatigue failures. In *American Society of Mechanical Engineers*, 78-PVP-26, pages 10-26, 1978.

[57] I. N. Rabotnov. *Rabotnov: Selected Works-Problems of the Mechanics of a Deformable Solid Body*. Moscow Izdatel Nauka, 1991.

[58] L. M. Kachanov. On the creep fracture time. *Izv. Akad. Nauk USSR*, 8:26-31, 1958.

[59] N. R. Hansen and H. L. Schreyer. A thermodynamically consistent framework for theories of elastoplasticity coupled with damage. *International Journal of Solids and Structures*, 31(3): 359-389, 1994.

[60] J. Janson and J. Hult. Fracture mechanics and damage mechanics-a combined approach. *J. Mec. Theor. Appl. 1*, pages S18-28, 1977.

[61] E. J. Barbero and K. W. Kelly. Predicting high temperature ultimate strength of continuous fiber metal matrix composites. *Journal of Composite Materials*, 27(12):1214-1235, 1993.

[62] K. W. Kelly and E. Barbero. Effect of fiber damage on the longitudinal creep of acfmmc. *International Journal of Solids and Structures*, 30(24):3417-3429, 1993.

[63] J. R. Rice. *Continuum Mechanics and Thermodynamics of Plasticity in Relation to Microscale Deformation Mechanisms*, chapter 2, pages 23-79. Constitutive Equations in Plasticity. MIT Press, Cambridge, MA, 1975.

[64] J. C. Simo and T. J. R. Hughes. *Computational Inelasticity*. Springer, Berlin, 1998.

[65] D. Krajcinovic, J. Trafimow, and D. Sumarac. Simple constitutive model for a cortical bone. *Journal of Biomechanics*, 20(8):779-784, 1987.

[66] D. Krajcinovic and D. Fanella. Micromechanical damage model for concrete. *Engineering Fracture Mechanics*, 25(5-6):585-596, 1985.

[67] D. Krajcinovic. Damage mechanics. *Mechanics of Materials*, 8(2-3):117-197, 1989.

[68] A. M. Neville. *Properties of Concrete, 2nd Edition*. Wiley, New York, NY, 1973.

[69] ACI, American Concrete Institute.

[70] B. W. Rosen. The tensile failure of fibrous composites. *AIAA Journal*, 2(11):1985-1911, 1964.

[71] W. Weibull. A statistical distribution function of wide applicability. *Journal of Applied Mechanics*, 18:293-296, 1951.

[72] B. W. Rosen. *Fiber Composite Materials*, chapter 3. American Society for Metals, Metals Park, OH, 1965.

[73] A. S. D. Wang. A non-linear microbuckling model predicting the compressive strength of unidirectional composites. In *ASME Winter Annual Meeting*, volume WA/Aero-1, 1978.

[74] J. S. Tomblin, E. J. Barbero, and L. A. Godoy. Imperfection sensitivity of fibermicro-buckling in elastic-nonlinear polymer-matrix composites. *International Journal of Solids and Structures*, 34(13):1667-1679, 1997.

[75] D. C. Lagoudas and A. M. Saleh. Compressive failure due to kinking of fibrous composites. *Journal of Composite Materials*, 27(1):83-106, 1993.

[76] P. Steif. A model for kinking in fiber composites. i. fiber breakage via micro-buckling. *International Journal of Solids and Structures*, 26(5-6):549-61, 1990.

[77] S. W. Yurgartis and S. S. Sternstein. *Experiments to Reveal the Role of Matrix Properties and Composite Microstructure in Longitudinal Compression Strength*, volume 1185 of *ASTM Special Technical Publication*, pages 193-204. ASTM, Philadelphia, PA, 1994.

[78] C. Sun and A. W. Jun. Effect of matrix nonlinear behavior on the compressive strength of fiber composites. In *AMD*, volume 162, pages 91-101, New York, NY, 1993. ASME, American Society of Mechanical Engineers, Applied Mechanics Division.

[79] D. Adams and E. Lewis. Current status of composite material shear test methods. *SAMPE Journal*, 31(1):32-41, 1995.

[80] A. Maewal. Postbuckling behavior of a periodically laminated medium in compression. *International Journal of Solids and Structures*, 17(3):335-344, 1981.

[81] L. M. Kachanov. Rupture time under creep conditions problems of continuum mechanics. *SIAM*, pages 202-218, 1958.

[82] E. J. Barbero and P. Lonetti. Damage model for composites defined in terms of available data. *Mechanics of Composite Materials and Structures*, 8(4):299-315, 2001.

[83] E. J. Barbero and P. Lonetti. An inelastic damage model for fiber reinforced laminates. *Journal of Composite Materials*, 36(8):941-962, 2002.

[84] P. Lonetti, R. Zinno, F. Greco, and E. J. Barbero. Interlaminar damage model for polymer matrix composites. *Journal of Composite Materials*, 37(16):1485-1504, 2003.

[85] P. Lonetti, E. J. Barbero, R. Zinno, and F. Greco. Erratum: Interlaminar damage model for polymer matrix composites. *Journal of Composite Materials*, 38(9):799-800, 2004.

[86] E. J. Barbero, F. Greco, and P. Lonetti. Continuum damage-healing mechanics with application to self-healing composites. *International Journal of Damage Mechanics*, 14(1):51-81, 2005.

[87] S. Murakami. Mechanical modeling of material damage. *Journal of Applied Mechanics*, 55:281-286, 1988.

[88] J. M. Smith and H. C. Van Ness. *Introduction to Chemical Engineering Thermodynamics*, 3rd Edition. McGraw-Hill, New York, NY, 1975.

[89] E. R. Cohen, T. Cvita, J. G. Frey, B. Holmstrm, K. Kuchitsu, R. Marquardt, I. Mills, F. Pavese, M. Quack, J. Stohner, H. L. Strauss, M. Takami, and A. J. Thor. *Quantities, Units and Symbols in Physical Chemistry: The IUPAC Green Book*, 3rd Edition, http:// old. iupac. org/ publications/books/ author/ cohen. html. RSC Publishing, 2007.

[90] L. E. Malvern. *Introduction to the Mechanics of a Continuous Medium*. Prentice Hall, Upper Saddle River, NJ, 1969.

[91] Y. A. Cengel and M. A. Boles. *Thermodynamics: An Engineering Approach*, 3rd Edition. McGraw-Hill, New York, NY, 1998.

[92] J. Lubliner. *Plasticity Theory*. Collier Macmillan, New York, NY, 1990.

[93] J. A. Nairn. The strain energy release rate of composite microcracking: a variational approach. *Journal of Composite Materials*, 23(11):1106-29, 11 1989.

[94] J. A. Nairn, S. Hu, S. Liu, and J. S. Bark. The initiation, propagation, and effectof matrix microcracks in cross-ply and related laminates. *In Proc. of the 1st NASA Advanced Comp. Tech. Conf.*, pages 497-512, Oct. 29-Nov. 1, 1990.

[95] S. Liu and J. A. Nairn. The formation and propagation of matrix microcracks in crossply laminates during static loading. *Journal of Reinforced Plastics and Composites*, 11(2):158-78, Feb. 1992.

[96] J. A. Nairn. Microcracking, microcrack-induced delamination, and longitudinal splitting of advanced composite structures. Technical Report NASA Contractor Report4472, 1992.

[97] J. A. Nairn and S. Hu. *Damage Mechanics of Composite Materials*, volume 9 of *Composite Materials Series*, chapter Matrix Microcracking, pages 187-244. Elsevier, 1994.

[98] J. A. Nairn. Applications of finite fracture mechanics for predicting fracture eventsin composites. In *5th Int'l Conf. on Deform. and Fract. of Comp.*, 18-19 March 1999.

[99] J. A. Nairn. *Polymer Matrix Composites*, volume 2 of Comprehensive CompositeMaterials, chapter Matrix Microcracking in Composites, pages 403-432. Elsevier Science, 2000.

[100] J. A. Nairn. Fracture mechanics of composites with residual stresses, imperfect interfaces, and traction-loaded cracks. In *Workshop 'Recent Advances in Continuum Damage Mechanics for Composites'*, volume 61, pages 2159-67, UK, 20-22 Sept. 2000. Elsevier.

[101] J. A. Nairn and D. A. Mendels. On the use of planar shear-lag methods for stresstransfer analysis of multilayered composites. *Mechanics of Materials*, 33(6):335-362, 2001.

[102] J. A. Nairn. *Finite Fracture Mechanics of Matrix Microcracking in Composites*, pages 207-212. Application of Fracture Mechanics to Polymers, Adhesives and Composites. Elsevier, 2004.

[103] Y. M. Han, H. T. Hahn, and R. B. Croman. A simplified analysis of transverse ply cracking in cross-ply laminates. *Composites Science and Technology*, 31(3):165-177, 1988.

[104] C. T. Herakovich, J. Aboudi, S. W. Lee, and E. A. Strauss. Damage in composite laminates: Effects of transverse cracks. *Mechanics of Materials*, 7(2):91-107, 11 1988.

[105] J. Aboudi, S. W. Lee, and C. T. Herakovich. Three-dimensional analysis of laminates with cross cracks. *Journal of Applied Mechanics, Transactions ASME*, 55(2):389-397, 1988.

[106] F. W. Crossman, W. J. Warren, A. S. D. Wang, and G. E. Law. Initiation and growthof transverse cracks and edge delamination in composite laminates. ii-experimental correlation. *Journal of Composite Materials Suplement*, 14(1):88-108, 1980.

[107] D. L. Flaggs and M. H. Kural. Experimental determination of the in situ transverse lamina strength in graphite/epoxy laminates. *Journal of Composite Materials*, 16:103-16, 03 1982.

[108] S. H. Lim and S. Li. Energy release rates for transverse cracking and delaminations induced by transverse cracks in laminated composites. *Composites Part A*, 36(11):1467-1476, 2005.

[109] S. Li and F. Hafeez. Variation-based cracked laminate analysis revisited and fundamentally

extended. *International Journal of Solids and Structures*, 46(20):3505-3515, 2009.

[110] J. L Rebiere and D. Gamby. A decomposition of the strain energy release rate associated with the initiation of transverse cracking, longitudinal cracking and delamination in cross-ply laminates. *Composite Structures*, 84(2):186-197, 2008.

[111] S. C. Tan and R. J. Nuismer. A theory for progressive matrix cracking in composite laminates. *Journal of Composite Materials*, 23:1029-1047, 1989.

[112] R. J. Nuismer and S. C. Tan. Constitutive relations of a cracked composite lamina. *Journal of Composite Materials*, 22:306-321, 1988.

[113] T. Yokozeki and T. Aoki. Stress analysis of symmetric laminates with obliquelycrossed matrix cracks. *Advanced Composite Materials*, 13(2):121-40, 2004.

[114] T. Yokozeki and T. Aoki. Overall thermoelastic properties of symmetric laminates containing obliquely crossed matrix cracks. *Composites Science and Technology*, 65(11-12):1647-54, 2005.

[115] Z. Hashin. Analysis of cracked laminates: a variational approach. *Mechanics of Materials*, 4:121:136, 1985.

[116] J. Zhang, J. Fan, and C. Soutis. Analysis of multiple matrix cracking in [±θ/90n]s composite laminates, part I, inplane stiffness properties. *Composites*, 23(5):291-304, 1992.

[117] P. Gudmundson and S. Ostlund. First order analysis of stiffness reduction due tomatrix cracking. *Journal of Composite Materials*, 26(7):1009-30, 1992.

[118] P. Gudmundson and S. Ostlund. Prediction of thermoelastic properties of composite laminates with matrix cracks. *Composites Science and Technology*, 44(2):95-105, 1992.

[119] P. Gudmundson and W. Zang. An analytic model for thermoelastic properties of composite laminates containing transverse matrix cracks. *International Journal of Solids and Structures*, 30(23):3211-31, 1993.

[120] E. Adolfsson and P. Gudmundson. Thermoelastic properties in combined bending and extension of thin composite laminates with transverse matrix cracks. *International Journal of Solids and Structures*, 34(16):2035-60, 06 1997.

[121] E. Adolfsson and P. Gudmundson. Matrix crack initiation and progression in composite laminates subjected to bending and extension. *International Journal of Solidsand Structures*, 36(21):3131-3169, 1999.

[122] W. Zang and P. Gudmundson. Damage evolution and thermoelastic properties of composite laminates. *International Journal of Damage Mechanics*, 2(3):290-308, 07 1993.

[123] E. Adolfsson and P. Gudmundson. Matrix crack induced stiffness reductions in[(0m/90n/+p/-q)s]m composite laminates. *Composites Engineering*, 5(1):107-23, 1995.

[124] P. Lundmark and J. Varna. Modeling thermo-mechanical properties of damaged laminates. In *3rd International Conference on Fracture and Damage Mechanics, FDM 2003*, volume 251-252 of *Advances in Fracture and Damage Mechanics*, pages 381-7, Switzerland, 2-4 Sept. 2003. Trans Tech Publications.

[125] M. Kachanov. *Elastic solids with many cracks and related problems*, volume 30 of *Advances in Applied Mechanics*, chapter X, pages 260-445. Academic Press, Inc., 1993.

[126] A. Adumitroiaie and E. J. Barbero. Intralaminar damage model for laminates subjected to membrane and flexural deformations. *Mechanics of Advanced Materials and Structures*, 2013.

[127] T. Yokozeki, T. Aoki, and T. Ishikawa. Transverse crack propagation in the specimen width direction of cfrp laminates under static tensile loadings. *Journal of Composite Materials*, 36(17):2085-99, 2002.

[128] T. Yokozeki, T. Aoki, and T. Ishikawa. Consecutive matrix cracking in contiguous plies of composite laminates. *International Journal of Solids and Structures*, 42(9-10): 2785-802, 05 2005.

[129] T. Yokozeki, T. Aoki, T. Ogasawara, and T. Ishikawa. Effects of layup angle and ply thickness on matrix crack interaction in contiguous plies of composite laminates. *Composites Part A(Applied Science and Manufacturing)*, 36(9):1229-35, 2005.

[130] E. J. Barbero, F. A. Cosso, and F. A. Campo. Benchmark solution for degradation of elastic properties due to transverse matrix cracking in laminated composites. *Composite Structures*, 98(4):242-252, 2013.

[131] S. Li, S. R. Reid, and P. D. Soden. A continuum damage model for transverse matrix cracking in laminated fibre-reinforced composites. *Philosophical Transactions of the Royal Society London, Series A(Mathematical, Physical and Engineering Sciences)*, 356(1746):2379-412, 10/15 1998.

[132] Janis Varna, Roberts Joffe, and Ramesh Talreja. A synergistic damage-mechanics analysis of transverse cracking [±θ/904]s laminates. *Composites Science and Technology*, 61(5):657-665, 2001.

[133] A. S. D. Wang, P. C. Chou, and S. C. Lei. A stochastic model for the growth of matrix cracks in composite laminates. *Journal of Composite Materials*, 18(3):239-54, 05 1984.

[134] S. Li, S. R. Reid, and P. D. Soden. Modelling the damage due to transverse matrix cracking in fiber-reinforced laminates. In *Proc. 2nd Int. Conf. on Nonlinear Mechanics (ICNP-2)*, pages 320-323. Peking University Press, 1993.

[135] D. H. Cortes and E. J. Barbero. Stiffness reduction and fracture evolution of oblique matrix cracks in composite laminates. *Annals of Solid and Structural Mechanics*, 1(1):29-40, 2010.

[136] E. J. Barbero and D. H. Cortes. A mechanistic model for transverse damage initiation, evolution, and stiffness reduction in laminated composites. *Composites Part B*, 41:124-132, 2010.

[137] E. J. Barbero, G. Sgambitterra, A. Adumitroaie, and X. Martinez. A discrete constitutive model for transverse and shear damage of symmetric laminates with arbitrary stacking sequence. *Composite Structures*, 93:1021-1030, 2011.

[138] G. Sgambitterra, A. Adumitroaie, E. J. Barbero, and A. Tessler. A robust three-node shell element for laminated composites with matrix damage. *Composites Part B: Engineering*, 42(1):41-50, 2011.

[139] A. M. Abad Blazquez, M. Herraez Matesanz, Carlos Navarro Ugena, and E. J. Barbero. Acoustic emission characterization of intralaminar damage in composite laminates. In *Asociación Española de Materiales Compuestos MATCOMP 2013*, Algeciras, Spain, July 2-5, 2013.

[140] E. J. Barbero and F. A. Cosso. Determination of material parameters for discrete damage mechanics analysis of composite laminates. *Composites Part B: Engineering*, 2013.

[141] H. T. Hahn. A mixed-mode fracture criterion for composite materials. *Composites Technology Review*, 5:26-29, 1983.

[142] Allan F. Bower. *Mechanics of Solids*. CRC, 2010. Free at http://solidmechanics.org/.

[143] Simulia. Abaqus analysis user's manual, version 6.10.

[144] J. Varna, R. Joffe, N. V. Akshantala, and R. Talreja. Damage in composite laminates with off-axis plies. *Composites Science and Technology*, 59(14):2139-2147, 1999.

[145] H. Chai, C. D. Babcock, and W. G. Knauss. One dimensional modeling of failure in laminated plates by delamination buckling. *International Journal of Solids and Structures*, 17(11):1069-1083, 1981.

[146] M.-K. Yeh and L.-B. Fang. Contact analysis and experiment of delaminated cantilever composite beam. *Composites: Part B*, 30(4):407-414, 1999.

[147] W. L. Yin, S. N. Sallam, and G. J. Simitses. Ultimate axial load capacity of adelaminated beamplate. *AIAA Journal*, 24(1):123-128, 1986.

[148] D. Bruno. Delamination buckling in composite laminates with interlaminar defects. *Theoretical and Applied Fracture Mechanics*, 9(2):145-159, 1988.

[149] D. Bruno and A. Grimaldi. Delamination failure of layered composite plates loaded in compression. *International Journal of Solids and Structures*, 26(3):313-330, 1990.

[150] G. A. Kardomateas. The initial post-buckling and growth behaviour of internal delaminations in composite plates. *Journal of Applied Mechanics*, 60(4):903-910, 1993.

[151] G. A. Kardomateas and A. A. Pelegri. The stability of delamination growth in compressively loaded composite plates. *International Journal of Fracture*, 65(3):261-276, 1994.

[152] I. Sheinman, G. A. Kardomateas, and A. A. Pelegri. Delamination growth during preand postbuckling phases of delaminated composite laminates. *International Journalof Solids and Structures*, 35(1-2):19-31, 1998.

[153] D. Bruno and F. Greco. An asymptotic analysis of delamination buckling and growthin layered plates. *International Journal of Solids and Structures*, 37(43):6239-6276, 2000.

[154] H.-J. Kim. Postbuckling analysis of composite laminates with a delamination. *Computer and Structures*, 62(6):975-983, 1997.

[155] W. G. Bottega and A. Maewal. Delamination buckling and growth in laminates-closure. *Journal of Applied Mechanics*, 50(14):184-189, 1983.

[156] B. Cochelin and M. Potier-Ferry. A numerical model for buckling and growth of delaminations in composite laminates. *Computer Methods in Applied Mechanics*, 89(1-3):361-380, 1991.

[157] P.-L. Larsson. On delamination buckling and growth in circular and annular orthotropic

[158] W. -L. Yin. Axisymmetric buckling and growth of a circular delamination in a compressed laminate. *International Journal of Solids and Structures*, 21(5):503-514, 1985.

[159] K. -F. Nilsson, L. E. Asp, J. E. Alpman, and L. Nystedt. Delamination buckling and growth for delaminations at different depths in a slender composite panel. *International Journal of Solids and Structures*, 38(17):3039-3071, 2001.

[160] H. Chai and C. D. Babcock. Two-dimensional modeling of compressive failure in delaminated laminates. *Journal of Composite Materials*, 19(1):67-98, 1985.

[161] J. D. Withcomb and K. N. Shivakumar. Strain-energy release rate analysis of plates with post-buckled delaminations. *Journal of Composite Materials*, 23(7):714-734, 1989.

[162] W. J. Bottega. A growth law for propagation of arbitrary shaped delaminations in layered plates. *International Journal of Solids and Structures*, 19(11):1009-1017.

[163] B. Storåkers and B. Andersson. Nonlinear plate theory applied to delamination in composites. *Journal of the Mechanics and Physics of Solids*, 36(6):689-718, 1988.

[164] P. -L. Larsson. On multiple delamination buckling and growth in composite plates. *International Journal of Solids and Structures*, 27(13):1623-1637, 1991.

[165] M. A. Kouchakzadeh and H. Sekine. Compressive buckling analysis of rectangular composite laminates containing multiple delaminations. *Composite Structures*, 50(3):249-255, 2000.

[166] J. M. Comiez, A. M. Waas, and K. W. Shahwan. Delamination buckling: Experiment and analysis. *International Journal of Solids and Structures*, 32(6/7):767-782, 1995.

[167] G. A. Kardomateas. Postbuckling characteristics in delaminated kevlar/epoxy laminates: An experimental study. *Journal of Composites Technology and Research*, 12(2):85-90, 1990.

[168] O. Allix and A. Corigliano. Modeling and simulation of crack propagation in mixedmodes interlaminar fracture specimens. *International Journal of Fracture*, 77(2):111-140, 1996.

[169] D. Bruno and F. Greco. Delamination in composite plates: Influence of shear deformability on interfacial debonding. *Cement and Concrete Composites*, 23(1):33-45, 2001.

[170] J. W. Hutchinson and Z. Suo. Mixed mode cracking in layered materials. *Advances in Applied Mechanics*, 29:63-191, 1992.

[171] N. Point and E. Sacco. Delamination of beams: An application to the dcb specimen. *International Journal of Fracture*, 79(3):225-247, 1996.

[172] J. G. Williams. On the calculation of energy release rate for cracked laminates. *International Journal of Fracture*, 36(2):101-119, 1988.

[173] A. Tylikowsky. Effects of piezoactuator delamination on the transfer functions of vibration control systems. *International Journal of Solids and Structures*, 38(10-13):2189-2202, 2001.

[174] C. E. Seeley and A. Chattopadhyay. Modeling of adaptive composites including debonding. *International Journal Solids and Structures*, 36(12):1823-1843, 1999.

[175] W. J. Bottega and A. Maewal. Dynamics of delamination buckling. *International Journal of Non-Linear Mechanics*, 18(6):449-463, 1983.

[176] G. Alfano and M. A. Crisfield. Finite element interface models for the delamination analysis of laminated composites: Mechanical and computational issues. *International Journal for Numerical Methods in Engineering*, 50(7):1701-1736, 2001.

[177] P. P. Camanho, C. G. D'avila, and M. F. Moura. Numerical simulation of mixed-mode progressive delamination in composite materials. *Journal of Composite Materials*, 37(16):1415-1438, 2003.

[178] Z. Zou, S. R. Reid, and S. Li. A continuum damage model for delaminations in laminated composites. *Journal of the Mechanics and Physics of Solids*, 51(2):333-356, 2003.

[179] X-P Xu and A. Needleman. Numerical simulations of fast crack growth in brittle solids. *Journal of the Mechanics and Physics of Solids*, 42:1397-1434, 1994.

[180] J. R. Reeder. A bilinear failure criterion for mixed-mode delamination in composite materials. *ASTM STP 1206*, pages 303-322, 1993.

[181] H. Chai. Three-dimensional fracture analysis of thin-film debonding. *International Journal of Fracture*, 46(4):237-256, 1990.

[182] J. D. Withcomb. Three dimensional analysis of a post-buckled embedded delamination. *Journal of Composite Materials*, 23(9):862-889, 1989.

[183] D. Bruno, F. Greco, and P. Lonetti. A 3d delamination modelling technique based on plate and interface theories for laminated structures. *European Journal of MechanicsA/Solids*, 24:127-149, 2005.

[184] J. G. Williams. The fracture mechanics of delamination tests. *Journal of Strain Analysis*, 24(4):207-214, 1989.

[185] W. -L. Yin and J. T. S. Wang. The energy-release rate in the growth of a onedimensional delamination. *Journal of Applied Mechanics*, 51(4):939-941, 1984.

[186] J. R. Rice. A path independent integral and the approximate analysis of strain concentrations by notches and cracks. *Journal of Applied Mechanics*, 35:379-386, 1968.

[187] Simulia. Abaqus analysis user's manual, version 6. 10. chapter 29. 5: Cohesive elements.

[188] C. G. D'avila, P. P. Camanho, and M. F. De Moura. Mixed-mode decohesion elements for analyses of progressive delamination. In *42nd AIAA/ASME/ASCE/AHS/ASC Structures, Structural Dynamics, and Materials Conference, April 16-19, 2001*, volume 3, pages 2277-88, Seattle, WA, 2001. AIAA.

[189] Simulia. Abaqus analysis user's manual, version 6. 10. chapter 11. 4. 3: Crack propagation analysis.

[190] E. J. Barbero and J. N. Reddy. Jacobian derivative method for three-dimensional fracture mechanics. *Communications in Applied Numerical Methods*, 6(7):507-518, 1990.

[191] M. Benzeggagh and M. Kenane. Measurement of mixed-mode delamination fracture toughness of unidirectional glass/epoxy composites with mixed-mode bending apparatus. *Composite Science and Technology*, 56:439, 1996.

[192] J. Reeder, S. Kyongchan, P. B. Chunchu, and D. R.. Ambur. Postbuckling and growth of de-

laminations in composite plates subjected to axial compression. In *43rd AIAA/ASME/ASCE/AHS/ASC Structures, Structural Dynamics, and Materials Conference*, Denver, Colorado, vol. 1746, p. 10, 2002.

[193] A. Caceres. *Local Damage Analysis of Fiber Reinforced Polymer Matrix Composites, Ph. D. dissertation*. PhD thesis, West Virginia University, Morgantown, WV, 1998.

[194] E. J. Barbero. Composite materials forum. http://forum.cadec-online.com/viewforum.php?f=4.

[195] G. Puri. *Python Scripts for Abaqus*. abaquspython.org, Atlanta, GA, 2011.

[196] M. Metcalf. A tool to convert f77 source code to f90. ftp://ftp.numerical.rl.ac.uk/pub/MandR/convert.f90.

内 容 简 介

本书共 10 章,详细论述了复合材料分析模型的建立方法、优点和局限性,以及分析结果的正确性,主要内容包括正交各向异性材料力学、有限元分析介绍、层合板的刚度和强度、稳定性分析、自由边应力、细观力学计算、连续损伤力学、离散损伤力学、分层等。本书提供了大量详细分析的实例,以帮助读者正确理解相关理论和正确使用有限元分析工具。

本书适合于航空、航天、土木、机械、车辆、船舶等专业的工程技术人员及高校师生阅读、参考。